EMULSIONS:

Theory and Practice

PAUL BECHER

Research Chemist, Atlas Powder Co.

American Chemical Society
Monograph Series

REINHOLD PUBLISHING CORPORATION
NEW YORK, USA
CHAPMAN & HALL, LTD., LONDON

Library of Congress Catalog Card Number: 57–14847

eat for Chemistry
Series:

REINHOLD PUBLISHING CORPORATION
Publishers of
*Chemical Engineering Catalog, Chemical Materials
Catalog, Progressive Architecture, Materials
in Design Engineering, Automatic Control,
Advertising Management of American Chemical
Society*

To The Memory Of
RAYMOND ELLER KIRK
Chemist, Teacher, Friend

Warm thyself by the fire of the wise.
—Sayings of the Fathers, II, 15

GENERAL INTRODUCTION

American Chemical Society's Series of Chemical Monographs

By arrangement with the Interallied Conference of Pure and Applied Chemistry, which met in London and Brussels in July, 1919, the American Chemical Society was to undertake the production and publication of Scientific and Technologic Monographs on chemical subjects. At the same time it was agreed that the National Research Council, in cooperation with the American Chemical Society and the American Physical Society, should undertake the production and publication of Critical Tables of Chemical and Physical Constants. The American Chemical Society and the National Research Council mutually agreed to care for these two fields of chemical progress. The American Chemical Society named as Trustees, to make the necessary arrangements of the publication of the Monographs, Charles L. Parsons, secretary of the Society, Washington, D. C.; John E. Teeple, then treasurer of the Society, New York; and Professor Gellert Alleman of Swarthmore College. The Trustees arranged for the publication of the ACS Series of (a) Scientific and (b) Technological Monographs by the Chemical Catalog Company, Inc. (Reinhold Publishing Corporation, successor) of New York.

The Council of the American Chemical Society, acting through its Committee on National Policy, appointed editors and associates (the present list of whom appears at the close of this sketch) to select authors of competent authority in their respective fields and to consider critically the manuscripts submitted. Since 1944 the Scientific and Technologic Monographs have been combined in the Series. The first Monograph appeared in 1921 and, up to 1957, 134 treatises have enriched the Series.

These Monographs are intended to serve two principal purposes: first, to make available to chemists a thorough treatment of a selected area in form usable by persons working in more or less unrelated fields to the end that they may correlate their own work with a larger area of physical science discipline; secondly, to stimulate further research in the specific field treated. To implement this purpose the authors of Monographs are expected to give extented references to the literature. Where the literature is of such volume that a complete bibliography is impracticable, the authors are expected to append a list of references critically selected on the basis of their relative importance and significance.

PREFACE

My wished end is, by gentle
concussion, the emulsion of truth.
—J. Robinson, Preface to Eudoxa (1658).

This book is an essentially self-contained discussion of modern emulsion theory and practices, with particular attention to the developments of the last fifteen to seventeen years. Although significant work of a period earlier than this has been cited, often in considerable detail, certain older work is now seen to be of limited value, and such papers have been given only superficial mention or passed over in silence.

Similarly, a great many patents are, for the most part, devoid of general scientific value, and the writer has felt that no useful purpose would be served by the inclusion of literally thousands of such citations. A careful selection of patent references, which can properly be considered as part of the literature of emulsions, has been made, however, and will give the reader a general concept of the trends in this field.

With few exceptions all original papers cited have been consulted by the writer. When, for reasons of language or inaccessibility, this has proven impossible, I have attempted to give the appropriate *Chemical Abstracts* reference. I have, in a few cases, given references to review articles in obscure journals, on the assumption that one man's obscure journal may be another's standby.

By the same token, references have been frequently made to secondary sources, i.e., books and encyclopedias, since it is felt that these may often be more accessible than primary sources. This was done only, however, when the writer felt that the secondary source accurately represented the original publication.

The major portion of this book was written while the author was in the employ of the Colgate-Palmolive Co., and I wish to acknowledge the cooperation of co-workers and superiors there in this work. In particular, I wish to express my profound gratitude for the help of Miss M. Matossian and her talented library staff, without whose help this book quite literally could not have been written.

I also should like to thank the following for permission to reproduce illustrations and, in some cases, for supplying useful data: Prof. E. G. Richardson and the *Journal of Colloid Science*; Abbé Engineering Co.; Brook-field Engineering Laboratories, Inc.; Brush Electric Co.; Central Scientific Co.; Chemicolloid Laboratories, Inc.; Cornell Machine Co.; Kinetic Dispersion Corp.; Manton-Gaulin Manufacturing Co.; Reinhold Publishing Corp.; Tri-Homo Corp.; Troy Engine & Machine Co.; U. S. Industrial Chemicals Co., Division of National Distillers & Chemical Corp.

I should also like to express my appreciation to Dr. W. A. Hamor, editor of the Monograph series, for his continuous interest in the progress of this work; to Mr. Stuart B. Flexner for his critical editing of the manuscript, and, finally, to my wife, who, in addition to providing the obvious coopera-tion necessary for writing a book in a house containing two small children, did most of the work involved in compiling Appendix B, and assisted in the preparation of the index.

PAUL BECHER

Willmete, New Jersey
October, 1957

CONTENTS

ix

CHAPTER 1

Introduction

The theory of emulsions has grown in a rather haphazard way. Emulsion theory is partly an outgrowth of classical colloid chemistry and partly a development of the ancient arts involved in the production of commercial emulsions. As a result, some of the terms used tend to be, at best, ambiguous. Thus, "emulsifier" refers to both a chemical entity and to a piece of machinery; and, in reference to these same machines, the term "homogenizer" is often loosely applied to what is properly a colloid mill.

Definition of Emulsion

In a useful discussion of some aspects of emulsion formulation, Sutheim[1] has pointed out the wide variability existing in the definition of the term "emulsion," and has illustrated this with a series of such definitions from random sources. Adapting and expanding Sutheim's listings, the following series of definitions is worthy of study.

1. An emulsion is a very fine dispersion of one liquid in another with which it is immiscible.[2]

2. An emulsion is a system containing two liquid phases, one of which is dispersed as globules in the other.[3]

3. Emulsions are mechanical mixtures of liquids that are immiscible under ordinary conditions, and which may be separated into layers on standing, heating, freezing, by agitation or the addition of other chemicals.[4]

4. An emulsion is a two-phase liquid system consisting of fairly coarse dispersions of one liquid in another with which it is not miscible.[5]

5. Emulsions are intimate mixtures of two immiscible liquids, one of them being dispersed in the other in the form of fine droplets.[6]

6. . . . emulsions are finely divided liquid to semi-solid substances. . . .[7]

7. Emulsions . . . are microscopically visible droplets of one liquid . . . suspended in another.[8]

8. Emulsions are stable and intimate mixtures of the oil or oily material with water.[9]

9. An emulsion . . . consists of a stable dispersion of one liquid in another liquid.[10]

All these definitions are of some value. It is, however, somewhat striking that only two make any reference whatever to stability (8 and 9), while only one makes any reference to instability (3). Only one (7) attempts to limit the size of the dispersed phase; that is, if we are to understand Mc-Bain's words "microscopically visible" literally, to mean visible under an optical microscope. It is also surprising to find that five of the definitions contain no reference to immiscibility (2, 6, 7, 8, and 9). Schulz's definition (6) is almost completely meaningless.

It should be pointed out that it is possible to overemphasize stability. For example, it is of value to consider certain types of solvent extraction systems as being extremely unstable emulsions; in the case of, e.g., petroleum emulsions, instability is the desideratum. On the other hand, stability is often the main consideration in industrial practise. A limitation on droplet size, except within fairly large limits, is important only if one wishes to make a sharp distinction between the phenomena of emulsification and solubilization.[11] This distinction is not always possible or, indeed, desirable.

A useful synthesis of the above quotations may be made as follows:

> An emulsion is a heterogeneous system, consisting of at least one immiscible liquid intimately dispersed in another in the form of droplets, whose diameter, in general, exceeds 0.1 μ. Such systems possess a minimal stability, which may be accentuated by such additives as surface-active agents, finely divided solids, etc.

Terminology of Emulsions

In discussing emulsions, it is necessary to be able to distinguish clearly each of the two phases present. The phase which is present in the form of finely divided droplets is called the *disperse* or *internal* phase; the phase which forms the matrix in which these droplets are suspended is called the *continuous* or *external* phase. The disperse phase may also be referred to as the *nondisperse* or *discontinuous* phase.

The existence of two distinct emulsion types was first pointed out by Wo. Ostwald. The distinction consists in noting which component is the continuous and which the disperse phase. Thus, taking the classic case of an emulsion of oil and water, there may exist either an oil-in-water (oil is the disperse phase) or a water-in-oil (water is the disperse phase) emulsion. These emulsion types are conveniently abbreviated O/W and W/O, respectively. This terminology may often be applied conveniently to emulsions in which the phases are not, strictly speaking, oil and/or water (e.g., metal emulsions). The terminology is somewhat inexact for the unusual case of multiple emulsions (cf. p. 149).

Surface-active or other agents which are added to an emulsion to increase its stability by interfacial action* are known as *emulsifiers* or *emulsifying agents*. Stability is also increased by the action of mechanical devices such as *simple stirrers, homogenizers*, or *colloid mills*. These three types of equipment are often generically called "emulsifiers." This, of course, may lead to confusion with *chemical* emulsifiers or emulsifying agents. Therefore, it might be appropriate to adopt the term *emulsator* to describe any mechanical aid to emulsification.

Plan of the Book

The next chapter presents a fairly complete, albeit brief, discussion of the facts and theories of surface chemistry which are relevent to the topic of emulsions. Chapter 3 concerns itself with the physical properties of emulsions as a function of emulsion composition. Chapter 4 covers the theories of emulsion stability and Chapter 5 deals with such manifestations of instability as inversion, creaming, and complete demulsification. Chapter 6 introduces the more practical aspects of the book, containing a discussion of the chemistry of emulsifying agents and of emulsifier efficiency as a function of composition. The seventh chapter treats of the technique of emulsification, including a discussion of the various types of emulsators. Chapter 8 covers the practical formulation of numerous emulsion types, and Chapter 9 deals with the commercial aspects of demulsification.

Two appendices include a discussion of measurements on emulsions and emulsion components, and a listing (as complete as possible) of commercially available emulsifiers.

Bibliography

1. Sutheim, G. M., in "Emulsion Technology," 2nd Ed., pp. 285–86, Brooklyn, Chemical Publishing Co., 1946.
2. Alexander, J., "Colloid Chemistry," p. 102, New York, D. Van Nostrand, 1924.
3. Clayton, W., "Theory of Emulsions," 4th Ed., p. 1, Philadelphia, The Blakiston Co., 1943 (quoting F. Selmi's definition of 1845).
4. Encyclopedia Brittanica, 14th Ed., **8,** p. 416.
5. Hatschek, E., "Introduction to the Physics and Chemistry of Colloids," Philadelphia, P. Blakiston's Son & Co., 1926.
6. Sutheim, G. M., "Introduction to Emulsions," p. 1, Brooklyn, Chemical Publishing Co., 1946.
7. Schulz, G., "Emulsions and Emulsifiers," p. 2, O. T. S. Report PB 98002, Department of Commerce, Washington, 1949.

---

* It is necessary to specify this, since other agents, e.g., added to increase the viscosity of the emulsion, also have the ultimate effect of increasing stability.

8. McBain, J. W., as quoted in McBain, M. E. L. and Hutchinson, E., "Solubilization," p. xi, New York, Academic Press, 1955.
9. "Emulsions," Carbide & Carbon Chemicals Corp., 6th Ed., p. 8, 1937.
10. Roberts, C. H. M., *J. Phys. Chem.* **36,** 3087 (1932).
11. Webster's New Collegiate Dictionary, G. & C. Merriam Co.
12. As for example, is made by McBain and Hutchinson, *op. cit.* pp. 189–190.

CHAPTER 2

Surface Activity

The vast increase in interfacial surface accompanying emulsification makes necessary a brief discussion of surface activity. For example, the emulsification of only ten cubic centimeters of oil to form droplets of radius 0.1 μ creates a total interfacial area of 300 square meters, an increase on the order of a millionfold. Under these circumstances the special properties of surfaces and (more specifically) interfaces become of paramount importance.

Surface Tension

Various phenomena associated with surface tension, e.g., capillary rise, have apparently been recognized for hundreds of years. For example, Partington[1] cites Leonardo da Vinci in this connection. For many years explanations of surface-tension effects depended on the assumed existence of a "contractile skin" on the surface of the liquid. While liquids certainly behave as though such a skin exists, and while the properties of the surface are different from the bulk, no recourse to such a picture is actually necessary.

As is well known, short-range attractive (van der Waals') forces exist between molecules. These are the forces which are responsible for the nonideality of the substance in the gaseous form, and, indeed, for its existence in the liquid form, since a truly ideal gas possesses no possibility of liquefaction.

In the bulk of a liquid, the molecules are sufficiently close so that the effect of attractive forces are considerable. Indeed, the attractive forces are sufficiently great to keep all but a small number of the molecules from escaping into the vapor state. Although these forces are relatively large in magnitude, they tend to balance out in the bulk of the liquid. On the other hand, the molecules in the surface region, not being completely surrounded by other (liquid) molecules, are subjected to an unbalanced attraction, which has the net effect of an attractive force directed inward normal to the surface (see Figure 2-1). The smaller the surface, the lower this net force. Thus, a condition of minimum surface leads to a lower energy, and we say that the surface of a liquid has a "tendency" to con-

5

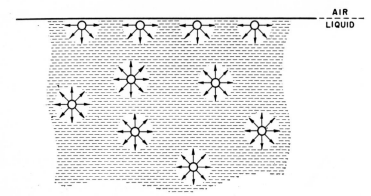

FIGURE 2-1. The forces acting on molecules on the surface and in the interior of a liquid.

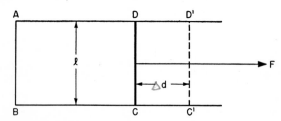

FIGURE 2-2. The physical definition of surface tension.

tract. This molecular view accounts, qualitatively at least, for the appearance of a contractile skin.

Physical Definition of Surface Tension. Imagine a small rectangular wire frame $ABCD$ with one side moveable (Figure 2-2). If the length AB (or CD) be l, and the length AD (or BC) be d, the surface area of a film of liquid contained in the frame will be $2ld$ (since there are *two* sides to the film).

If the moveable side CD is moved through the distance Δd to a new position $C'D'$ by the exertion of a force F, the work done on the liquid

$$w = F\Delta d \qquad (2.1)$$

The force F is balanced by a counter force operating along the length CD. If γ is defined as the force in dynes per centimeter acting along this length, then the force opposing the expansion of the film is $2\gamma l$, and Eq. 2.1 reduces to

$$w = 2\gamma l\Delta d = \gamma\Delta S,$$

where ΔS ($= 2l\Delta d$) is the increase in surface.

Thus,

$$\gamma = \frac{w}{\Delta S} \tag{2.2}$$

In other words, γ may be defined as *the work in ergs necessary to generate one square centimeter of surface*. This is the physical definition of surface tension.

Units of Surface Tension. The definition of surface tension given above suggests that the appropriate units for surface tension should be ergs per square centimeter. However, since the surface tension can also be defined in terms of a force acting along a 1-cm length of surface, it is also appropriate to use *dynes per centimeter*. Since an erg is equal to a dyne-centimeter it is clear that the choice of units has no effect on the numerical value of the surface tension. The unit usually employed is dynes/centimeter. Table 2-1 gives typical surface-tension values for some pure liquids and oils of interest.

Surface Tension and Vaporization. The elementary molecular theory and the physical definition of surface tension presented above may be correlated. When the surface of a film is increased by extending the wire frame, more molecules must be brought into the higher energy condition in the surface. This is analogous to the phenomena accompanying vaporization. Stefan[2] was apparently the first to suggest that there should be a relationship between the surface tension (or more correctly the *specific cohesion*; $a^2 = 2\gamma/d$, where d = density) and the latent heat of vaporization. Partington[3] gives an extensive list of theoretical and empirical equations relating these two properties.

It is also not surprising to note that the vapor pressure of small droplets (where the surface-to-mass ratio is high) is very different from that of the bulk phase. It can be shown[4] that the ratio of the vapor pressure p' of the droplet can be related to that of the bulk liquid p by the expression

$$RT \ln \frac{p'}{p} = \frac{2\gamma M}{rd}, \tag{2.3}$$

where M is the molecular weight of the vapor, d the density of the liquid, and γ, R, and T have their usual meanings. In Table 2-2 the ratio p'/p is given for water at 20°C as a function of droplet radius. These data show that the effect of droplet radius on vapour pressure is a sizeable one.

Surface Free Energy. Since the surface tension of most liquids decreases with rising temperature,* it follows that there should be an ab-

---

* The exceptions consist of certain liquid metals and do not concern the present discussion.

TABLE 2-1. SURFACE TENSION OF SEVERAL PURE SUBSTANCES AT 20°C.
(DYNES/CM)*

Mercury	485.0	Chloroform	27.13
Water	72.80	Carbon Tetrachloride	26.66
Acetylene Tetrabromide	49.67	Ethyl Caproate	25.81
Nitrobenzene	43.38	Methyl Propyl Ketone	24.15
Nitromethane	36.82	Diisoamyl	22.24
Bromobenzene	36.26	n-Octane	21.77
Chloracetone	35.27		
Oleic Acid	32.50	n-Hexane	18.43
Carbon Disulfide	31.38	Ethyl Ether	17.10
Benzene	28.86	Castor Oil†	39.0
Caprylic Acid	28.82		
Toluene	28.4	Olive Oil†	35.8
n-Octyl Alcohol	27.53	Cottonseed Oil†	35.4
		Liquid Petrolatum†	33.1

* International Critical Tables, except as indicated.
† Halpern, A., *J. Phys. Chem.* **53**, 896 (1949).

TABLE 2-2. EFFECT OF DROPLET SIZE ON VAPOR PRESSURE (WATER, 20°C)

Radius. cm	p'/p
10^{-4}	1.001
10^{-5}	1.011
10^{-6}	1.114
10^{-7}	2.95

sorption of heat when the surface is expanded. This may be elaborated by the following thermodynamic argument:[5]

Consider a single surface film of area S with a surface tension γ at an absolute temperature T, and let the following cyclic process* be carried out:

i. Increase the area S to $S + dS$ (T = const.). The work done *by* the system is $-\gamma dS$ (cf. Eq. 2.2).

* The analogy with the Carnot cycle of classical thermodynamics should be noted.

ii. Lower the temperature to $T - dT$. The surface tension changes to $\gamma - d\gamma \equiv \gamma - (d\gamma/dT)\, dT$.

iii. At constant temperature $(T - dT)$, decrease the area from $S + dS$ to S. The work done in this step is then $(\gamma - d\gamma)S \equiv [\gamma - (d\gamma/dT)dT]dS$.

iv. Raise the temperature to T, whereby the system returns to its original state.

Work is done only in steps i. and iii. of the above cycle, and the net work performed is:

$$-\gamma dS + [\gamma - (d\gamma/dT)dT]\, dS \equiv -(d\gamma/dT)dTdS.$$

Now if l_s is the heat absorbed per unit increase of surface at constant temperature (the latent heat of extension of the film), then the heat absorbed at the higher temperature, in step i, is $l_s dS$. By the second law of thermodynamics

$$-(d/dT)dTdS = l_s dS(dT/T).$$

Hence:

$$l_s = -T(d\gamma/dT) \tag{2.4}$$

Equation 2.4 was first stated by Lord Kelvin. By virtue of the decrease of surface tension with temperature, l_s must always be positive.

Furthermore, if u is the increase in the energy of the film as a result of a unit increase in area, then

$$l_s = u - \gamma \tag{2.5}$$

(since the heat absorbed must be equal to the energy increase plus the work done), and

$$u = \gamma - T(d\gamma/dT) \tag{2.6}$$

from which it follows that γ is the *surface free energy* of the film.

Effect of Temperature on Surface Tension. As pointed out in the previous section, the surface tension of most liquids decreases with increasing temperature. This is a reasonable corollary to the molecular picture of surface tension. The increased kinetic energy imparted to the surface molecules by the rise in temperature will tend to overcome the net attractive forces of the bulk liquid. Furthermore, as the temperature of the liquid approaches the critical, the cohesive forces between the molecules approaches zero. It is to be expected, therefore, that the surface tension will vanish at the critical temperature.

Eötvös[6] proposed the relation

$$\gamma(Mv)^{2/3} = k(T_c - T) \tag{2.7}$$

where M is the molecular weight, v the specific volume, T_c the critical temperature of the liquid, T the temperature, and k is a universal constant. It will be noted that the equation has the desired property of causing the surface tension to vanish at T_c, and can be derived from quite reasonable assumptions. The constant k is found to have the approximate value 2.2 for a large number of liquids.

The fit of the above equation with experimental data is somewhat inexact, and Ramsay and Shields[7] suggested the modification

$$\gamma(Mv)^{2/3} = k(T_c - T - d) \qquad (2.8)$$

where d is an additional constant having the value of approximately 6.0. If Eq. 2.8 is not to be regarded as purely empirical, it must be interpreted as implying that the surface tension becomes zero at an absolute temperature about six degrees below the critical. It is a little difficult to understand the physical significance of this, but the theoretical researches of Meyer[8] indicate that the behavior of liquids in the critical region is not as sharply defined as the Eötvös equation would lead us to suppose.

Interfacial Tension

The previous discussion has considered the properties of a surface existing between a liquid and a gas, the gas being either air or the saturated vapor of the liquid. Of considerably more importance to the theory of emulsions are the boundary tensions existing between two liquids, and, to a slightly lesser extent, those existing between a liquid and a solid. Such a boundary tension is referred to as an *interfacial tension*. (As implied above, surface tension is actually an interfacial tension between a liquid and a gas.)

Liquid-Liquid Interfaces. When two immiscible liquids are placed in contact, an interface results. Returning to the kinetic-molecular picture of the surface discussed above (p. 5), the net attractive forces operating on a molecule in the interface will be somewhat different than in the case of a simple surface, since there will undoubtedly be some van der Waals' interaction with the surface molecules of the second liquid. Equally certain, however, is the fact that this attraction will be of a different order of magnitude than that of the bulk molecules, and that an imbalance of forces will exist, with all the consequent physical effects (see Figure 2-3). It should be recognized that what is true of one liquid is equally true, *pari passu*, of the other. Hence, it is to be expected that the value of interfacial tension will usually lie between the individual surface tensions of the two liquids. Table 2-3 gives the values of the observed interfacial tensions against water for a number of liquids.

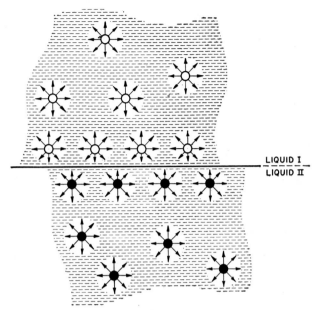

FIGURE 2-3. Forces acting on molecules at the interface and in the interior of a pair of liquids.

TABLE 2-3. INTERFACIAL TENSION OF LIQUIDS AGAINST
WATER AT 20° C. (DYNES/CM)*

Mercury	375.0	Chloroform	32.80
n-Hexane	51.10	Nitrobenzene	25.66
n-Octane	50.81	Ethyl Caproate	19.80
Carbon Disulfide	48.36	Oleic Acid	15.59
Diisoamyl	46.80	Ethyl Ether	10.70
Carbon Tetrachloride	45.0	Nitromethane	9.66
Brombenzene	39.82	n-Octyl Alcohol	8.52
Acetylene Tetrabromide	38.82	Caprylic Acid	8.22
Toluene	36.1	Chloracetone	7.11
Benzene	35.0	Methyl Propyl Ketone	6.28
		Olive Oil†	22.9

* International Critical Tables, except as indicated.
† Sutheim, G. M., "Introduction to Emulsions," p. 25, Brooklyn, Chemical Publishing Co., Inc., 1946.

Adhesion and Cohesion. In order to understand the relationship between surface and interfacial tensions, the concepts of adhesion and cohesion are introduced here. Consider, as in Figure 2-4A, a cylinder of liquid 1 sq cm in cross section. If the cylinder is pulled apart at the point S, thus creating 2 sq cm of new surface, the work done is 2γ, which is defined as

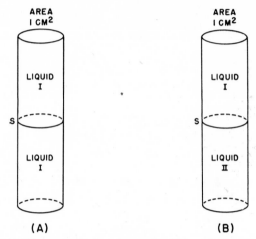

(A)   **(B)**

FIGURE 2-4. (A) When a cylinder of pure liquid is pulled apart, work equal to the work of cohesion W_c is required. (B) When cylinders of two different liquids are pulled apart at their mutual interface, work equal to the work of adhesion W_a is required.

the *work of cohesion*, W_c, of the liquid. Now consider *two* liquid cylinders in contact, with an interface of 1 sq cm (Figure 2-4B); the interfacial tension at this interface being γ_{12}. If these cylinders are pulled apart, two new surfaces with the individual surface tensions γ_1 and γ_2 are created. The net work thus required, or the *work of adhesion* between the two surfaces, W_A, is then given by

$$\text{Immiscible} \qquad W_A = \gamma_1 + \gamma_2 - \gamma_{12} \qquad (2.9)$$

This equation is due to Dupré.[9] In principle, if one could experimentally determine the work of adhesion, the equation could be used to calculate interfacial tension. In point of fact, however, the direct determination of the work of adhesion is quite impossible, and Dupré's equation is used for the calculation of adhesion from the observed surface and interfacial tensions.

The work of adhesion, and in particular, its sign, is of importance in the theory of wetting. It will be referred to subsequently in the discussion of the stabilization of emulsions by finely divided solids (*infra* pp. 125–131).

Antonoff's Rule. Dupré's equation applies only to liquids that can be considered completely immiscible. To be sure, no liquids are *completely* immiscible; but in systems where miscibility is appreciable a rule first proposed by Antonoff[10] would appear to apply. In this case, according to Antonoff, the interfacial tension is related to the individual surface tensions by the following equation:

$$\text{miscible} \qquad \gamma_{12} = \gamma_1 - \gamma_2 \qquad (2.10)$$

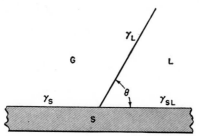

FIGURE 2-5. Interfacial tensions at the solid-liquid interface.

The surface tensions of this equation are not, however, those of pure liquids, but are those of one liquid saturated with the other. The rule seems to be observed in a large number of cases, but large positive deviations apparently occur in other instances.[11] Thus Antonoff's rule has been the subject of polemic discussion.*

Solid-liquid Interfaces. If a liquid is placed in contact with a plane surface of solid, three interfaces exist: i.e., between the solid and gas (or vapor of the liquid), between the liquid and gas (or vapor), and between the solid and liquid. There thus exist three corresponding surface (or interfacial) tensions: γ_S, γ_L, and γ_{SL}. The three phases meet at a point such that there exists a *contact angle* θ between the liquid and solid (see Figure 2-5). If θ is finite, the liquid forms a droplet on the surface of the solid; if $\theta = 0$, the liquid spreads on the solid, i.e., it "wets" it. By an argument due to Gauss[13] it can be shown that the relation between the interfacial tensions and the angle of contact must be

$$\gamma_S - \gamma_{SL} = \gamma_L \cos \theta \qquad (2.11)$$

Dupré's equation (Eq. 2.9) applies equally in the case of a solid-liquid interface, and thus

$$W_A = \gamma_S + \gamma_L - \gamma_{SL} \qquad (2.12)$$

Combining Eqs. 2.11 and 2.12, it follows that

$$W_A = \gamma_L (1 + \cos \theta) \qquad (2.13)$$

where W_A is the work of adhesion between the liquid and solid. This relation is ascribed to Young,[14] who reportedly derived it in words in 1805; Harkins[15] denies that this relation can be found in Young's works in any form, and Adam[16] also ascribes it to Dupré.

For emulsion theory, a more interesting case consists of the situation

---

* See, for example, Partington's comment.[12]

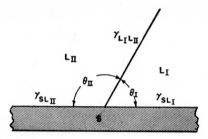

FIGURE 2-6. Interfacial tensions at the solid-liquid I-liquid II interface.

arising at the mutual triple interface between *two* immiscible liquids and a solid, i.e., the vapor phase is replaced by a second liquid (Figure 2-6). In this case the three interfacial tensions are: $\gamma_{SL_{I}}$, $\gamma_{SL_{II}}$, and $\gamma_{L_{I}L_{II}}$. Equation 2.11 applies to the system in the form

$$\gamma_{SL_I} - \gamma_{SL_{II}} = \gamma_{L_I L_{II}} \cos \theta_I,$$

the acute angle being defined as the angle of contact. (A precisely analogous equation is obtained if one makes the other choice.) The application of this relation and its consequences for emulsions stabilized with finely divided solids is discussed in Chapter 4.

Harkins[17] has pointed out that the quantity γ_S in, for example, Eq. 2.11 is not the surface tension of the solid (a quantity which, in any case, is not determinable[18]) but is actually the interfacial tension existing between the solid and a film of the vapor of the liquid (or of the supervening gas). This would introduce an additional term in Eq. 2.13, which can be obtained from the adsorption isotherm of the gas or vapor on the solid. In most cases the correction is slight.

Surface Films

A special type of interface between immiscible liquids occurs when a small amount of a water-insoluble material is allowed to spread on a water surface. Dupré's equation applies; but spreading will only occur if the work of adhesion between the two liquids is greater than the work of cohesion of the water-insoluble liquid.

Spreading Coefficient. This statement defines the quantity which Harkins[18] terms the *spreading coefficient*:

$$S = W_A - W_C \qquad (2.15)$$

Thus, if $S > 1$, one liquid will spread on the other; when $S < 1$, the super-

TABLE 2-4. SPREADING COEFFICIENT FOR LIQUIDS ON WATER AT 20° C*

Ethyl Ether	45.0	Benzene	8.94
Methyl Propyl Ketone	42.37	Diisoamyl	3.76
n-Octyl Alcohol	36.75	Nitrobenzene	3.76
Caprylic Acid	35.76	n-Hexane	3.27
Chloracetone	30.42	Carbon Tetrachloride	1.14
Ethyl Caproate	27.19	n-Octane	0.22
Nitromethane	26.32	Brombenzene	−3.28
Oleic Acid	24.71	Carbon Disulfide	−6.99
Olive Oil	14.1	Acetylene Tetrabromide	−15.74
Chloroform	12.87	Mercury	−787.0

* Calculated from data of Tables 2-1 and 2-3. The positive value for carbon tetrachloride is anomalous, as this liquid *does not* spread on water.

natant liquid will simply form floating lenslike drops. This, for example, is what occurs when a highly refined liquid petrolatum is placed on pure water. While it is far from a necessary condition, it has been noted that substances with polar groupings are more likely to spread on water than nonpolar compounds. To give a particularly striking example, n-octyl alcohol on water at 20° C exhibits a spreading coefficient of 35.74, whereas n-octane shows an S value of only 0.22. The hydrocarbon will spread, but has only a very slight tendency to do so. Table 2-4 gives values of S for a number of liquids on water.

Now, if the amount of material allowed to spread on the water surface be small, the film thus formed has unusual properties. Historically, the first exact work on such films was done by Benjamin Franklin in 1765. Franklin spread a film of olive oil on a pond in Clapham Common, and, by a simple calculation, arrived at the conclusion that the film was one ten-millionth of an inch, or 25 Å, thick.[19] In view of later work, this is a most remarkably accurate determination.

Monomolecular Films. The first modern examination of the properties of such films was begun by A. Pockels[20] (1891). Miss Pockels studied the behavior of films of olive oil upon the surface of water in a trough. She found that when the film was compressed between barriers the surface tension decreased rapidly after a definite surface area per unit mass was reached. Rayleigh[21] studied similar films by this method, and arrived at the important conclusion that the films formed in this manner were one molecule thick, i.e., were *monomolecular films*.

Langmuir,[22] using the methods of Pockels and Rayleigh, introduced an important technical refinement to this type of measurement. Rather than determining the surface tension of the film, he introduced the use of a fixed *floating* barrier which, by means of a suitable mechanical linkage, measured the *force* being exerted on the film by another (movable) barrier. Further improvements in the device are due to Adam, and McBain and Wilson.[22a] The specific form developed by these last workers has been dubbed the *PLAWM* apparatus, this designation arising from the initial latters of the names *P*ockels, *L*angmuir, *A*dam, *W*ilson, and *M*cBain.

Additional refinements required for making more elaborate measurements are due to Harkins and his coworkers.[22b] Figure 2-7 shows a simple, commercially available form of the Langmuir device, which is generally referred to as the *hydrophil balance*. The mechanical linkage to the barrier, enabling one to measure the force by means of a torsion balance, is clearly shown.

A much more elaborate model, developed by Harkins, is shown in Figure 2-8; measurements are made in a thermostat, and results of high precision are obtainable. As shown, this precision film balance is contained in a liquid-filled thermostat, permitting controlled, accurate work at temperatures varying from as low as −30° to 70°C.

Figure 2-9 is a schematic representation of the type of data obtained if one uses such a device to measure the area of a film of, say, stearic acid as a function of the compressive force. Initially (large area), only a slight

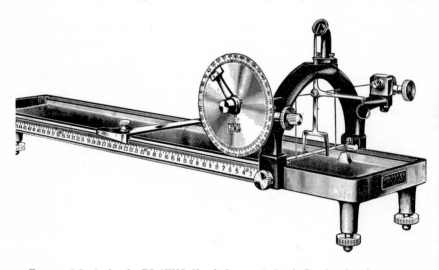

FIGURE 2-7. A simple *PLAWM* film balance. A fixed, floating barrier measures the force being exerted on the film by a movable barrier. *Courtesy: Central Scientific Co.*

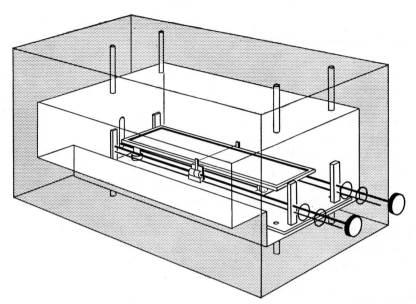

FIGURE 2-8. A diagrammatic representation of a precision film balance, according to Harkins. The balance is contained in a liquid-filled thermostat, permitting accurate work at temperatures from −30° to 70°C.

increase in force is required to decrease the area (section *ab* of the curve); then there is a sudden large increase in force corresponding to a very slight decrease in area (section *bc* of the curve); this is followed by a flat portion of the curve in which further decrease in area of the film is reflected by no increase in the force exerted on the barrier.*

This behavior may be explained as resulting from the fact, recognized by Rayleigh, that these films are monomolecular. Thus, the initial portion *ab* of the curve of Figure 2-9 arises from the steady crowding together of the molecules of the fatty acid until, at *b*, they are packed as tightly as is possible. Any further compression will, of course, result in a large pressure being exerted on the barrier (section *bc*); finally, the film will buckle and collapse, so that no further pressure can be exerted. As a matter of fact, just prior to the film's collapse, it is often possible to observe strain lines and striations in it.

---

* Strictly speaking, this type of curve is obtained only with so-called "condensed" films; at higher temperatures the films behave like two-dimensional gases and give force-area curves similar to the *P-V* curves of real gases. These "expanded" films need not be considered in detail here; Harkins[23] has identified five types of expanded films.

The area corresponding to the densest packing of the monomolecular film can be obtained by extrapolating the sharply rising part of the force-area curves to zero force (indicated by the dotted line intersecting the abscissa at O in Figure 2-9). For a pure substance we can calculate the number of molecules, n, actually present in a film containing a given weight of material, from the relation $n = (w/M)N$ where w is the weight of the film, M the molecular weight, and N Avogadro's number. Thus it should be possible to calculate the area occupied by each molecule; or, putting it differently, it should be possible to compute the cross-sectional area of the molecule.

Table 2-5, adapted from Adam,[24] shows the results of such measurements for a large number of long-chain compounds of the general formula $C_nH_{2n+1}X$ in terms of the nature of the end-group X. For substances of this sort it is found that the hydrocarbon chain has to contain in excess of 10 to 17 carbon atoms in order to obtain condensed films at the temperature of measurement.

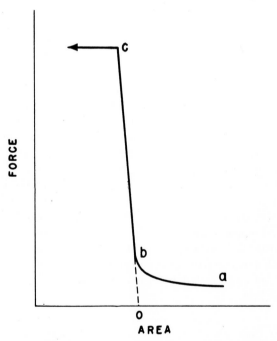

FIGURE 2-9. Typical force-area curve obtained with a monomolecular film of, e.g., saturated fatty acids. The extrapolation of the steeply rising portion of the curve to the abscissa at O permits the calculation of the area per molecule.

TABLE 2-5. AREA PER MOLECULE FOR CONDENSED
FILMS OF COMPOUNDS $C_nH_{2n+1}X$

Series	End Group X	Area/molecule ($Å^2$)
Fatty Acids	—COOH	20.5
Dibasic Esters	—COOC$_2$H$_5$	20.5
Amides	—CONH$_2$	20.5
Methyl Ketones	—COCH$_3$	20.5
Triglycerides (area per *chain*)	—COOCH$_2$	20.5
Esters of Saturated Acids	—COOR	22.0
Alcohols	—CH$_2$OH	21.6
Phenols, and other simple *p*-substituted benzene compounds	—⟨benzene ring⟩OH	24.0
	—⟨benzene ring⟩OCH$_3$	24.0
	—⟨benzene ring⟩NH$_2$	24.0

The interesting thing about these data is the remarkable consistency observed in the values of the cross-sectional areas of the great variety of compounds in a given series. The fact that 20.5 sq Å is found for such a large number of compounds, and that, for example, the *p*-substituted benzene compounds all show an area per molecule of 24.0 sq Å indicates such a high order of consistency that it must be readily explicable.

It will be recognized that all of these molecules have in common what may be termed a polar-nonpolar structure. That is, the molecule may be considered as being made up of two distinct sections; one possessing polar* characteristics, the other, nonpolar. The polar portions of these molecules are characterized by a fair degree of water solubility. For example, acetic acid (which is the lowest carboxylic acid possessing any hydrocarbon character) is infinitely soluble in water. On the other hand, the nonpolar (in this case, hydrocarbon) portions of these molecules are distinguished by their distinct nonsolubility in water. For this reason, the polar portion of the molecule is commonly referred to as the *hydrophilic* (water-loving) group; the nonpolar as *hydrophobic* (water-hating), or better as *lipophilic* (oil-loving). Molecules of this type have been given the useful designation

---

* The term "polar" when used in the sense implied here, as in most discussions of surface activity, is apparently not capable of exact definition. What is implied is that, although the molecule considered as a whole possesses no net dipole moment, the polar moecules (or groups) tend to be more soluble in polar solvents, such as water, while nonpolar molecules (or groups) tend to be more soluble in typical (nonpolar) organic solvents, such as benzene. A polar group is thus usually considered to possess a high degree of water solubility, and *vice versa*.

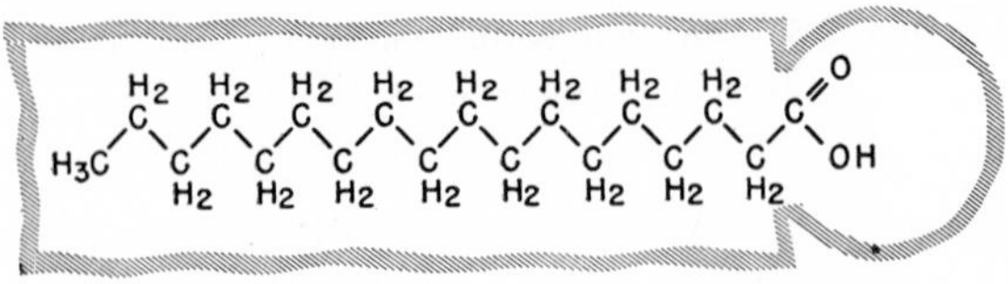

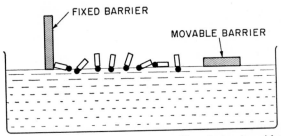

FIGURE 2-10. The structure of stearic acid considered as an amphiphilic compound. The section in the rectangle is the nonpolar, hydrophobic, hydrocarbon tail; the section in the circle is the polar, hydrophilic, carboxylic head.

FIGURE 2-11. Fully expanded monomolecular film of stearic acid (point *a*, Figure 2-9).

amphipathic by Hartley,[25] which Winsor[26] has modified to *amphiphilic* as being more descriptive.

The explanation of the data of Table 2-4 is considerably facilitated by the adoption of the useful convention, introduced by Harkins, of representing such amphiphilic molecules by the symbol ♀, where the circular portion of the symbol represents the hydrophilic polar "head" of the molecule, and the rectangle represents the lipophilic nonpolar "tail." Figure 2-10 indicates schematically how this symbolism applies to stearic acid.

Using this convention, Figure 2-11 indicates schematically the situation existing at point *a* on the curve of Figure 2-9, i.e., when the film of stearic acid is fully expanded. The molecules lie helter-skelter over the water surface, but there is a distinct tendency towards *orientation*, that is, the majority of the acid molecules lie in such a manner that each polar carboxylic head is "dissolved" in the water, while each nonpolar hydrocarbon tail either sticks out or droops over onto the surface of the water. As the molecules are crowded together upon compression of the film, however, the amount of surface becomes limited, and the polar groups occupy the available surface area owing to their water solubility. Finally, when the molecules are tightly packed together (*b* on the force-area curve, Figure 2-9), the situation is represented by Figure 2-12. Now the molecules are arranged so that the polar heads are all in the water surface, while the tails

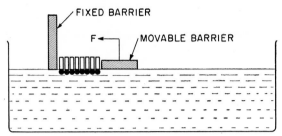

FIGURE 2-12. Fully compressed monomolecular film of stearic acid (point *b*, Figure 2-9).

have been forced to occupy a position above the surface. Each molecule is literally "standing on its head."

From this, it is quite clear that the figure of 20.5 sq Å for hydrocarbon derivatives is actually the cross-sectional area of the cylinder occupied by the hydrocarbon chain. Similarly, the value of 24.0 sq Å for benzene derivatives must correspond to the cross-sectional area of the benzene ring. The intermediate values found for alcohols and esters may possibly arise from hydrogen bonding effects.

Inasmuch as the total volume of the film is known, and its area can be accurately measured, its thickness can also be calculated. For example, a value of 27.5 Å is obtained for the length of the hydrocarbon chain of stearic acid. This is in agreement with values calculated for the known carbon-carbon bond distances in saturated hydrocarbon chains. The agreement with Franklin's estimate, made nearly two hundred years ago, is also striking.

Surface Properties of Solutions

The preceding section considered the situation existing at the interface between two immiscible pure liquids. From the point of view of the theory of emulsions, however, it is of considerably greater interest to consider the effect of "impure" liquids (i.e., of the effect of solute) on surface and interfacial properties.

That small amounts of solute can have violent effects on these properties is shown by a demonstration due to Hardy.[27] A large lens of *Liquid Petrolatum, Squibb* (or any other liquid with a large negative spreading coefficient) is formed on the surface of a sheet of water. A drop of oleic acid is placed upon the center of the lens; after a short time considerable agitation is noted close to this point, and then the lens appears to shatter, with almost explosive violence, into small fragments which are projected to the edges of the container.

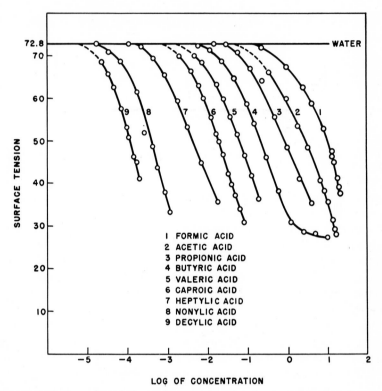

FIGURE 2-13. Lowering of the surface tension of water, as a function of concentration, by the homologous series of lower fatty acids.

Surface Tension of Solutions. That the presence of a solute should have an effect on the surface tension is perhaps not surprising; what is surprising is the tremendous variability of the effects observed. For instance, Figure 2-13 shows the effect on the surface tension of water by *lower* members of the series of homologous fatty acids, as a function of concentration. In all cases, a lowering is observed, but the effect is strongly dependent on the molecular weight. This is in accord with Traube's rule, according to which the concentrations for equal lowering of the surface tension in dilute solutions decrease by a third for each additional CH_2 in an homologous series.[28]

This, of course, does not exhaust the possible variations in surface tension. McBain, Ford, and Wilson[29] have classified the three main types of surface-tension curves, which are represented schematically in Figure 2-14. The Type I curve, showing decreasing surface tension, as typified by the fatty acid curves of Figure 2-15, is by far the most common. The Type II

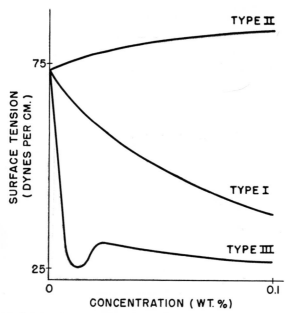

FIGURE 2-14. Schematic representation of the principal types of surface tension concentration curves, after McBain, Ford, and Wilson.[29]

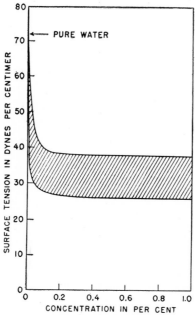

FIGURE 2-15. The range of surface tensions found in solutions of most surface-active compounds.[30]

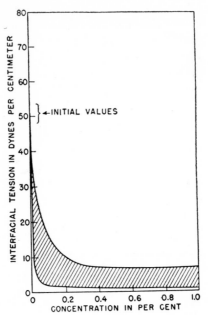

FIGURE 2-16. The range of interfacial tensions found in solutions of most surface-active compounds.[30]

curve, in which there is a slight but definite increase in surface tension, is found in the solutions of strong electrolytes, and for certain compounds (e.g., sugars) containing a large number of hydroxyl groups.

From the point of view of this discussion, perhaps the most interesting compounds are those which give Type III curves, namely most soaps and detergents, i.e., compounds belonging to the class which we have termed *amphiphilic*. For reasons which will be apparent in the next section, these compounds are also generally referred to as *surface-active* compounds or "surfactants."* It should also be pointed out that the minimum in the Type III curve is probably an artifact (cf. p. 27, below).

Figure 2-15, adapted from Fischer and Gans,[30] shows the range of surface tensions found in solutions of most surface-active materials.

Interfacial Tensions of Solutions. The interfacial tensions which occur in solutions giving rise to surface-tension curves of the Types I and II variety are relatively uninteresting. The striking fact about Type III systems is that the interfacial tensions become so small. Figure 2-16 gives the range of interfacial tensions observed for the majority of solutions of amphiphilic compounds.

---

* This word was coined several years ago. It serves no useful purpose whatever, and is etymologically indefensible.

A striking illustration of the range of possible interfacial tension values is given by Harkins and Zollman.[31] A paraffin oil exhibited an interfacial tension of 40.6 dynes/cm toward water; this was lowered to 31.05 dynes/cm when the aqueous phase was made 0.001 M in oleic acid. When this was neutralized by the equivalent amount of sodium hydroxide (forming the corresponding soap) the interfacial tension fell to 7.2 dynes/cm. When the aqueous phase was then made 0.001 M with respect to sodium chloride, the interfacial tension then fell to less than 0.01 dynes/cm. When the paraffin oil was replaced by olive oil, the interfacial tension dropped to 0.002 dynes/cm, the lowest interfacial tension ever measured!

The effect of the addition of the sodium chloride is remarkable; however, it is perfectly normal. The addition of an electrolyte to solutions of amphiphilic compounds generally results in such striking decreases in surface and interfacial tensions.[31]

Gibbs' Adsorption Equation. It has already been pointed out that the phenomenon of surface (and interfacial) tension may be explained on a molecular basis by the statement that the van der Waals field of force acting on a molecule at the surface of a liquid is different from the forces acting on a similar molecule in the bulk of the liquid. If this consideration is extended to solutions, it will be clear that the molecules for which the interaction energy is lower than average will tend to accumulate in the surface. This will have the effect of keeping the free energy of the system at a minimum.

The consequences of this accumulation were derived by a thermodynamic argument of Gibbs[32] in 1876. Gibbs defined a quantity Γ, called the *surface excess* (it could equally well be termed *surface deficiency*, since it may have negative values), as the difference in the surface and bulk concentrations of a molecular species.*

By free energy considerations, and by the use of the Gibbs-Duhem equation, it can be shown[33] that

$$S^s \partial T + s \partial \gamma + \sum n_i^s \partial \mu_i^s = 0 \qquad (2.16)$$

where S^s is the surface entropy, s the surface area, n_i^s the number of molecules of species i in the surface, and μ_i^s the surface chemical potential of molecular species i. At constant temperature, Eq. 2.16 reduces to:

$$s \partial \gamma + \sum n_i^s \partial \mu_i^s = 0 \qquad (2.17)$$

Dividing through by the surface area, s

$$\partial \gamma + \frac{n_1^s}{s} \partial \mu_1^s + \frac{n_2^s}{s} \partial \mu_2^s + \cdots = 0 \qquad (2.18)$$

---

* The term "surface" is purposely left more or less undefined here; it is exactly defined in the original paper by Gibbs.

The various fractions n_i^s/s are obviously the surface excess Γ_2 (as defined above) for each molecular species. Hence

$$\partial\gamma + \Gamma_1\partial\mu_1^s + \Gamma_2\partial\mu_2^s + \cdots = 0 \tag{2.19}$$

By this definition it will be noted that the "surface excesses" are really the excess amounts per unit *area* of the surface, and are thus not concentrations in the conventional sense.

For a two-component system (e.g., consisting of a solvent and a single solute) 2.19 becomes

$$\partial\gamma + \Gamma_1\partial\mu_1^s + \Gamma_2\partial\mu_2^s = 0 \tag{2.20}$$

Now, in a system at equilibrium (at constant temperature, pressure, and surface area) it can be shown that the surface chemical potential of any constituent is equal to that in the bulk phase, i.e., $\mu^s = \mu$. Eq. II.20 may then be written

$$\partial\gamma + \Gamma_1\partial\mu_1 + \Gamma_2\partial\mu_2 = 0, \tag{2.21}$$

where μ_1 and μ_2 refer to the chemical potentials *in the solution.*

Since the definition of the surface as given by Gibbs, although precise, is arbitrary, it may be defined (in the case of dilute solutions) in such a way that the surface excess of the solvent Γ_1 is effectively zero. This then leads to

$$\partial\gamma + \Gamma_2\partial\mu_2 = 0$$

$$\Gamma_2 = -\left(\frac{\partial\gamma}{\partial\mu_2}\right)_T \tag{2.22}$$

The chemical potential for any component of a system, for example the solute, may be represented by $\mu^o + RT \ln a$, where μ^o is a constant for a given substance at constant temperature and a is the *activity* of the solute. Substituting into Eq. 2.22

$$\Gamma_2 = -\frac{1}{RT}\left(\frac{\partial\gamma}{\partial \ln a_2}\right)_T$$

$$= -\frac{a_2}{RT}\left(\frac{\partial\gamma}{\partial a_2}\right)_T \tag{2.23}$$

In dilute solutions, the concentration may be substituted for the activity, so that

$$\Gamma_2 = -\frac{c}{RT}\left(\frac{\partial\gamma}{\partial c}\right)_T, \tag{2.24}$$

which is the commonly quoted form of the Gibbs equation.

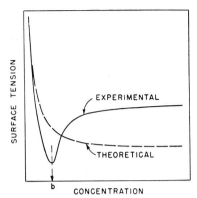

FIGURE 2-17. Surface tension-concentration curve for type III compounds. The solid curve is observed for many materials, but the rise in surface tension after the minimum corresponds to a theoretically impossible change in adsorption. The theoretically expected curve is shown by the dashed line. This curve is realized when highly purified materials are used.

Interpretation of Gibbs' Equation. It is interesting to reexamine the generalized surface tension-concentration curves of Figure 2-14 in the light of Gibbs' adsorption equation (Eq. 2.24). Clearly, when the slope of the surface tension-concentration curve $(\partial\gamma/\partial c)_T$ is negative, the surface excess Γ is positive, i.e., the concentration of solute in the surface is greater than the bulk concentration, and vice versa. Thus it is seen that solutes which give Type II curves will be *negatively adsorbed*; in other words, the surfaces of such solutions will be richer in solvent than the bulk solution. On the other hand, Type I and Type III curves indicate that the solute concentration is in excess in the surface. This is markedly the case for Type III curves where, at low concentrations, the slope of the curve is strongly negative.

One complication results from the application of Eq. 2.24 to the Type III curve. Following the rapid drop to the minimum value (b in Figure 2-17), the surface-tension rises, finally more or less flattening out. In the region in which the surface tension is rising the slope is positive, and Gibbs' equation requires that the solution should suddenly change from a state of strong positive adsorption to an equally strong negative one. However, experiments with solutions in this concentration range indicate that the adsorption remains positive.

It would appear that this curve represents a failure of the Gibbs' equation; the theoretically expected curve is indicated by the dashed line in Figure 2-17.* However, this is not the case. Fatty alcohol sulfates carefully

---

* The flattening at higher concentrations is not a consequence of the Gibbs' equation, but is not inconsistent with it. The cause of this effect is considered later.

prepared by Miles and Shedlovsky[34] gave surface tension-concentration curves without minima; addition of quite small amounts of fatty alcohol (the usual impurity) caused the minimum to return. Bulkeley and Bitner[35] found no minimum with sodium oleate solutions when care was taken to exclude carbon dioxide. It would appear that the minimum is the result of a complex interaction which occurs when more than one surface-active species is present. It has also been suggested that the minimum is due to the presence of a submerged electrical double layer[36] (cf. p. 110); but the data of J. V. Robinson,[37] on the creation of such minima with nonionic detergents, shows that this cannot be valid.

The high surface concentrations obtained with Type III compounds justify the designation "surface-active" so often applied to them.

Surface Tension as a Function of Time. Since surface tension is an orientation effect, it is to be expected that a freshly created surface will not have exactly the same properties as one which has been allowed to come to equilibrium. This is true even of pure liquids; it must be even more significant in solutions of surface-active materials where migration of solute molecules to the surface (in accordance with the Gibbs equation) must take a finite time. With pure liquids, of course, this time is very short; it can, however, be determined by use of the vibrating-jet technique and is found to be less than 0.003 seconds for most liquids.

On the other hand, Rayleigh[38] demonstrated that surfaces of the age 0.01 second obtained from 2.5 per cent solutions of sodium oleate exhibited a surface tension of the order of that of pure water, while the same solution when measured in a capillary rise apparatus gave a value of 25 dynes/cm.

This effect is also observed in interfacial tensions as well. Table 2-6 shows values of the interfacial tension existing between benzene and a 0.001 M aqueous solution of sodium oleate at 20° C, as a function of time.[39]

The effect of additives on the time of attainment of equilibrium has been studied by several workers.[34, 40]

Experimental Verification of Gibbs' Equation. Numerous attempts have been made to establish the validity of Eq. 2.24 experimentally. In order to do this one requires some definition of what constitutes the "surface" of the liquid. McBain[41] has assembled an impressive array of data

TABLE 2-6. INTERFACIAL TENSION BETWEEN BENZENE AND 0.001 M
SODIUM OLEATE AS A FUNCTION OF TIME[31]

Time (min)	0.0	0.5	1.0	2.0	3.0	5.0	7.5	10.0
γ_i	(35.0)*	14.1	12.9	11.9	11.0	10.7	10.7	10.7

* Estimated.

which implies that the effective depth of the liquid surface is moderately large, perhaps of the order of 1000 Å, and certainly of the order of several hundred Ångstroms. If this be the case, experimental verification of the Gibbs equation, although difficult, is certainly possible.

The first method devised[42] consists of passing a large number of liquid hydrocarbon drops or bubbles of gas up through the solution in such a way that the surface of the drops or bubbles become fully saturated with the adsorbed material. The drops are discharged at the top of the column in such a way that the adsorbed material is released into the solution. As a consequence, the solution becomes more concentrated at the top of the column than at the bottom. This difference in concentration, together with the estimated total area of the bubbles, allows one to calculate the amount adsorbed, i.e., Γ. Unfortunately, most of the data collected by this technique shows adsorptions several times greater than that required by Eq. 2.24; the technique, however ingenious, does not seem to be foolproof.

A more direct attack on the problem was the so-called "microtome method" used by McBain and coworkers.[43] An extremely precise and ingenious apparatus was devised whereby the surface of the solution was skimmed by a knife blade travelling at high speed (about 35 feet per second). The knife carried with it a small cylindrical vessel which collected the solution cut off the surface. The concentration of this surface layer was then compared with that of the bulk solution by means of an interferometer. In all the cases examined by this technique, including both positively and negatively adsorbed substances, the results obtained agreed with those predicted by the Gibbs' equation. Recent work by Cockbain[43a] on the adsorption of sodium dodecyl sulfate at the n-decane-water interface has only served to emphasize the validity of Gibbs' relation.

It should also be pointed out that Gibbs' equation permits the calculation of the area per molecule of adsorbed films of certain fatty acids. Such calculated areas agree well with those obtained by surface film measurements with the hydrophil balance (p. 16).[44]

Bulk Properties of Solutions of Surface-Active Compounds

In the previous section the effect of solutes, and, in particular, surface-active solutes on the surface properties of their solutions have been considered. In the present section, the effect of surface-active materials on the bulk properties of their solutions will be discussed.

The modern investigations into the properties of such solutions may well be considered to have begun with the publications of McBain and coworkers about 1911. It had long been known that dilute solutions of soaps behaved like normal solutions of weak electrolytes; however, anomolous behavior occurred at higher concentrations. For example, the conductivity of such solutions was found to increase to a point equivalent to that shown by strong electrolytes; whereas the colligative properties (i.e., osmotic pressure, freezing point lowering, vapor pressure lowering, etc.) fell far below those calculated from the ideal laws, even when allowance was made for ionization effects. In the following pages, the magnitude of these deviations will be indicated, and their explanation, in terms of McBain's concept of *colloidal electrolytes*, introduced.

Electrical Conductivity. An early, classic investigation by McBain, Laing, and Titley[45] on the electrical conductivity of the potassium salts of the fatty acids illustrates the behavior most aptly. In Figure 2-18 the equilivalent conductances of these salts is plotted as a function of volume normality, data being given for 90° and 18° C. As can be seen, the soaps below the laurate behave in a fairly normal way. For the C_{12} and higher soaps, however, the curve drops off rapidly in equivalent conductance at low concentrations, passes through a minimum, and then rises again. The similarity between this curve and the surface tension-concentration curve of Type III (Figure 2-14) is immediately apparent.

Because of the possibility of hydrolysis of the soaps at low concentrations, however, there is a certain element of ambiguity in such data. On

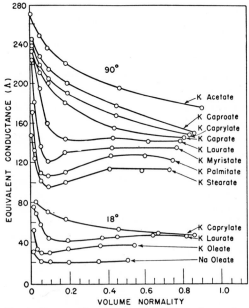

FIGURE 2-18. Equivalent conductances of the potassium salts of the fatty acids as a function of concentration at 90° and 18°C [45]

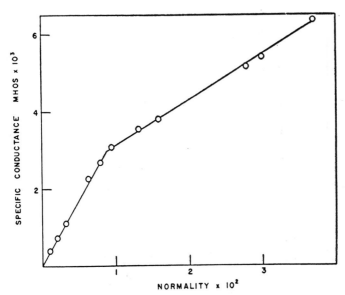

FIGURE 2-19. Specific conductance of aqueous solutions of lauryl sulfonic acid as a function of concentration.[46]

the other hand, nonideal behavior has also been observed in the conductance of surface-active electrolytes where hydrolysis is not a problem. This is shown in striking manner in Figure 2-19, where the specific conductance of lauryl sulfonic acid is plotted against the normality of its aqueous solutions.[46] The dotted line indicates the theoretical behavior expected. Similar results have been found with long-chain amine salts,[47] alkyl sulfuric esters,[48] alkane sulfonates,[46, 49] and many other types of agents.

A property related to conductivity is the ionic transport number. Discontinuities and abnormal behavior found with conductance is reflected in this property. Figure 2-20 gives the values of the cationic transport number of laurylamine hydrochloride as a function of concentration.[47]

Colligative Properties. As has been pointed out, the conductivity values obtained for the solutions of surface-active agents show a deviation towards *increased* conductivity. On the other hand, the colligative, or osmotic, properties show a decided *decrease*.

The various colligative properties may be described in terms of the *osmotic coefficient g*. For example, it is defined as the ratio of the observed lowering of the freezing point of a solution to that which would be expected for an ideal fully dissociated electrolyte forming two ions:

$$g = i/2 = 1 - j = \theta/(2 \times 1.858m), \tag{2.25}$$

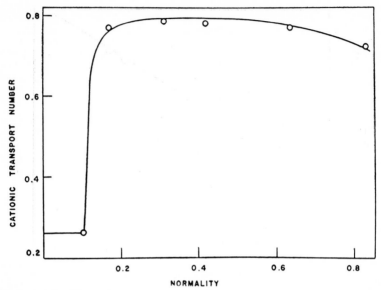

FIGURE 2-20. The cationic transport number of laurylamine hydrochloride as a function of concentration.[47]

where g is the osmotic coefficient, i is van't Hoff's coefficient, j is the Lewis-Randall function, m is the molality, 1.858 is the molal freezing point lowering for an ideal solute in water, and θ is the observed lowering of freezing point.

Thus, Figure 2-21 shows the osmotic coefficient of sodium tetradecyl sulfate as a function of the square root of the concentration.[50] The straight line at low concentrations is what is predicted for ideal behavior; the rapid drop at values of the square root of the molality higher than about 0.15 is characteristic of surface-active compounds. The sulfosuccinate esters give similar results[50] when studied by freezing point methods.

By the nature of the colligative properties of solutions, similar effects should be observed in, for example, the dependence of the osmotic pressure on concentration. The measurement of the osmotic pressure of solutions of surface-active compounds presents considerable experimental complications, however, because of the difficulty of finding a suitable semi-permeable membrane. Hess and Suranyi[51] have surmounted these difficulties to a considerable extent. Figure 2-22 shows their results for sodium dodecyl sulfate solutions.

McBain and Salmon[52] have also investigated the dew point lowering, of solutions of a series of sodium and potassium salts of the fatty acids, with similar departures from ideality being found.

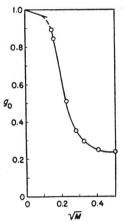

FIGURE 2-21. The osmotic coefficient g of sodium tetradecyl sulfate as a function of the square root of the concentration.[50] The straight line at low concentrations corresponds to ideal behavior.

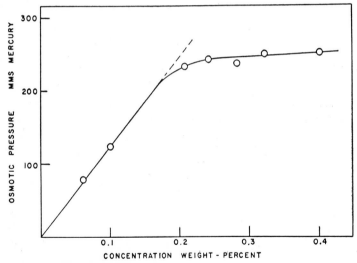

FIGURE 2-22. The osmotic pressure of aqueous solutions of sodium dodecyl sulfate as a function of concentration.[51]

Solubility. The solubility of surface-active compounds as a function of temperature has been the subject of a large number of investigations.[53, 54] In these, it is found that low solubilities occur until a particular temperature, characteristic for each compound, is reached, whereupon there is a sudden large increase in solubility. Figure 2-23 illustrates this, using the data of Tartar and Wright[54] on the solubility of the higher alkyl sulfonates.

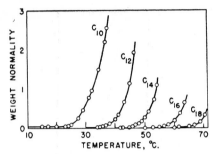

FIGURE 2-23. The solubility of the higher alkyl sulfonates as a function of temperature.[54]

Relation of Bulk to Surface Properties

Enough has been said in the previous section to indicate that the properties of solutions of surface-active compounds are unusual. There are certain other properties of these solutions, i.e., optical and solubilization phenomena, which could also have been adduced for this purpose. It will be more satisfactory, from a heuristic point of view, to introduce these properties in connection with the explanation of the general relation between surface and bulk properties.

Colloidal Electrolytes; Micelle Theory. As McBain[55] has pointed out, theoretical colloid chemistry at the turn of the century consisted largely of a rather naive formalism, and the properties, per se, of colloids were rather nebulously described. It was in this atmosphere that McBain, principally, introduced the then revolutionary concept of *colloidal electrolytes* in an effort to explain some of the anomalous data described in the previous section.

Initially, to state the hypothesis in a simple form, McBain assumed that these phenomena occur as a result of the spontaneous formation of particles of colloidal dimensions from, e.g., ions; that the colloidal particles are in true reversible equilibrium with the ions; and, therefore, that colloidal electrolytes are truly stable in the strictest thermodynamic sense.*

The stable colloidal particles have a self-organizing structure in that the polar groups are exposed to the water, while the hydrophobic groups are in contact. This minimizes the interfacial energy, and leaves the maximum number of water molecules in mutual contact. Such a structure has been termed a *micelle*. In a qualitative sense, it is clear that the micelle will

* I find I must quote McBain in this context.[55]

"So novel was this finding that when in 1925 some of the evidence for it was presented to the Colloid Committee for the Advancement of Science in London, it was dismissed by the Chairman, a leading international authority, with the words, 'Nonsense, McBain.' "

account for the data presented in the previous section. That is, the association, and hence reduction in the total number, of particles (e.g., ions or undissociated molecules) will account for the nonideal behavior with respect to colligative and similar properties. On the other hand, the micelles will be more conductive than in the unassociated form, since their mobility as calculated from Stokes' law would increase by a factor of $m^{2/3}$ for a micelle containing m ions.

Quantitatively, however, calculations of the extent of aggregation based on colligative properties and on conductance do not agree particularly well. This led McBain to postulate the existence of micelles consisting of undissociated molecules. It is possibly not necessary to make this assumption, but micelle formation with nonionic amphiphilic compounds is known to occur, and this, of course, would lead to the existence of uncharged colloidal aggregates. Thus the existence of such micelles cannot properly be ruled out, even with ionic materials.[56]

Structure of Micelles. Up to this point, nothing has been said about the structure of these colloidal electrolyte aggregates, or micelles. It is precisely this micelle structure which is most open to discussion, although a great deal of theoretical and experimental work has been performed in an attempt to explain it. The first point, which has tacitly been made above, is that all departures from nonideality found in solutions of surface-active materials are owing to the formation of micelles. Since these solutions are ideal in behavior at low concentrations, and since the change in behavior is found to be rather abrupt, it is obvious that the formation of micelles must occur at some specific concentration. As a natural consequence of this, all of the departures of the various properties from ideality would occur at the same concentration for a given substance. That this is indeed the case is shown strikingly in the data of Preston[57] for sodium dodecyl sulfate, as shown in Figure 2-24. It should be pointed out, however, that the agreement is not always this good, especially between the results of different workers. This can reasonably be ascribed to differences in the purity of materials, which apparently exert a significant effect, and to slight differences in technique.

Although the results portrayed in Figure 2-24 are at first sight somewhat startling, Jones and Bury[58] have pointed out that they are a natural result of the

> "laws of mass action, when applied to the association of simple molecules to form complex molecules containing a *large* number of single molecules"

The operative word in the foregoing quotation is, of course, *large*.

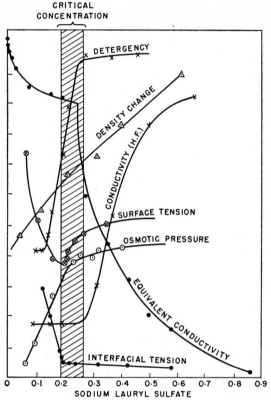

FIGURE 2-24. The data of Preston[57] showing the functional dependance of the colligative properties on the concentration of sodium dodecyl sulfate. It will be noted that the various properties undergo a more-or-less discontinuous change within a narrow range of concentrations.

The following thermodynamic argument illustrates this. In a solution of surface-active molecules in which aggregates have begun to form there exists an equilibrium between associated and unassociated form

$$(A)_m \leftrightharpoons mA$$

where A represents the single ionic or molecular species, and m the number of such ions or molecules associated into a micelle. Now (using concentrations rather than activities), if c represents the total solute concentration, c_i that portion of the solute in the form of individual molecules, and c_m the portion united into micelles (expressed in equivalent concentrations), then the equilibrium constant is given by

$$K = (c_i)^m / c_m \tag{2.26}$$

The concentration of molecules in micellar form is thus given by

$$c_m = (c_i)^m / K \qquad (2.27)$$

It was pointed out by Grindley and Bury[59] that if c_i was much smaller than $K^{1/m}$ then the right-hand side of Eq. 2.27 will, for large m, be very small, i.e., the amount of solute in the form of micelles will be negligible. It is only when c_i becomes at all comparable with $K^{1/m}$ that the micellar concentration becomes significant; and when c_i exceeds this value, c_m increases rapidly with c.

This is actually the behavior found experimentally; the assumption that m must be large is thus justified. Davies and Bury[60] have designated the concentration at which the micellar concentration becomes appreciable as the "critical concentration for micelles." This is perhaps more conveniently (although less exactly) termed the *critical micelle concentration*, and abbreviated C.M.C.* This is, of course the concentration at which the sudden changes in the properties occur, e.g., in Figure 2-24.

Davies and Bury also point out that the larger the value of m the more abrupt will be the change in slope of the plot of the property under consideration as a function of the concentration. For infinite m the break will become a true discontinuity, such as that observed for a complete change of phase of the solute, i.e., saturation of the solution.

Insertion of reasonable values into Eq. 2.27 leads to the conclusion that m must lie between 50 and 100. Evidence will be adduced defining m a little more exactly; evidence will also be cited indicating much smaller values.

Such a micelle, containing 50–100 ions, can be regarded as a single large polyvalent ion. By modern electrolyte theory this polyvalent ion can be considered to be surrounded by a cloud of *gegen-ions*, i.e., ions of opposite charge to those of the micelle. The complex consisting of the micelle plus the attendant gegen-ions is more properly to be considered as constituting the thermodynamic micelle. In this case, certain modifications in Eq. 2.27 are required.[61] If p is the number of gegen-ions associated with an m-ion micelle, the C.M.C. is given by

$$c_m = \frac{(c_i p / m)^p (c_i)^m}{K}, \qquad (2.28)$$

---

* There is a material widely used in detergents and, to a smaller extent, in emulsion practise known as carboxy methyl cellulose. This is usually abbreviated CMC, leading to confusion with the above cited abbreviation for critical micelle concentration. This confusion is further confounded by the practise of some authors of omitting the periods in the latter abbreviation.

thus introducing the concentration of gegen-ions. The change in properties at the critical micelle concentration can be shown to be more abrupt for any given value of m if this relation is used, rather than Eq. 2.27.

These considerations have led us to a theoretically reasonable value of the size of a micelle; in order to draw any conclusions as to the micellar shape additional experimental data must be quoted.

Solubilization. In a sense it is almost inappropriate to introduce the phenomenon of solubilization in such an off-hand manner, as it were, as an appendage to a discussion of micelle structure. This is a large topic in its own right, and M. E. L. McBain and E. Hutchinson[62] and P. A. Winsor[62] have devoted monographs to it. The latter, indeed, chooses to regard emulsions as a special intermediate stage in an intermicellar equilibrium.

On the other hand, from the purely heuristic point of view, something of a case can be made for a concept entirely opposed to that of Winsor, i.e., looking on solubilization as representing a rather special case of emulsion formation. Solubilization is defined by McBain[63] as

"a particular mode of bringing into solution substances that are otherwise insoluble in a given medium."

If one is content to consider the word "solution" in the foregoing quotation rather loosely, this could be considered as defining an emulsion (although McBain himself specifically differentiates between the two). On the other hand, certain systems which have been called "transparent emulsions" in the literature are almost certainly examples of solubilization.

At least as long ago as 1846 it was recognized that soap solutions have the power of increasing the solubility of certain substances; an extensive listing of the early literature on this phenomenon is given by M. E. L. McBain and Hutchinson.[64]

A good example of the magnitude of the effect is given in Figure 2-25, showing the effect on the solubility of benzene, ethylbenzene, butylbenzene, and naphthalene of various concentrations of potassium laurate.[65] It should be emphasized that this is true solubility, in the sense that a homogeneous system is formed.

It was early recognized that the phenomenon of solubilization is connected in some way with the existence of micelles; indeed, a useful method of determining the C.M.C. depends on this phenomenon. Shepard and Geddes[66] found that the ultraviolet absorption spectrum of a cationic dye was affected by the presence of cationic detergents. Consideration of this led Corrin[67] to the conclusion that titration of a cationic dye with an *anionic* detergent would lead to a color change at the critical micelle concentration. The color change is caused by solubilization, and apparently solubilization cannot occur until micelles are formed. Table 2-7[68] gives values of the

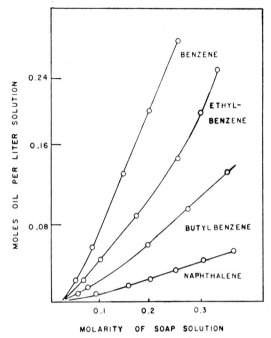

FIGURE 2-25. Solubilization of benzene, ethylbenzene, butylbenzene, and naphthalene by potassium laurate.[65]

TABLE 2-7. TITRIMETRIC DETERMINATION OF C.M.C. WITH VARIOUS DYES[68]

Dye	Potassium Laurate	Sodium Dodecyl Sulfate	Sodium Decyl Sulfonate	Decyltrimethyl Ammonium Bromide
Pinacyanol Chloride	0.0235	0.00602	0.0400	
Rhodamine 6G	.0234	.00612	.0387	
Sky Blue FF				.0.0643
Eosin				.0635
Fluorescein				.0610
2,6-Dichlorphenolindophenol				.0602

C.M.C. of various surface-active compounds obtained by the dye titration method. The generally good agreement between different dyes argues well for the validity of the method and for the validity of the assumed relation between solubilization and micelle formation.

The phenomena of solubilization becomes clear when reference is made to the statement given earlier (p. 34) concerning the self-organizing character of the micelles. Since the hydrophobic portions of the various molecules constituting the micelle are in contact, it is reasonable to assume that the micelle solubilizes the hydrophobic solute by the simple process of "dissolving" it in the hydrophobic portion of the micelle. This process will, of course, have a significant effect on the shape and size of the micelle, and will, in turn, give a valuable clue as to the original form of the micelle itself.

X-Ray Effects Connected With Micelles. When a narrow beam of essentially monochromatic x-rays is passed through a layer of about 1 mm of a soap solution, it is found that diffraction takes place above certain concentrations. Below these concentrations the solution appears isotropic.[69] Since diffraction is evidence of an ordered structure, the diffraction is

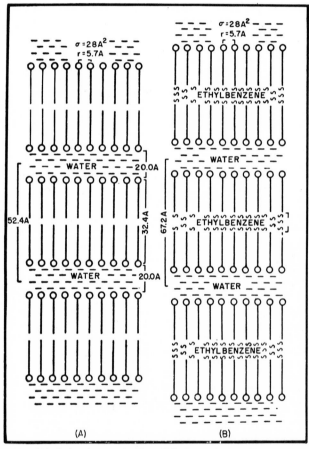

FIGURE 2-26. (A) A laminar micelle, with interplanar distances as calculated from X-ray data. (B) The effect of solubilization of ethylbenzene on the dimensions of the micelle as calculated from X-ray data. The solubilized molecules are "dissolved" in the hydrophobic interior of the micelle.[71]

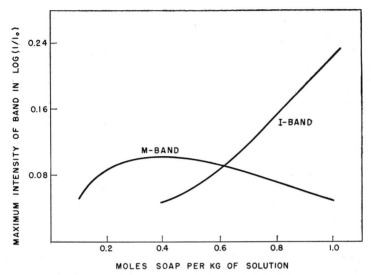

FIGURE 2-27. X-ray density distributions obtained by Mattoon, Stearns, and Harkins[70] on solutions of potassium laurate. The I-band spacings are believed to correspond to the intermicellar distance, while the M-band corresponds to the head-to-head distance in the micelle (cf. Figure 2-26).

considered to be due to the formation of micelles. The observed diffraction patterns, interpreted by the Bragg diffraction law, led to the conclusion that the micelle was a laminar structure, very much as illustrated in Figure 2-26A. The soap molecules are pictured as organized in double layers (tail-to-tail) with the hydrophilic heads facing out into the aqueous phase. This picture was principally advanced by McBain, who had hypothesized such a structure before x-ray data were available.

The data obtained by x-ray measurement is illustrated in Figure 2-27, showing the results obtained by Mattoon, Stearns, and Harkins[70] for solutions of potassium laurate. Two distinct *bands,** designated the I- and M-bands, are observed. The spacings observed are believed to correspond to the intermicellar distances (I-band) and the head-to-head distance in the micelle (M-band). It will be noted that these bands show their maximum intensities at different concentrations, which leads to the existence of *two* critical micelle concentrations from x-ray data.

The effect of solubilizate on the x-ray diffraction patterns was studied by Harkins, Mattoon, and Corrin,[71] using ethylbenzene as a typical material, and the change in x-ray spacings was consistent with the picture

---

* In crystalline materials, x-ray diffraction yields sharply defined lines. In partially-ordered materials, however, e.g., high-polymeric substances, "crystalline" solutions, etc., diffuse bands are observed; this is also the case in micellar solutions.

indicated in Figure 2-26B, i.e., with the previously suggested "solution" of the hydrophobic solute in the interior of the micelle.

Unfortunately, satisfactory as is this interpretation of the x-ray data in terms of a micelle structure, the laminar micelle leaves much to be desired. One of its difficulties is that there is no simple way for the micelle to end. In principle, there is no reason why the lamellae should not be infinite; indeed, this is thermodynamically necessary. This last statement follows from the fact that if the lamellae are *not* infinite there will still exist a definite interface between the hydrophobic portion of the surface-active compound and the aqueous phase at the edge of the micelle. Of course, the larger the micelle the smaller the total hydrophobic interfacial area for the whole system becomes, and hence the free energy is minimized. Actually, of course, free energy becomes inappreciably small for a very large micelle; but this micelle would have to be much larger than seems possible, according to present experiments.

This argument requires a considerably different shape for the micelle. Ignoring for the moment the requirements of the x-ray data, it is clear that a spherical grouping would satisfy most of the geometric and thermodynamic requirements for micelle structure. Such a micelle has been proposed by Hartley[25] and is illustrated in Figure 2-28. It will be noted that no high degree of internal order in the micelle is hypothesized. On the other hand, attempts have been made to preserve the lamellar structure by combining it with the minimal interface of the Hartley micelle; this is done by the assumption of something approaching a disc-shaped structure.

An extension of this concept has been carried through by Debye,[72] using a thermodynamic argument. In this theory, one starts by assuming the micelles to be disc-shaped platelets. As each molecule adds to the micelle, a certain amount of hydrocarbon-chain adhesional energy is liberated, the

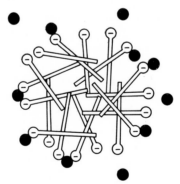

FIGURE 2-28. A spherical micelle, according to Hartley.[25] This sort of micelle is much less "organized" than the laminar micelle.

total amount being proportional to the size of the micelle. Thus, the work of assembling the micelle is

$$W = N^{3/2} w_e - N w_m ,$$

where N is the number of surface-active molecules in the micelle, w_m is the work of hydrocarbon-chain adhesion per ion, and w_e is a constant which relates to the work done against the electrostatic repulsion of the ionic molecular heads. When W is plotted as a function of N, a curve is obtained with a minimum, and it is considered that this value of N represents a stable micellar quantity. It is interesting to point out that Debye had been led to assume a sausage-shaped micelle for laurylamine hydrochloride on the basis of light-scattering data.[73]

Recently, Reich[74] has pointed out that Debye's thermodynamic analysis contains some errors. A more refined analysis leads to the conclusion that a Hartley micelle of reasonable size is to be preferred from considerations of entropy.

This can be rendered consistent with the x-ray data if one interprets the observed bands as representing scattering data rather than Bragg diffraction.[75] If this is done, results not inconsistent with sphericity are obtained. In this connection, the observations on the x-ray diffraction of spherical emulsion particles by Schulman and coworkers[76] may be cited. In systems of this sort, diffraction (or better, scattering) patterns similar to those observed in solutions of colloidal electrolytes are obtained. Also, if Winsor's[26] concept of an intermicellar equilibrium is considered as bridging the gap between emulsions and solubilization, spherical micelles become almost a necessity.

For the purposes of the present work the precise shape and size of the micelle is of small significance. The fact that micelles exist and are of some reasonable size is what is important. The precise nature of the micelle is still very much an open question, and it is, of course, possible (indeed, probable) that no single simple structure will be found to satisfy all the phenomena associated with micelle formation. In this connection, the recent observation of micelle formation in nonaqueous systems by Kaufman and Singleterry[77] may be of signal importance.

Bibliography

1. Partington, J. R., "Advanced Treatise on Physical Chemistry," **2**, p. 134, London, Longmans Green & Co., 1951.
2. Stefan, J., *Ann. Physik* **29**, 655 (1886).
3. Partington, J. R., *op. cit.*, pp. 148–157.
4. Adam, N. K., "The Physics and Chemistry of Surfaces," 3rd Ed., p. 14, London, Oxford University Press, 1941.
5. Partington, J. R., *op. cit.*, pp. 138–139.

6. Eötvös, R., *Ann. Physik* **27**, 448 (1886).
7. Ramsey, R. and Shields, J., *Trans. Roy. Soc. (London)* **A184**, 647 (1893).
8. Mayer, J. E. and Mayer, M. G., "Statistical Mechanics," pp. 309–314, New York, John Wiley & Sons, Inc., 1940.
9. Dupré, A., "Theorie Mechanique de la Chaleur," p. 393 (1869).
10. Antonoff, G., *J. Chim. Phys.* **5**, 372 (1907).
11. Adam, N. K., *op. cit.*, p. 214.
12. Partington, J. R., *op. cit.*, p. 170 (esp. footnote 7).
13. Partington, J. R., *op. cit.*, p. 163.
14. Young, T., *Proc. Roy. Soc. (London)* **95**, 65 (1805).
15. Harkins, W. D., "The Physical Chemistry of Surface Films," p. 280, New York, Reinhold Publishing Corp., 1952.
16. Adam, N. K., *op. cit.*, p. 179.
17. Harkins, W. D., *op. cit.*, p. 282.
18. Harkins, W. D., *op. cit.*, p. 23.
19. McBain, J. W., "Colloid Science," p. 59, Boston, D. C. Heath & Co., 1950.
20. Pockels, A., *Nature* **43**, 437 (1891).
21. Rayleigh, L., *Phil. Mag.* **48**, 331 (1899).
22. Langmuir, I., *J. Am. Chem. Soc.*, **39**, 1848 (1917).
22a. Adam, N. K., *op. cit.*, pp. 17–33.
22b. Harkins, W. D., *op. cit.*, pp. 121–128.
23. Harkins, W. D., *op. cit.*, pp. 106–113.
24. Adam, N. K., *op. cit.*, pp. 50–51.
25. Hartley, G. S., "Paraffin-Chain Salts," p. 45, Paris, Hermann et Cie., 1936.
26. Winsor, P. A., "Solvent Properties of Amphiphilic Compounds," p. 2, London, Butterworth Scientific Publications, 1954.
27. Hardy, W. B., *Proc. Roy. Soc. (London)* **A86**, 634 (1911); **A88**, 303 (1913).
28. Traube, I., *Ann.* **265**, 27 (1891); Rehbinder, P. A., *Z. physik. Chem.*, **111**, 447 (1924).
29. McBain, J. W., Ford, T. F., and Wilson, D. A., *Kolloid-Z.* **78**, 1 (1937).
30. Fischer, E. K. and Gans, D. M., *Ann. N. Y. Acad. Sci.* **49**, 371 (1946).
31. Harkins, W. D. and Zollman, H., *J. Am. Chem. Soc.* **48**, 69 (1926); Harkins, W. D., *op. cit.*, p. 90.
32. Gibbs, J. W., *Trans. Conn. Acad. Sci.* **3**, 391 (1876) "Collected Works," **1**, p. 230, New Haven, Yale, 1928.
33. A condensed proof is given by Adam, N. K., *op. cit.*, pp. 107–113; cf. also Moilliet, J. L. and Collie, B., "Surface Activity," pp. 54–62, New York, D. van Nostrand Co., Inc., 1951; Glasstone, S., "Thermodynamics for Chemists," pp. 241–245, New York, D. van Nostrand Co., Inc., 1947.
34. Miles, G. D. and Shedlovsky, L., *J. Phys. Chem.* **48**, 57 (1944).
35. Bulkeley, R. and Bitner, F. G., *Bur. Standards J. Research* **5**, 951 (1930).
36. Wohl, K., *Annals N. Y. Acad. Sci.* **46**, Art. 6, 204 (1946).
37. Robinson, J. V., Abstracts 124th A.C.S. Meeting (Sept. 1953).
38. Rayleigh, L., *Nature* **41**, 566 (1890).
39. International Critical Tables, Vol. IV. The data are there referred to a paper by Harkins and McLaughlin, but this reference is incorrect, and I have not been able to find the original source.
40. Burcik, E. J. and Newman, R. C., "A Fundamental Study of Foams and Emulsions," OTS Report PB 111420, Dept. of Commerce, Washington, D.C.
41. McBain, J. W., *op. cit.*, pp. 68–79.
42. Adam, N. K., *op. cit.*, pp. 113–115.

43. McBain, J. W. and Humphreys, C., *J. Phys. Chem.* **36**, 300 (1932); McBain, J. W. and Swain, R. L., *Proc. Roy. Soc. (London)* **A154**, 608 (1936); McBain, J. W. and Wood, L. A., *ibid.* **A174**, 286 (1940).

43a. Cockbain, E. G., *Trans. Faraday Soc.* **50**, 874 (1954).

44. Adam, N. K., *op. cit.*, pp. 115–121.

45. McBain, J. W., Laing, M. E. and Titley, A. F., *J. Chem. Soc.* **115**, 1279 (1919).

46. McBain, M. E. L., Dye, W. B., and Johnston, S. A., *J. Am. Chem. Soc.* **61**, 3210 (1939).

47. Ralston, A. W. and Hoerr, C. W., *J. Am. Chem. Soc.* **64**, 97, 772 (1942); **65**, 976 (1943).

48. Lottermoser, A. and Puschel, F., *Kolloid-Z.* **63**, 175 (1933).

49. Wright, K. A., Abott, A. D., Sivertz, V., and Tartar, H. V., *J. Am. Chem. Soc.* **61**, 549, 552 (1939).

50. McBain, J. W. and Bolduan, O. E. A., *J. Phys. Chem.* **47**, 94 (1943).

51. Hess, K. and Suranyi, L. A., *Z. physik. Chem.* **A184**, 321 (1939).

52. McBain, J. W. and Salmon, C. S., *J. Am. Chem. Soc.* **43**, 426 (1920).

53. Adam, N. K. and Pankhurst, K. G. A., *Trans. Faraday Soc.* **42**, 523 (1946).

54. Tartar, H. V. and Wright, K. A., *J. Am. Chem. Soc.* **61**, 539 (1939).

55. McBain, J. W. in Alexander, J. (Ed.), "Colloid Chemistry," **5**, pp. 102–103, New York, Reinhold Publishing Corp., 1944.

56. See Moilliet, J. L. and Collie, B., *op. cit.*, pp. 21–24, for a well-argued contrary view.

57. Preston, W. C., *J. Phys. and Colloid Chem.* **52**, 84 (1948).

58. Jones, F. E. and Bury, C. R., *Phil. Mag.* **4**, 841 (1927).

59. Grindley, J. and Bury, C. R., *J. Chem. Soc.* **1929**, 679.

60. Davies, D. G. and Bury, C. R., *J. Chem. Soc.* **1930**, 2263.

61. Moilliet, J. L. and Collie, B., *op. cit.*, p. 15.

62. McBain, M. E. L. and Hutchinson, E., "Solubilization," New York, Academic Press Inc., 1955; Winsor, P. A., *op. cit.*

63. McBain, M. E. L. and Hutchinson, E., *op. cit.*, p. xi.

64. McBain, M. E. L. and Hutchinson, E., *op. cit.*, pp. 1–13.

65. Stearns, R. S., Oppenheimer, H., Simon, E., and Harkins, W. D., *J. Chem. Phys.* **15**, 496 (1947).

66. Shepard, S. E. and Geddes, A. L., *J. Chem. Phys.* **13**, 63 (1945).

67. Corrin, M. L., Klevens, H. B., and Harkins, W. D., *J. Chem. Phys.* **14**, 480 (1946).

68. Harkins, W. D., *op. cit.*, p. 302.

69. Hess, K., *Fette u. Seifen* **49**, 81 (1942); cf. Harkins, W. D., *op. cit.*, pp. 329–330, for additional references.

70. Mattoon, R. W., Stearns, R. S., and Harkins, W. D., *J. Chem. Phys.* **15**, 209 (1947); **16**, 644 (1948).

71. Harkins, W. D., Mattoon, R. W., and Corrin, M. L., *J. Am. Chem. Soc.* **68**, 220 (1946); *J. Colloid Sci.* **1**, 105 (1946).

72. Debye, P., *J. Phys. Chem.* **53**, 1 (1949).

73. Debye, P. and Anacker, E. W., *J. Phys. and Colloid Chem.* **51**, 18 (1947).

74. Reich, I., *J. Phys. Chem.* **60**, 257 (1956).

75. Corrin, M. L., *J. Chem. Phys.* **10**, 844 (1948).

76. Schulman, J. H. and Riley, D. P., *J. Colloid Sci.* **3**, 383 (1948); Schulman, J. H. and Friend, J. A., *ibid.* **4**, 497 (1949).

77. Kaufman, S. and Singleterry, C. R., *J. Colloid. Sci.* **10**, 139 (1955).

CHAPTER 3

Physical Properties of Emulsions

Practical considerations of organization and length dictate separate discussions of the physical properties (gross and semi-microscopic) of emulsions and of the problem of emulsion stability. In reality, however, they cannot be considered separately. Emulsions are essentially unstable heterogeneous systems; they are partly dispersions, partly colloids. The properties of an emulsion often depend largely on its composition and on its mode of preparation. These physical properties are also the very considerations which govern the stability of the system.

Thus, in the present chapter, certain of the various physical properties of emulsions will be discussed. Additional detail, as required by the nature of the treatment, will be found in the next chapter, as part of the review of emulsion theory.

Particle Size and Size Distribution

In the emulsion definition given earlier (p. 2) it was stated that the internal phase of an emulsion was dispersed in droplets of a diameter greater than 0.1 μ. In fact, very few emulsion droplets are smaller than 0.25 μ in diameter, and the largest found are about a hundred times greater. Table 3-1, reproduced from Sutheim,[1] indicates the range of emulsion droplet diameter sizes in comparison with other naturally occurring and synthetic materials.

The fact that a range of particle size is quoted implies that even in a single emulsion the droplet diameters may be far from uniform. That this is the case is illustrated in Figure 3-1, showing the distribution of particle sizes encountered in a thoroughly homogenized asphalt emulsion.[2] More diffuse distributions are illustrated later, e.g., in Table 7-1, p. 211, where the effect of the mode of preparation on this property is discussed.

As will be shown later, droplet-size distribution with a maximum of low diameter droplets, and (of lesser importance in this connection) with this maximum sharply defined, apparently represents a situation of maximum stability, all other things being equal. Thus, changes in the droplet-size distribution curve with time, leading to a more diffuse distribution and with the maximum at higher diameters, is a measure of the instability of an emulsion.

TABLE 3-1.
RELATIVE PARTICLE SIZES[1]

Range of Dimensions	Visibility	Description of State	Examples in Nature	μ	Examples in Emulsions and Paints	μ
10⁻¹ cm. = 1 mm.	Plain visibility with the naked eye		Frog's egg	1000		
10⁻² cm.	Limit of visibility with the naked eye (approx. 50μ)	Coarse dispersions	Fine sand	500		
			Ameba	100	Mesh of No. 325 screen	44
			Potato starch	45-110	Coarse pigments	30
					Coarse emulsions	5-25
10⁻³ cm.	Plain visibility with the microscope	Fine dispersions	Corn starch	15-20	Mesh of No. 1,250 screen	10
			Red blood corpuscle	7.5-8.5	Butterfat particles in milk	5-10
			Rice starch	3-7		
10⁻⁴ cm. = 1μ	Limit of microscopic resolving power (approx. 0.25 μ)	EMULSIONS	Average bacteria (typhoid)	1	Finest earth colors	2-3
					Butterfat particles in homogenized milk	1-2
					Oil droplets in good emulsion vehicle	0.5-1.5
			Wavelength of visible light	0.4-0.65	Precipitated pigments	0.3-0.8
			Colloid gold particles	0.2	Finest emulsion droplets	0.25
			Smallest bacteria	0.1	Lampblack	0.1
10⁻⁵ cm.		Colloid dispersions		mμ		mμ
10⁻⁶ cm.	Limit of ultramicroscopic resolving power (5-10 mμ)		Filterable virus (probable)	30	Colloidal carbon black	10-30
	Field of the electron microscope	Molecular range	Largest molecules e.g., protein (probable)	6		
10⁻⁷ cm. = 1 mμ			Palmitic acid (length of chain)	2.4		
10⁻⁸ cm. = 1 Å	Below any visibility	Atomic range	Methyl group	0.14		
			Hydrogen (H_2)	0.1		

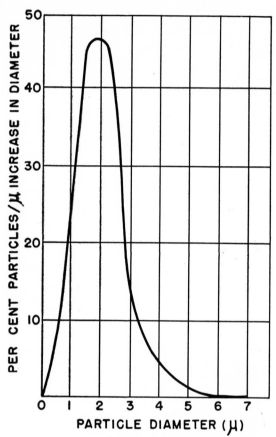

FIGURE 3-1. The particle-size distribution in a well-homogenized asphalt emulsion.[2]

The manner in which these distributions change with time is illustrated in Figure 3-2, derived from the data of Fischer and Harkins.[3] In this study, octane emulsions were stabilized with 0.005 M sodium oleate and with 0.005 M cesium oleate. As can be seen, within seven days a radical change occurred in the particle-size distribution; the two emulsifying agents give equivalent results. Similar emulsions, stabilized with soaps at the concentration of 0.1 M, remained unchanged after four years in sealed glass tubes. The reasons for this difference in stability as a function of emulsifier concentration will be considered more fully below (cf. p. 89), but it may briefly be noted that it is related to the fraction of the interfacial area which is considered to be covered by the stabilizing agent.

Opposite effects are also observed i.e., a diffuse distribution in an emulsion may change to a sharp distribution. Thus, Berkman and Egloff[4] cite the data of Wright to the effect that oil-field emulsions, upon ageing, become coarser and more uniform. Similar observations were made by King and Mukherjee[5] on emulsions stabilized with high molecular weight colloids, e.g., gelatin.

Phenomena of this sort are observed during the actual preparation of emulsions, and have been termed "limited coalescence" by Wiley.[6] It had been suggested earlier, e.g., by Lewis,[7] that there was a definite critical value for particle size in emulsions; but this certainly is not always the case. It seems, however, in many instances, to be true for emulsions stabilized by finely-divided solids.[6]

Although the lower limit for emulsion droplet radius has been defined as 0.1 μ, Bowcott and Schulman[8] have found that in transparent emulsions (either O/W or W/O) stabilized by soap or long-chain fatty alcohols the droplet diameters are of the order of 100 to 500 Å. Evidence for these

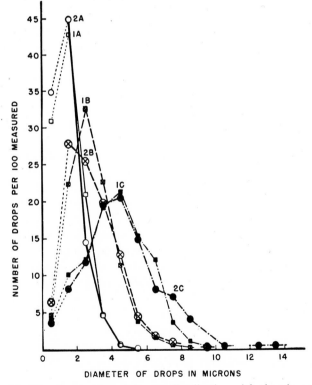

FIGURE 3-2. The change of particle-size distribution with time in octane emulsions stabilized with sodium and cesium oleate at concentrations of 0.005M. A radical change in distribution is observed within seven days.[3]

dimensions is provided by phase diagrams that show the composition of the continuous phase and the interphase. Sedimentation rate measurements on one of the systems studied are in agreement with the quoted range of droplet diameters.

Particle size has been related to the mode of preparation of the emulsion by Jürgen-Lohmann[9] and Kiyama, Kinoshita and Suzuki;[10] these results will be discussed in detail in Chapter 5.

Isakovich[11] has considered the problem of the propagation of sound in emulsions as related to the droplet dimensions. In an emulsion of one liquid in another, "thermal" dispersion of the sound velocity, i.e., with Newtonian isothermal compressions and rarefactions, can exist on a microscopic scale between the components of the emulsion. Macroscopically, however, the phenomenon is still Laplacian, that is, with adiabatic compressions and rarefactions. The range of this "Laplacian-Newtonian" velocity of sound is at lower frequencies. Transition to a "Laplacian-Laplacian" velocity, in which the compressions and rarefactions are Laplacian on both the microscopic and macroscopic scales, occurs at a critical frequency at which the length of the temperature wave is of the same order of magnitude as the particle dimensions of the emulsion.

Damping of sound in emulsions can be considerable even at moderate frequencies. For example, in a 10 per cent emulsion of benzene in water, with droplet diameters of the order of 5 μ, the damping coefficient in 1.5 $\times$ 10^6 hertz is approximately 1.5 $\times$ 10^{-2}, or about 100 times as great as in pure benzene.

Jellinek[12] has undertaken a systematic treatment of the distribution functions of emulsions and their average quantities. In this analysis, Jellinek has considered the distribution not only of particle diameters, but also of the surface areas and volumes of the droplets. The quantities $S_{n,l}$, $S_{n,a}$, and $S_{n,v}$ represent the total number of particles of a given diameter (l), surface area (a), or volume (v), expressed as integral distribution functions. Similar expressions are derived for total surface areas ($S_{A,l}$, $S_{A,a}$, and $S_{A,v}$) and the total volumes ($S_{V,l}$, $S_{V,a}$, and $S_{V,v}$). From these, derivative functions may be obtained for the rate of change of number, area, and volume with changing droplet l, a, or v, as well as the average number, area, and volume as a function of these parameters. All these functions are interrelated, and the corresponding weight distribution can be obtained by multiplying the volume distribution by the droplet density.

From these basic equations, relations are derived for the specific surface area (i.e., surface area s of one gram of disperse phase), the number of droplets (N_g) per one gram of disperse phase, the inhomogeneity (u) of the emulsion, etc. Thus, if l_m is the diameter of the maximum number of drop-

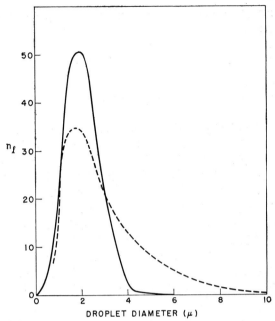

FIGURE 3-3. The droplet-size distribution in an emulsion as calculated from Eq. 3.1, with $l_m = 2\mu$ and $N = 50,000$ (solid curve). The dashed line is the experimentally derived curve for a mineral oil-in-water emulsion.[13]

lets (i.e., the diameter at the maximum of the distribution curve), the rate of change of the number of droplets with diameter l is

$$n_l = (2N/l_m{}^3)e^{-(2/3)(l^3/l_m{}^3)}l^2 \qquad (3.1)$$

The volume distribution with respect to diameter is similarly given by

$$V_l = (N/3l_m{}^3\pi)e^{-(2/3)(l^3/l_m{}^3)}l^5 \qquad (3.2)$$

The quantity termed by Jellinek the inhomogeneity $u = 0.14l_m{}^2$; the specific surface $S = 4.76/\rho l_m$; and the number of droplets per gram $N_g = 0.41/\rho l_m{}^3$. In Eqs. 3.1 and 3.2, N is the total number of droplets in the emulsion. Figure 3-3 shows the distribution of Eq. 3.1 for the case $l_m = 2\mu$, and $N = 50,000$. On the same coordinates is plotted the distribution curve for a mineral oil-in-water emulsion, experimentally derived from measurements made on 50,000 drops (dashed line).[13] The theoretical curve is in good agreement with experiment at low diameters, but deviates considerably at higher values. The general shape is correct, however; and better agreement would, no doubt, be secured if higher values of N had been employed.

TABLE 3-2. EFFECT OF PARTICLE SIZE ON EMULSION APPEARANCE[16]

Particle Size	Appearance
Macro globules	Two phases may be distinguished
Greater than 1 μ	Milky white emulsion
1 to approx. 0.1 μ	Blue-white emulsion
0.1 to 0.05 μ	Gray semitransparent (dries bright)
0.05 μ to smaller	Transparent

A large number of distribution curves were experimentally determined by Cooper,[14] who gives a rather complete statistical analysis of such curves in general. Russ[15] gives a succinct discussion of the general topic.

Griffin[16] has indicated the effect of droplet size on the appearance of the emulsion; his comments are summarized in Table 3-2.

From certain theoretical points of view it is convenient to regard solvent extraction systems as extremely unstable emulsions. Langlois, Gullberg and Vermeulen[17] have devised experimental techniques for determining the distribution curves of such unstable systems, and these techniques have application in more stable emulsions. This and other methods of determining particle-size distributions of emulsions are discussed in the Appendix (p. 329).

Brownian Motion. In connection with emulsion particle size, a few words on the subject of the Brownian motion are in order. When viewed under the microscope, the particles which make up fine suspensions are seen to undergo a continuous, random, zig-zag motion. This is called Brownian motion, after the botanist Brown, who apparently was the first to observe this effect (1828). It is now known that the movement is caused by the actual bombardment of the particles by the molecules of the suspending medium.

The theory of the Brownian motion is outside the scope of this work, but certain experimental observations of Exner[18] on the velocity of motion as a function of particle diameter are pertinent. In a study of Brownian motion of a suspension of gamboge particles in water, the following observations were made:

Diameter (μ)	Velocity (μ/sec)
4	Not visible
3	Just visible
1.3	2.7
0.9	3.3
0.4	3.8

It thus appears that for the majority of the droplets in an average emulsion, no Brownian motion would occur. However, it might be expected to be appreciable for smaller droplets.

This has a consequence insofar as stability is concerned: the Brownian motion increases the probability of inter-particle collisions, and hence of coagulation. To be sure, once coagulation has reached a certain point, this effect will not be important. However, the Brownian motion certainly has a negative effect on the stability of small emulsion droplets.

Concentration

Two separate concentration terms are of interest. The first of these concerns the relative amounts of the two phases making up the emulsion. While any system of concentration may be employed, e.g., weight per cent, molarity (for pure substances), etc., for theoretical purposes a convenient measure of concentration is the *volume per cent* or *volume fraction* of the disperse or internal phase. This is especially true when discussing problems relating to viscosity and to inversion.

The second concentration term which should be defined in discussing emulsions is the concentration of the emulsifying agent. It is not particularly significant what form this takes, weight per cent being perhaps as usual as any. In reading the literature, however, care must be taken to distinguish between a concentration of emulsifying agent based upon the *total* emulsion and one based on the volume or weight of *one phase*. For example, an emulsion might be described as being made up by dispersing ten parts by weight of an oil in 90 parts by weight of a 0.1 per cent sodium oleate solution. Obviously, the concentration of the emulsifier (sodium oleate) in the final emulsion is less than 0.1 per cent.

The effect on emulsion stability of each of these concentration terms will be discussed subsequently.

Optical Properties of Emulsions

A useful summary of some the older data on the optical properties of emulsions is given by Clayton;[19] however, it must be admitted that this aspect has been slighted by most researchers. The semi-qualitative effect of particle size on the appearance of emulsions has been referred to (Table 3-2). An important paper by Bailey, Nichols, and Kraemer[20] dealt with this relation between optical properties and particle size, and reviewed the attempts up to that time (1936) to relate light-scattering with the properties of suspensions. Recent work on light scattering (in the high polymer field) has revived interest in this technique. In a recent experimental arrangement, Lothian and Chappel[21] reviewed the functional relationship between the values of K, the total scattering coefficient, and the wavelength of the incident light, the refractive index, absorption and radius of the particles, and the angular size of the detector used. Exact methods of calculation are discussed and optimum conditions for measurement are suggested.

Langlois, Gullberg, and Vermeulen[17] have determined the *interfacial area* in emulsions by optical measurements. Photoelectric measurements of light transmission relative to clear fluids were carried out with an optical probe which was inserted in the emulsion. The relative light transmission was found to be related to the interfacial area by the equation

$$I_0/I = 1 + \beta A, \tag{3.3}$$

where I_0/I is the ratio of the light intensity transmitted by the clear liquid to that transmitted by the emulsion of which it is the external phase; A is the total interfacial area; and β is a constant which is a function of the ratio n_d/n_c (the refractive indices of the disperse and continuous phases, respectively). The relation between β and the refractive index ratio was determined experimentally and is presented graphically.

Transparent and Chromatic Emulsions. Normally, emulsions are opaque, cream-colored liquids. However, as has been pointed out earlier (p. 52), emulsions in which the particle size is quite small may be transparent. X-ray studies on such emulsions have yielded valuable information on the structure of the interfacial phase.[22] Transparency may also result in emulsions which have disperse and continuous phases of the same refractive indices. If the two liquids have the same refractive indices but different optical dispersive powers, then not a transparent, but rather a highly colored ("chromatic") emulsion will result. This is actually a special case of the so-called Christiansen effect.[23]

According to Francis,[23a] brilliant colors may be observed in such emulsions at a light barrier when viewed by transmitted light. The color appears at the barrier and is a function of the temperature and composition of the emulsion. The series of colors observed is complementary to the rainbow colors and includes pink and purple (but not green). This phenomenon is used in the observation of refractive indices of substances as liquids above their normal boiling points, the location of tie lines for binodal curves of certain ternary systems, and for the analysis of glycerol and other compounds.

Viscosity. The resistance to flow of emulsions is probably one of the most important of their gross properties, both from the practical and theoretical point of view. Practical considerations arise from the fact that a commercial emulsion may be marketable only at a specific viscosity. The attainment of the proper viscosity, while maintaining stability and other desirable properties, may be a problem of no small magnitude.

From the theoretical point of view, viscosity measurements, allied with hydrodynamic theory, are capable of giving considerable information about the structure of emulsions, and are often a clue to their stability.

This being the case, it is appropriate to preface the present section with a general discussion of the types of viscous flow which may be expected in emulsion systems. The viscosity (or, more correctly, coefficient of viscosity) is defined as the shearing stress τ exerted across an area when there is unit velocity gradient normal to the area.

$$\tau = \eta \, \frac{du}{dy} = \eta D, \qquad (3.4)$$

for flow in the x direction. If the shear is σ, it may be shown that D is the time rate of change of shear and Eq. 3.4 may be written

$$\tau = \eta \, \frac{d\sigma}{dt} \qquad (3.5)$$

In most simple liquids, τ is proportional to $d\sigma/dt$ (or D) as long as laminar flow takes place. In other words the constant η is independent of the velocity gradient; the liquid is Newtonian.*

For most emulsions, however, Newtonian flow is the exception rather than the rule, and η is a function of the rate of shear. Figure 3-4 shows the four types of flow behavior which fluids may exhibit: Newtonian, plastic, pseudoplastic, and dilatant. Newtonian flow has been defined above.

In the case of *plastic fluids* (or Bingham plastics) the substance has some inherent structure which completely resists shearing forces up to a magnitude τ_y, the so-called *yield strength* or *yield value*. For a plastic liquid this is defined by the intercept of the flow curve with the shear stress axis (cf. Figure 3-4). When the yield strength is reached the structure breaks down completely, and (after a short curved region) a straight line relationship ensues. The slope of this line is defined as the *coefficient of rigidity*. Drilling muds, in the field of emulsions, are examples of liquids with these flow properties.

In the case of *pseudo-plastic flow*, no yield value is exhibited, but the viscosity coefficient is clearly functionally dependent on the shear rate. In such liquids some sort of intermolecular structure is apparently built up under the influence of the shearing force.

---

* Newton's definition may be of interest: "The resistance which arises from the lack of slipperiness of the parts of the liquid, other things being equal, is proportional to the velocity with which the parts of the liquid are separated from one another."[24]

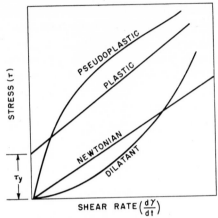

FIGURE 3-4. The four types of flow which may be exhibited by fluids: Newtonian, plastic, pseudoplastic, and dilatant.

Dilatant flow is the opposite of pseudo-plastic flow; the viscosity coefficient increases with rate of shear. Dilatancy is apparently quite rare in emulsion systems.

In addition to the above rather readily defined types of flow there exist so-called *thixotropic* and *rheopectic* systems. In these, the flow properties depend not only on the rate of shear, but also on the length of time the shearing stress is applied. In thixotropic flow, viscosity decreases with time, and *vice versa* for rheopectic. Mayonnaise is an example of an emulsion which may exhibit thixotropic properties; rheopectic flow, like dilatant flow, is rare.

In non-Newtonian systems, viscosity is a function of the rate of stress. Thus, a series of different viscosity values can be obtained by measurements on the same liquid. It is most inaccurate to refer to this measurement as the viscosity of the liquid; the very useful designation "apparent viscosity" is widely used. ("Differential" and "peculiar" viscosity are terms which have also been proposed.[25]) Thus, if the shear rate is stated in reporting such data, the quantity is unequivocally defined and is useful for control purposes. On the other hand, such values are of questionable significance for theoretical considerations. Unfortunately, in reporting the data on emulsion viscosity, it has not always been made clear that measurements were made under Newtonian conditions, or at least, in the linear portion of the pseudo-plastic curve.

Turning now to emulsions specifically, Sherman[26] has listed six factors which may affect the viscosity (or apparent viscosity) of an emulsion:

1. Viscosity of the external phase (η_0).
2. Viscosity of the internal phase (η_i).
3. Volume concentration of the disperse phase (ϕ).
4. Nature of the emulsifying agent and the film precipitated at the interface.
5. Electroviscous effect.
6. Particle-size distribution.

The subsequent discussion will be based on this valuable analysis.

Viscosity of External Phase. Virtually all theoretical or, for that matter, empirical, treatments of emulsion viscosity consider the viscosity of the continuous or external phase η_0 to be of prime importance in defining the viscosity of the final emulsion. Most equations indicate a direct proportionality between the viscosity of the emulsion and that of the external phase. Most of the published equations may be put in the form

$$\eta = \eta_0(x), \tag{3.6}$$

where x represents the summation of all the other properties which may affect the viscosity. Sherman[26] makes the valuable point that, in many emulsions, the emulsifying agent is dissolved in the external phase, and hence η_0 is the viscosity of this solution, rather than of the pure liquid. This is especially important when colloidal stabilizing agents (such as gums) are added, since they have a marked effect on the viscosity.

Viscosity of Internal Phase. The effect of the viscosity of the internal phase has not had much consideration. Taylor,[27] extending the hydrodynamic considerations of the classical Einstein relation (cf. below), by assuming that any interfacial film which existed merely transmitted tangential stress from one phase to the other, arrived at the relation

$$\eta = \eta_0 \left\{ 1 + 2.5\phi \left(\frac{\eta_i + \frac{2}{5}\,\eta_0}{\eta_i + \eta_0} \right) \right\}, \tag{3.7}$$

where the symbols have the meanings assigned above.

In a study of the viscosity of emulsions of milk fat in skim milk, Leviton and Leighton[28] modified Taylor's equation by introducing a power series in the volume fraction of the disperse phase ϕ

$$\ln \frac{\eta}{\eta_0} = 2.5 \left(\frac{\eta_i + \frac{2}{5}\,\eta_0}{\eta_i + \eta_0} \right) (\phi + \phi^{5/3} + \phi^{11/3}) \tag{3.8}$$

in order to make it applicable at higher concentrations. At low values of ϕ, Eq. 3.8 reduces to Taylor's equation.

Recently, a more elaborate hydrodynamic analysis has been carried out by Oldroyd,[29] resulting in an extremely complex equation. Oldroyd carried through a calculation of the elastic properties of a dilute emulsion of one incompressible viscous liquid in another, arising from the interfacial tension existing between the two phases. The relaxation time and retardation time for the system vary directly as the droplet diameter and inversely as the interfacial tension. The effect of slip at the interface which might be associated with the presence of an interfacial film of a third component introduced as a stabilizer (cf. below) was also calculated.

In Eq. 3.7 the term involving the viscosity of the internal phase

$$\left(\frac{\eta_i + \tfrac{2}{5} \eta_0}{\eta_i + \eta_0} \right)$$

is introduced to correct for currents set up in the dispersed droplet under the shearing flow force, under the supposition that the droplet is fluid. In this connection, Bond and Newton[30] showed that, provided the radius of the droplet were less than a certain critical value, spherical drops surrounded by a *more* viscous medium behave as rigid spheres. The conclusions drawn by Richardson[31] and Broughton and Squires[32] from data on stable emulsions leads to the conclusion that, under the conditions studied by these workers at least, the particles of the internal phase were small enough to behave like rigid spheres. Toms,[33] on the basis of his examination of a large number of O/W emulsions, concluded that, while η_i was of no importance, the chemical nature of the internal phase was quite significant. Toms' data will be discussed in greater detail below, in connection with the effect of emulsifying agents.

The investigations described above have involved oil-in-water emulsions. Sherman[34] has recently carried out a parallel investigation of W/O emulsions with ϕ values in the range of 70 per cent. Such emulsions exhibit plasticity, and yield values were obtained. Table 3-3 gives data on the viscosity and yield values of emulsions of solutions of mannitan monoöleate and of sorbitan sesquioleate (as emulsifying agents) in a mineral oil which had a viscosity of 0.250 poises at 21.0° C. Sherman also concluded that η_i was of no consequence, but that the chemical nature of the disperse phase could be of considerable significance. Thus, while emulsions prepared with equivalent proportions of propylene glycol, glycerine, or sorbitol syrup dissolved in the aqueous phase had similar consistencies when stabilized with nonionic agents, on the inclusion of 2 per cent of colloidal carbon as an additional stabilizer the viscometric constants determined for the propylene glycol system were found to be of lower magnitude than for the other two systems. This is interpreted as arising from an alteration in the wetting equilibrium of the carbon particles, presumably caused by the propylene glycol solution.

TABLE 3-3. VISCOSITY AND YIELD VALUES OF W/O EMULSIONS*[34]

Percentage Concentration of Emulsifier	Viscosity (poises)		Yield Value	
	Mannitan Monoöleate	Sorbitan Sesquioleate	Mannitan Monoöleate	Sorbitan Sesquioleate
0.5	3.09	liquid	579	—
1.0	—	1.92	—	681
1.1	3.11	—	681	—
2.0	3.18	3.78	920	834
4.0	3.51	3.83	954	919
5.0	3.89	—	988	—
6.0	4.20	3.73	1022	971
7.0	3.99	—	1022	—
8.0	3.99	3.73	1022	1260

* $\phi = 0.66$; temperature of determination = $21.0 \pm 0.1°$ C.

Summing up, hydrodynamic theory seems to indicate that the viscosity of the internal phase can be significant *if* the droplets behave as liquids. If the conditions are such that the droplets behave as rigid spheres, the chemical nature of the disperse phase is likely to have a greater effect. This is, in a sense, fortunate, since a good deal can be learned about the flow of suspensions in general, and emulsions in particular, by studies on model systems consisting of suspensions of rigid spheres (cf. below).

Concentration of Internal Phase. Principal attention of workers in field of emulsion viscosity has been devoted to the effect of the volume concentration of the disperse phase. As has been pointed out earlier (p. 53), the convenient theoretical concentration term is volume per cent or volume fraction. The symbol ϕ, as used in the following, is usually taken to mean the latter, although some workers prefer the percentage form of expression.

The classic equation relating the viscosity of a suspension with that of the suspending liquid and the volume fraction is that of Einstein[35]

$$\eta = \eta_0(1 + 2.5\phi), \qquad (3.9)$$

which is derived on hydrodynamic grounds. Unfortunately, it is a limiting law. In the form stated above it is probably not valid for values of ϕ greater than 0.02, and is therefore of extremely limited usefulness. When, however, the law is properly expressed in its limiting form (at infinite dilution) it is quite exact:

$$\left[\frac{(\eta/\eta_0) - 1}{\phi}\right]_{\phi \to 0} = \left[\frac{\eta_{sp}}{\phi}\right]_{\phi \to 0} = 2.5 \qquad (3.10)$$

where the fraction η/η_0 is also called the relative viscosity η_r, and η_{sp} is the so-called specific viscosity. Subsequent workers have attempted to modify Eq. 3.9 by introducing a power series in the volume concentration (e.g., Eq. 3.8, above). These take the general form

$$\eta = \eta_0(1 + 2.5\phi + a\phi^2 + b\phi^3 + \cdots), \qquad (3.11)$$

where a and b are constants. Guth, Gold, and Simha[36] were able to calculate the value of a on theoretical grounds, and arrived at the equation

$$\eta = \eta_0(1 + 2.5\phi + 14.1\phi^2) \qquad (3.12)$$

which increases the validity of the equation to a value of ϕ of about 0.06. It may be of interest to point out that Eilers,[37] in an investigation of emulsions of paraffin oil and of bitumen, found that his data was fitted by the equation

$$\eta_{sp} = 2.5\phi + 4.94\phi^2 + 8.78\phi^3 \qquad (3.13)$$

which will be recognized as being of the form of Eq. 3.11. Sherman[38] has investigated the fit of the Einstein equation at infinite dilution (Eq. 3.10) and the Guth-Simha equation (Eq. 3.12) for a W/O emulsion at various emulsifier concentrations, with the results shown in Table 3-4. As can be seen, in the limiting form, there is reasonable agreement with theory for the Einstein equation, the theoretical value of the constant being 2.5. On the other hand, the data support the suggestion that the interaction between the droplets (which is measured by the square power term in ϕ in the Guth-Simha equation) is less than their theory requires.

Recent work by Oliver and Ward[39] shows an interesting facet of equations of the power-series form. These workers, analyzing data on model emulsions consisting of suspensions of rigid spheres of varying size distributions, and measured with varying types of viscometers, discovered that a straight line relationship existed if one plotted $1 - 1/\eta_r$ versus ϕ for values of the volume concentration up to 20 per cent (or slightly higher). Algebraically expressed, this yields the result

$$\eta_r = \frac{1}{1 - k\phi} = 1 + k\phi + k^2\phi^2 + k^3\phi^3 + \cdots, \qquad (3.14)$$

which again is of the form of Eq. 3.11, except that a functional relationship is supposed to exist between the constants. Table 3-5 details the value of the constant k calculated by Oliver and Ward from their own data and that of others. As can be seen, the results are scattered rather closely around the Einstein value of 2.5 for the first power term. On the other hand, the coefficient of the square term would be 6.25, in considerable disagreement with the Guth-Simha figure of 14.1.

TABLE 3-4. EINSTEIN AND GUTH-SIMHA CONSTANTS FOR A W/O EMULSION[38]

Concentration Emulsifier (%)	$\left[\dfrac{\eta_{sp}}{\phi}\right]_{\phi \to 0}$ (Einstein)	$\left[\dfrac{\eta_{sp} - 2.5\phi}{\phi^2}\right]$ (Guth-Simha)
1.0	2.27	9.71
2.0	2.60	2.27
3.5	2.27	6.66
5.0	2.75	0.00

TABLE 3-5. OLIVER-WARD CONSTANTS FOR MODEL EMULSIONS[39]

Type Viscometer	Size Ratio of Suspended Particles	Range of Concentration (vol. %)	k
Capillary	*	0–15+	2.41[40]
Rotating cylinder	2.0:1	0–15+	2.57[41]
Rising sphere	1.8:1	0–20+	2.41[42]
Rising sphere	1.2:1	0–20+	2.77[42]
Rotating cylinder	1.4:1	0–20	2.56[43]
Capillary	1.4:1	0–35	2.42[44]
Capillary	1.2:1	0–25	2.38[45]
Capillary	5.0:1	25–35	2.34[46]
Capillary	1.6:1	0–30	2.45[39]

* Killed yeast cells; size ratio not known. All other experiments based on materials which were perfect spheres.

An equation of a completely different form has been suggested by Richardson[31] on theoretical grounds. He suggests that, at any particular rate of shear, the relation between emulsion viscosity and volume concentration is exponential, i.e.,

$$\eta = \eta_0 e^{k\phi} \quad \text{or} \quad \ln \eta/\eta_0 = k\phi \tag{3.15}$$

Working with oil-in-water emulsions stabilized with sodium oleate, triethanolamine oleate, and saponin, Broughton and Squires[32] modified Eq. III.15 by the addition of a constant term

$$\ln \eta/\eta_0 = k\phi + a \tag{3.16}$$

In a study of nitrocellulose lacquer emulsions stabilized with sodium polyvinyl acetate-phthalate, Simpson[47] found Eq. 3.16 to apply. In these systems, Newtonian flow is exhibited only below certain critical volume concentrations; above these a yield value exists.

Generally speaking, emulsions containing more than 50 per cent of disperse phase often show considerable non-Newtonian behavior. For such systems, Hatschek[48] has derived the well-known relationship

$$\eta = \eta_0 \left(\frac{1}{1 - \phi^{1/3}}\right) \tag{3.17}$$

This relationship has been widely quoted as being appropriate for emulsions, but it must be used with caution. Hatschek pointed out that it only applies in the linear portion of shear-flow curves, but it has also been applied in concentration ranges where it has no theoretical justification (although, to be sure, it has been found to fit the data reasonably well in such regions for a number of cases).

Sibree,[49] in an extensive series of investigations of paraffin-in-water emulsions, found it necessary to modify Hatschek's equation

$$\eta = \eta_0 \left[\frac{1}{1 - (h\phi)^{1/3}} \right], \tag{3.18}$$

where the constant h was termed the *volume factor* by Sibree. Sibree found that h was very nearly 1.3 for a number of oil-in-water emulsions of a variety of particle sizes. Table 3-6 reproduces some of Sibree's data. From these data it is possible to draw the conclusions that the droplets behave as though they were about 30 per cent greater in volume than would be expected from the volume concentration. Gabriel,[50] in an investigation of the viscosity of asphalt emulsions, has suggested that this may be ascribable to "hydration" of the oil globules, but this is a somewhat improbable mechanism, especially in view of the fact the Broughton and Squires[32] found values of h of *less* than unity in the concentration range $\phi = 0.40$ to 0.75. The latter workers also observed that h sometimes decreased in value with increasing concentration.

Sibree himself pointed out that since all his emulsions were prepared with the same continuous phase, and stabilized with sodium oleate, the

TABLE 3-6. VALUES OF SIBREE'S CONSTANT h FOR COARSE AND FINE O/W EMULSIONS[49]

Type of Emulsion	Oil Phase	η/η_0(obs.)	ϕ (obs.)	ϕ(calcd. from Eq. 3.18)	h ($= \phi$ calcd./ϕ obs.)
Coarse	Limpid	7.2	0.50	0.639	1.28
(Particle size: 220 μ to 20 μ)	paraffin	16.4	0.60	0.828	1.38
		26.4	0.682	0.890	1.30
		78	0.78	0.962	1.28
Coarse	Viscous	9.3	0.48	0.711	1.48
(Particle size: 340 μ to 20 μ)	paraffin	18	0.576	0.786	1.36
		33	0.656	0.912	1.39
		79	0.753	0.962	1.28
Fine	Limpid	8.6	0.50	0.690	1.38
(Particle size: None larger than	paraffin	12.8	0.55	0.784	1.42
10 μ)		19.0	0.576	0.850	1.48
		190	0.692	0.984	1.42
		900	0.72	0.997	1.40

possibility existed that h might depend on emulsifier concentration. Other workers have investigated the effect of emulsifier concentration and type on h and on the constants of the various equations proposed; these will be discussed below (p. 67).

Eilers[37] made a study of the applicability of the equation of Bredeé and de Booys[51]

$$\eta_r = \left[\frac{1 + 2.5\phi}{6(1 - \phi)} \right]^6 \tag{3.19}$$

to emulsions consisting of nonplastic, nonsolvated, and truly spherical particles possessing a high inherent viscosity. For this study bitumen emulsions were used, prepared according to technical methods, using a solution of a potash rosin soap (containing excess caustic potash) as the continuous phase. One emulsion was prepared from an asphalt-free, so-called albino bitumen that was low in water-soluble materials; another emulsion was prepared from a commercial Venezuelan bitumen. Viscosity measurements were carried out in a capillary viscometer, and Eq. 3.19 was found to be applicable for values of ϕ between zero and 0.65. For higher concentrations, however, the equation gives values which are too low. Taking into account the effect of packing on flow at higher concentrations, and the fact that agreement with Einstein equation is necessary at low concentrations, Eilers proposed a modification of Eq. 3.19

$$\eta_r = \left[\frac{1 + 2.5\phi}{2(1 - a\phi)} \right]^2, \tag{3.20}$$

where $a = 1.35$. Values of the viscosity calculated by this equation are somewhat lower than those observed experimentally; good agreement was obtained with the first emulsion when a is 1.28, and in the case of the second emulsion when $a = 1.30$. The numerical agreement with Sibree's constant is presumably accidental.

In a recent investigation by Maron and Madow,[52] on latex emulsions in the concentration range zero to 0.60, it was also found necessary to make allowance for the crowding effect of the dispersed globules on each other with increasing concentration. This was achieved by the following relation

$$\log \eta/\eta_0 = xZ, \tag{3.21}$$

where $Z = y\phi(1 - y\phi)$ and x and y are constants.

It should also be pointed out that a large number of other equations connecting viscosity and concentration have been derived; these are not all derived with emulsion systems in mind, but some of them may be applicable in particular cases. Generally, they are all reducible to the power series form of Eq. 3.11. Among these may be cited two by Arrhenius[53] (one of Arrhenius' equations is identical with Richardson's[31]), and equations Fikentscher,[54] Papkov,[55] Houwink,[56] Fikentscher and Mark,[57] Sakurada,[58] and Baker and Mardles.[59]

Interfacial Film and Emulsifying Agents. Turning now to the effect of the interfacial film and of the emulsifying agent which constitutes that film, it is interesting to note that only in recent years has this aspect of emulsion viscosity received any attention. Toms[33] made an intensive investigation of a series of oil-in-water emulsions of n-hexane, cyclohexane, xylene, toluene, decalin, benzene, tetralin, aniline, chlorobenzene, nitrobenzene, and chloroform, stabilized by sodium laurate and oleate, and potassium laurate, myristate, and oleate. Toms data were used to calculate the volume factor h of Eq. 3.18, and the results of a large number of his measurements are given in Table 3-7.

As can be seen, while the average value of h is not far from the theoretical 1.3, the individual variations are considerable, and appear to be attributable to both the variation in emulsifying agent and internal phase. Variation in emulsifier concentration (at least, in the range studied) appears to have less effect. Toms explained these data on the basis of the mutual solubilities of organic liquids, soaps, and water. According to this view, the variation in viscosity (and hence in h) might be due to three causes:

TABLE 3-7. VALUES OF VOLUME FACTOR h FOR O/W EMULSIONS OF VARIOUS ORGANIC LIQUIDS[33]

Internal Phase	Na Soaps				K Soaps					
	Laurate		Oleate		Laurate		Myristate		Oleate	
	(*)	(†)	(*)	(†)	(*)	(†)	(*)	(†)	(*)	(†)
n-Hexane	1.30	1.28	1.20	1.16	1.30	1.26	1.29	1.27	1.20	1.16
Cyclohexane	1.30	1.31	1.19	1.20	1.30	1.29	1.31	1.31	1.20	1.17
Xylene	1.33	1.33	1.24	1.20	1.32	1.31	1.34	1.34	1.24	1.19
Toluene	1.31	1.34	1.15	1.14	1.32	1.30	1.32	1.33	1.18	1.16
Decalin	1.35	1.34	1.30	1.30	1.36	1.34	1.34	1.34	1.32	1.29
Benzene	1.27	1.29	1.12	1.12	1.28	1.26	1.25	1.25	1.12	1.11
Tetralin	1.35	1.35	1.32	1.31	1.36	1.36	1.36	1.40	1.32	1.31
Aniline	1.30	1.25	1.40	1.42	1.26	1.18	1.37	1.30	1.38	1.41
Chlorbenzene	1.32	1.34	1.16	1.15	1.32	1.33	1.34	1.34	1.17	1.15
Nitrobenzene	—	1.29	1.29	1.26	—	1.32	1.32	1.34	1.30	1.28
Chloroform	—	1.28	1.10	1.11	1.24	1.25	1.27	1.23	1.12	1.13

* 0.001 mole soap per 100 cc emulsion
† 0.002 mole soap per 100 cc emulsion

1. Part of the soap stabilizer may enter the disperse globules and form a gel of solvated soap; this would bring about a change in rigidity of the interfacial film and alter the deformability of the globules.

2. The soap at the interface may alter the degree of mutual dispersion of the two liquids and thus effectively change the volume ratio ϕ.

3. A transfer of organic liquid across the interface might lead to peptization of the soap micelles in the aqueous external phase (thus changing the viscosity of the external phase η_0).

The above work concerned itself with O/W emulsions, as has most of the work in this field; recently, however, Sherman[38] has carried through an extensive investigation of the effect of the interfacial film on the viscosity of W/O emulsions. Some of Sherman's data is reproduced in Table 3-8, exhibiting the effect of emulsifier concentration on the emulsion viscosity. This property does have a striking effect, as can be seen by reference to η_{sp}/ϕ column of Table 3-8 which lists the Einstein constants calculated from the viscosity data. These data also illustrate how the fit of the Einstein equation becomes progressively worse as the concentration of disperse phase increases.

The information conveyed in the last column of the table is also quite important. As has been pointed out, Sherman was concerned primarily with W/O emulsions; however, at fairly high volume concentrations his emulsions exhibited the phenomenon of *inversion*. That is to say, the continuous and disperse phases exchanged, and the W/O emulsion inverted to O/W. This is a fairly general phenomenon, and is discussed in more detail on pp. 138–149. In this case, however, we are interested in the effect of inversion on the viscosity. Figure 3-5 shows the sort of viscosity-concentration relation which occurs when inversion takes place, the sharp drop in viscosity occurring at the inversion point. This striking effect is due, of course, to the fact that once inversion takes place, the volume concentration drops sharply, e.g., if inversion takes place (as indicated in Figure 3-5) at a $\phi = 0.74$ for a O/W emulsion, the corresponding ϕ for the inverted emulsion is 0.26. It should be pointed out that η_0 will also change, but this is likely to have a lesser effect. It should be clear that the portion of the curve to the right of the inversion point is actually the low concentration section of the inverted emulsion.

The choice of $\phi = 0.74$ as the inversion point in the schematic curve of Figure 3-5 is not fortuitous; a rather simple theory of the inversion process (discussed below, p. 141) requires that inversion shall always take place at this concentration. This is, in fact, far from true, as is well illustrated by Sherman's data. Table 3-9 summarizes the effect of emulsifier concentra-

TABLE 3-8. EFFECT OF EMULSIFIER CONCENTRATION AND VOLUME CONCENTRATION ON VISCOSITY OF W/O EMULSIONS[38]

Emulsifying agent (%)	Disperse phase (wt. %)	ϕ	Reciprocal mobility of emulsion at 21.0 ± 0.1 C., η (poises)	η_0 at 21.0 ± 0.1 C., (poises)	$\dfrac{\eta_{sp}}{\phi}$	Type of emulsion
1.0	10.0	0.0884	0.304	0.218	4.445	W/O
	20.0	0.1792	0.375	0.219	3.967	
	30.0	0.2727	0.500	0.220	4.653	
	40.0	0.3687	0.656	0.223	5.282	
	50.0	0.4669	0.937	0.225	6.784	
	60.0	0.5676	1.171	0.226	7.356	
	62.5	0.5941	2.695	0.228	18.190	
	65.0		0.247			O/W
	70.0		0.188			
2.0	10.0	0.0900	0.341	0.270	2.900	W/O
	20.0	0.1819	0.420	0.274	2.930	
	30.0	0.2762	0.523	0.278	3.183	
	40.0	0.3729	0.703	0.283	3.984	
	50.0	0.4713	0.789	0.287	3.708	
	60.0	0.5726	2.256	0.291	11.790	
	70.0		4.097			
	76.0		5.526			
	80.0		0.079			O/W
3.5	8.8	0.0849	0.353	0.285	2.815	W/O
	19.3	0.1744	0.456	0.290	3.286	
	27.5	0.2509	0.592	0.296	3.994	
	40.0	0.3707	0.844	0.301	4.874	
	50.0	0.4690	1.070	0.307	5.309	
	60.0	0.5704	2.359	0.311	11.550	
	70.0	0.6739	4.753	0.317	20.770	
	80.0	0.7799	12.050	0.323	46.610	
	82.5		0.260			O/W
5.0	10.0	0.0904	0.375	0.299	2.798	W/O
	20.0	0.1828	0.463	0.308	2.757	
	30.0	0.2773	0.551	0.315	2.705	
	40.0	0.3739	0.624	0.323	2.493	
	50.0	0.4726	0.789	0.331	2.934	
	60.0	0.5736	2.780	0.339	12.535	
	70.0		4.670			
	75.0		13.150			
	77.5		0.225			O/W
	80.0		0.151			

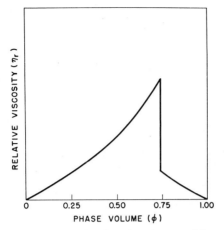

FIGURE 3-5. The dependence of the viscosity of an emulsion on the concentration of the internal phase. The sharp drop in viscosity at $\phi = 0.74$ corresponds to inversion of the emulsion. The portion of the curve to the right of the inversion point is the low-concentration section of the inverted emulsion.

TABLE 3-9. EFFECT OF EMULSIFIER CONCENTRATION ON INVERSION[26]

Emulsifier concentration (%)	Max. apparent viscosity (poises)	Value of ϕ at inversion
1.0	2.70	0.594
2.0	5.53	0.732
3.5	12.05	0.780
5.0	13.15	0.723

tion on the inversion concentration and the maximum viscosity attained prior to inversion.[26] The significance of these data will be considered more fully in the subsequent discussion.

Sherman[38] has also used his data to calculate the volume factor h (Eq. 3.18). These results are detailed in Table 3-10, and may be fruitfully compared with those of Toms (Table 3-7).

TABLE 3-10. EFFECT OF EMULSIFIER CONCENTRATION ON VOLUME FACTOR h[38]

Emulsifier Concentration (%)	$\eta_{obs.}$	$\eta_{calc.}$	$\phi_{obs.}$	$\phi_{calc.}$	h
1.0	1.171	1.315	0.5676	0.5253	0.93
	2.695	1.435	0.5941	0.7665	1.29
2.0	2.256	1.712	0.5726	0.6607	1.15
3.5	2.359	1.818	0.5704	0.6539	1.15
	4.753	2.577	0.6739	0.8128	1.21
	12.050	4.053	0.7799	0.9230	1.18
5.0	2.780	2.008	0.5736	0.6773	1.18

Neogy and Ghosh[60, 61] have recently made an exhaustive investigation of xylene-in-water emulsions stabilized by three cationic emulsifying agents (cetyl dimethylbenzylammonium chloride, cetyl pyridinium bromide, and cetyl trimethylammonium bromide) and of a water-in-benzene emulsion stabilized with magnesium oleate. These emulsions were non-Newtonian, and yield values were determined. In the case of the O/W emulsions it was found that the viscosity (as expected) increases with increasing concentration of disperse phase, while the yield value rises abruptly at 70–80 per cent of disperse phase. The type of emulsifying agent has apparently little effect in these cases.

On the other hand, in the case of the W/O emulsions, phase reversal occurred at a concentration of water between 50 and 60 per cent (cf. Table 3-9), and is marked by a sudden decrease in viscosity. It was found that the oil-in-water emulsions followed Sibree's equation (Eq. 3.18) fairly well, with best agreement being obtained at high concentrations ($\phi > 0.5$). The water-in-oil emulsions follow this equation only at 70 per cent water and differ widely in all other cases.

These authors used their data to test the various viscosity equations as a function of type of emulsifier and of emulsion type. Table 3-11 gives the values of the observed viscosities and those calculated by the Einstein (Eq. 3.9) and Hatschek (Eq. 3.17) equations for a xylene-in-water emulsion stabilized by cetyl pyridinium bromide. Table 3-12 gives similar data for a water-in-benzene emulsion stabilized by magnesium oleate.[60]

TABLE 3-11. CALCULATED AND OBSERVED VISCOSITIES OF XYLENE-IN-WATER EMULSIONS STABILIZED BY CETYL PYRIDINIUM BROMIDE[60]

ϕ	$\eta_{calc.}$		$\eta_{obs.}$
	Einstein	Hatschek	
0.20	1.273	2.043	2.43
0.30	1.485	2.565	3.09
0.40	1.696	3.223	4.42
0.50	1.908	4.113	6.93
0.60	2.120	5.420	11.23
0.70	2.332	7.573	26.31
0.80	2.544	11.83	62.10

TABLE 3-12. CALCULATED AND OBSERVED VISCOSITIES OF WATER-IN-BENZENE EMULSIONS STABILIZED BY MAGNESIUM OLEATE[60]

ϕ	$\eta_{calc.}$		$\eta_{obs.}$
	Einstein	Hatschek	
0.20	0.991	1.827	4.05
0.30	1.156	2.449	6.69
0.40	1.322	3.374	8.85
0.50	1.487	4.939	13.49
0.60	1.652	8.324	17.12

Table 3-13. Calculated and Observed Values of the Logarithm of the Relative Viscosity for Xylene-in-Water Emulsions Stabilize by Cetyl Pyridinium Bromide[60]

| | | log η_r (calc.) | |
ϕ	log η_r (obs.)	Sibree (h = 1.3)	Richardson (k = 1.84)
0.20	0.457	0.442	0.368
0.30	0.559	0.569	0.552
0.40	0.717	0.708	0.736
0.50	0.912	0.873	0.920
0.60	1.122	1.100	1.105
0.70	1.492	1.510	1.289

Table 3-14. Calculated and Observed Values of the Logarithm of the Relative Viscosity for Water-in-Benzene Emulsions Stabilized by Magnesium Oleate[60]

| | | log η_r (calc.) | | |
| | | | Richardson | |
ϕ	log η_r (obs.)	Sibree (h = 1.3)	(k = 2.82)*	(k = 1.43; a = 0.56)
0.20	0.787	0.442	0.564	0.846
0.30	1.005	0.569	0.845	0.989
0.40	1.127	0.708	1.127	1.132
0.50	1.310	0.873	1.410	1.275
0.60	1.413	1.100	1.692	1.418
0.70	1.564	1.510	1.974	1.561

* Calcd. from ϕ = 0.40 emulsion.

Similarly, Tables 3-13 and 3-14 give calculated and observed values of the logarithm of the relative viscosity for the same emulsions, calculated from the Sibree equation (Eq. 3.18) and Richardson's simple (Eq. 3.15) and modified (Eq. 3.16) equations. These equations give a reasonable fit, but the two-constant form of the Richardson equation is required to give the best fit for the magnesium-oleate stabilized emulsion. Neogy and Ghosh[61] state that only this latter equation can be expected to fit emulsion viscosity data to an accuracy of 5 per cent.

These authors summarize their results (of which only a small portion was cited above) as follows:

1. The viscosity of emulsions depends on the nature of the emulsifier used.

2. Emulsions stabilized with saponin (a naturally-occurring emulsifier) have a higher viscosity than those stabilized with soap.

3. Laurate- and caprate-stabilized emulsions have viscosities slightly lower than those stabilized by oleate and myristate.

4. Barring saponin, the differences in the viscosities diminish as the concentration of disperse phase is increased.

5. Sodium caprate is a much weaker emulsifier than the oleate, myristate, or laurate.

6. Emulsions stabilized with the cationic cetyl salts are least viscous, practically spontaneously formed, and quite stable.

7. The common view that an increase in viscosity increases stability by hindering the coalescence of drops cannot be justified.

8. The variation of viscosity with the nature of the emulstifier is explained by the assumption that the nature of the *interfacial film* which stabilizes the emulsion plays the vital part in viscosity.

9. Substitution of molar concentrations for the more commonly employed percentage concentrations does not affect the above conclusion (8).[61]

Generally speaking, these conclusions are consistent with those drawn by other workers, although the conclusion that high viscosity does not increase stability is probably not of general validity, even if correct in the systems studied by Neogy and Ghosh.

Other Observations on Interfacial Films. To the extent that the viscosity of the emulsion is a function of the type and concentration of emulsifying agent employed, it is also a function of the pH. This arises from the fact that the emulsifier may be effective over only a relatively narrow range of hydrogen-ion concentration. This is particularly true of cationic and anionic agents, and may be true for ester-type nonionics under strongly alkaline conditions. Sherman[62] has investigated this and has found, in the cases investigated, that inversion, with a sharp decrease in viscosity, occurred at a pH of about 9.0 for water-in-oil emulsions stabilized with three different nonionic agents.

In addition to these experimental studies, the effect of the interfacial film of emulsifying agent on flow properties has been investigated by, among others, Mardles and De Waele.[63] These workers have found that the rheological properties of emulsions and suspensions are affected by the character of the interfacial films. The stability of these systems is frequently controlled by rheological rather than hydrodynamic factors; this is especially true in concentrated emulsions. That this is the case is verified by the recent investigations of Oldroyd,[64] who studied the elastic and viscous properties of a dilute emulsion of one incompressible viscous liquid in another, when subjected to small variable rates of strain, by including the possibility of an interfacial film being everywhere present between the two phases. This film is assumed to resist deformation as a result of internal friction or of elasticity.

It was found that when the interfacial film is assumed to be ideally elastic, the viscosity in slow steady flow is the same as a suspension of solid spheres. On the other hand, if the film is purely elastic, it has no effect whatever on the type of elasto-viscous behavior of the emulsion. Therefore, it is possible, in principle, to distinguish between "solid" and "liquid" films by their effect on the viscosity; with the proviso, however, that the droplets must be above a certain minimum size.

The effect of surface-active materials on the viscosity and other mechanical properties of the oil-water interface has recently been studied by Criddle and Meader.[65] Some older, but nonetheless important, work has been done in this field by Serallach, Grinnel Jones, and Owen.[66] In this work, the strength of interfacial films stabilized by naturally occurring gums, e.g., acacia, irish moss, was studied.

Electroviscous Effect. The fifth phenomenon in Sherman's list (p. 57) which may be considered to affect the viscosity of emulsions is the so-called *electro-viscous* effect. This was first investigated theoretically for lyophobic colloids by Smoluchowski.[67] In this work it was shown that lyophobic particles bearing an electric charge would show a viscosity *exceeding* that of a similar system of uncharged particles. Taking this effect into account, Smoluchowski modified Einstein's equation to

$$\frac{\eta - \eta_0}{\eta_0} = 2.5\phi\left[1 + \frac{1}{\eta_0\kappa a^2}\left(\frac{\epsilon\zeta}{2\pi}\right)^2\right], \tag{3.22}$$

where a = the radius of the particles, κ = the specific conductivity of the suspension, ϵ = the dielectric constant of the dispersion medium, and ζ = the electrokinetic potential of the charged particles (the so-called *zeta potential*). The origin of the charges in emulsion systems, and their contributions to other emulsion properties, will be discussed in greater detail in subsequent sections of this and the following chapter. As stated above, however, the contribution of the effect of the charges will always be to increase the viscosity, since the electro-viscous term (i.e., the second term in the bracket in Eq. 3.22) will always be positive, irrespective of the sign of the charge.

It should be noted that, at infinite dilution, Smoluchowski's equation reduces to the Einstein relation, (as ϕ approaches zero, the specific viscosity approaches 2.5). This is due to the fact that the electro-viscous effect arises from interactions between the charges on adjacent drops. As the system grows more dilute, the distance between the charges increases, and the interaction, of course, falls off rapidly with an inverse power of the

TABLE 3-15. ELECTRO-VISCOUS EFFECT IN OIL-IN-WATER EMULSIONS[68]

Per cent, by weight, of emulsifier	Per cent, by weight, of oil	ϕ	η_r	K	$K_{\phi \to 0}$	V
5	5	0.0565	1.17	3.01	2.6–2.7	1.04–1.08
	10	0.112	1.40	3.57		
	20	0.222	2.12	5.04		
	30	0.328	3.52	7.70		
10	5	0.0559	1.21	3.77	3.0–3.1	1.20–1.24
	10	0.111	1.51	4.60		
	20	0.219	2.56	7.13		
	30	0.325	5.64	14.3		
17	5	0.0549	1.25	4.56	3.4–3.5	1.36–1.40
	10	0.109	1.68	6.23		
	20	0.216	3.94	13.6		
	30	0.321	22.15	66.1		
35	5	0.0531	1.35	6.59	4.8–5.0	1.92–2.00
	10	0.106	1.96	9.04		
	20	0.210	6.50	26.2		
	30	0.313	92.8	294		

interparticle distance. The electro-viscous term of Eq. 3.22 thus becomes zero.

In a recent paper, van der Waarden[68] has pointed out that, for certain colloid systems which have been investigated, Eq. 3.22 predicts values which are too high. Van der Waarden investigated the effect in an oil-in-water emulsion stabilized by sodium naphthalene sulfonates of high purity. Table 3-15 gives some of van der Waarden's data; the quantity K in the fifth column of the table is defined as $K = (\eta_r - 1)/\phi = \eta_{sp}/\phi$, or 2.5 times the bracket in Eq. 3.22. The sixth column gives this value at infinite dilution, as obtained by extrapolation, and V gives the value of K at infinite dilution divided by 2.5. This last quantity reflects the fact that the droplets of the emulsion behave as if their volume were V times the real volume. It is interesting, in connection with the earlier discussion, to note that $K_{\phi \to 0}$ approaches 2.5 (and V approaches 1) as the emulsifier concentration approaches zero.

From these, and other data, van der Waarden concluded that there was an apparent increase in the volume of the disperse phase: a negatively charged layer being built up at the interface, with a positively charged diffuse layer around the droplets. From the experimental data it was calculated that the disperse phase droplets are surrounded by a rigid layer 30–35 Å thick, irrespective of the actual droplet size.

Particle Size Distribution. The final property which may affect viscosity is the actual distribution of particle sizes among the droplets of the disperse phase. The types of distribution curves which may be expected have been discussed earlier (p. 46). On the basis of the various equations relating concentration and viscosity it does not appear probable that the particle size distribution affects the viscosity.

It is a well-known experimental fact, however, that when crude emulsions are homogenized, and thus subjected to a radical change in particle size distribution, a marked increase in viscosity often occurs. Sherman[26] ascribes this to a reduction in particle size, resulting in increased interfacial area and mutual interaction between the globules. In the case of an emulsion with a large value of ϕ, that is, a closely-packed system, the increased adsorption of the emulsifying agent at the interface under these conditions will probably exert a noticeable influence, as indicated above.

For example, Lyttelton and Traxler[69] have reported on the separation of asphalt emulsions into fractions with the same value of ϕ, but with widely differing viscosities. They found that the more viscous fraction was also the more homogeneous. Similar results are due to Richardson,[70] who examined the viscosities of a number of emulsions of differing size distributions at differing rates of shear. This was done by the use of a falling-ball technique; Figure 3-6 reproduces some of Richardson's data. Here the

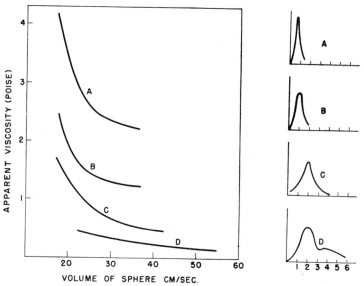

FIGURE 3-6. The viscosities of a series of emulsions as a function of the rate of shear. The particle-size distributions corresponding to the emulsions are shown in the subsidiary curves.[70]

apparent viscosities of four emulsions are shown as a function of their rate of shear; the distribution curves for the emulsions are also given.

From his studies, Richardson concluded that the apparent viscosity of emulsions having the same concentration and the same size distribution about a mean globule diameter is inversely proportional to this mean globule diameter. The observed viscosity variations, of a concentrated emulsion, with concentration and rate of shear can be explained in terms of the work done in distorting the globules and sliding them past each other.

Ward and Whitmore,[71] in a study of the viscosity of suspensions of rigid spheres, found that the relative viscosity was independent of the suspending liquid and of the *absolute* size of the spheres at a given concentration. However, relative viscosity is a function of sphere size distribution.

More recently, Roscoe[72] has obtained two modified forms of the Einstein equation, one of which applies to systems in which the particle size distribution is broad

$$\eta = \eta_0(1 - \phi)^{-2.5}, \tag{3.23}$$

and one for sharply-distributed emulsions

$$\eta = \eta_0(1 - 1.35\phi)^{-2.5} \tag{3.24}$$

Equation 3.23 is applicable over all concentration ranges. Equation 3.24 is appropriate at high concentrations; below $\phi = 0.05$ the Einstein Equation applies to such sharply-distributed systems. Figure 3-7 shows the fit of the

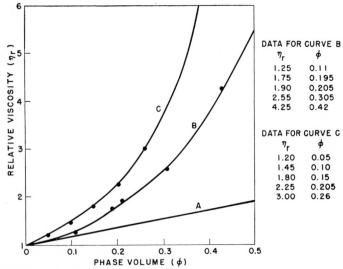

FIGURE 3-7. Fit of experimental data on the viscosity of emulsions with theoretical equations. Curve A, Eq. 3.9; B, Eq. 3.23; C, Eq. 3.24.[72]

curves to experimental data for two typical emulsions. Curve B is drawn according to Eq. 3.23, and curve C according to 3.24; curve A is the Einstein relation (Eq. 3.9).

Dielectric Constant

Relatively speaking, the dielectric constant of emulsions has been studied but little. This is somewhat surprising, since it could conceivably give considerable information about related electrical properties. Clayton[73] gives considerable data on earlier investigations of this property. All of these investigations are based on the assumption that the dielectric constant of a suspension or emulsion must be a linear function of the dielectric constants of the two components, and of their volume concentrations. Actually, this cannot be the case, as was pointed out some time ago by Fricke and Curtis.[74]

These writers point out, for example, that if a dielectric dispersed in water is subjected to an electric field, a part of the current passes through the system at the interfaces and become partly polarized. Thus, the dielectric properties of such systems are quite different from what would be expected from the contributions of the individual phases.

A most important contribution to this problem has recently been made by Kubo and Nakamura.[75] These workers have considered theoretically the dielectric constant of a dispersion in which spherical particles of dielectric constant ϵ_1 are dispersed in a medium of dielectric constant ϵ_0. The following assumptions were made:

1. Each particle of the disperse phase is spherical.
2. The particles have no *net* charge.
3. The radii of the spheres are very small compared to the dimensions of the whole heterogeneous system.
4. The radii of the spheres are very great in comparison to molecular dimensions, so that the disperse phase and dispersion medium can be considered as homogeneous media characterized by their dielectric constants alone.

Using these assumptions, Kubo and Nakamura arrive at the following equation for the relation between the dielectric constant ϵ of the dispersion and the volume fraction ϕ of the disperse phase.

$$\{3\epsilon_1/[(2+c)\epsilon_1 + (1-c)]\} \log [(\epsilon_1 - \epsilon)/(\epsilon_1 - \epsilon_0)]$$

$$- \{[(2+c)\epsilon_1 - 2(1-c)]/[(2+c)\epsilon_1 + (1-c)](2+c)\} \quad (3.25)$$

$$\log \{[(2+c)\epsilon + (1-c)/(2+c)\epsilon_0 + (1-c)]\} = \log (1-\phi),$$

where $c = 1 - (4\pi/9\sqrt{3})$. The original paper has curves showing values of the dielectric constants of both W/O and O/W emulsions as a function of ϕ, as calculated from Eq. 3.25.

Frenkel[75a] has shown that the electric moment p induced in a spherical drop of radius a and dielectric constant ϵ, placed in a medium of dielectric constant ϵ_0, and subjected to an external field E_0 is given by

$$p = E_0 a^3 (1 - \alpha)/(1 + \alpha),$$

where α is a certain function of γa, and $\gamma^2 = \kappa^2 + (i\omega/D)$. The quantity κ is the reciprocal radius of the ionic atmosphere* around the droplet, ω is the frequency of the applied field, and D is the diffusion coefficient of the ions in the conducting medium. Fradkina[75b] has shown that this leads to a macroscopic dielectric constant

$$\bar{\epsilon} = \epsilon_0 [1 + 3\phi(1 - \alpha)/(1 + 2\alpha)], \tag{3.26}$$

a much simpler result than that obtained by Kubo and Nakamara.[75] Fradkina has also shown that, for not too dilute aqueous of electrolytes dispersed in petroleum, $\gamma a \gg 1$, $\alpha = 0$, and

$$\bar{\epsilon} = \epsilon_0 (1 + 3\phi) \tag{3.27}$$

That is, the mean macroscopic dielectric constant depends solely on the water content of the emulsion and not on the electrolyte content of the water phase or on the degree of dispersion. Equation 3.27 can also be derived from the assumption that the polarization of the droplets is that of ideally conducting spheres.

If, on the other hand, the spheres have a dielectric constant ϵ_1, Eq. 3.27 becomes

$$\bar{\epsilon} = \epsilon_0 [1 + 3\phi(\epsilon_1 - \epsilon_0)/(\epsilon_1 + 2\epsilon_0)] \tag{3.28}$$

For pure water ($\epsilon_1 = 80$) dispersed in petroleum ($\epsilon_0 = 2$) this becomes

$$\bar{\epsilon} = \epsilon_0 (1 + 2.778\phi)$$

With the mutual perturbation of the droplets taken into account, one has

$$\bar{\epsilon} = \epsilon_0 [1 + 3\phi/(1 - \phi - 1.65\phi^{10/3})] \tag{3.29}$$

Values computed with Eq. 3.29, up to $\phi = 0.30$, differ by not more than one per cent from those obtained with four terms of the development of the formula

$$\bar{\epsilon} = \epsilon_0 (1 + 2\phi)/(1 - \phi), \tag{3.30}$$

* For a more precise definition of this quantity, see p. 112.

derived from Wagner's[75c] expression for the electrical conductivity of a system of isotropic spheres. In the case of ideally conducting emulsion drops, i.e., water with a salt content of several tenths of a per cent or higher, the effective electric field is reduced by a factor of $(1 - \phi)$, slightly modifying Eq. 3.30:

$$\bar{\epsilon} = \epsilon_0(1 + \phi)/(1 - 2\phi) \qquad (3.31)$$

This last equation was tested on W/O emulsions of $2N$ aqueous solutions of sodium chloride in petroleum, up to values of $\phi = 0.30$, with satisfactory agreement between the observed and calculated values of the dielectric constant. Interestingly, the dielectric constants show an upward trend with time.

Electrical Conductivity

This property, related somewhat to the previous one, has been somewhat more extensively investigated. This is because, properly used, it may afford a convenient way of distinguishing between O/W and W/O emulsions (cf. p. 327).

Emulsions in which water is the continuous medium may be expected to show a high conductivity; emulsions in which the oil phase is the continuous medium may be expected to show little, or no, conductance.

For example, in an investigation of inversion (p. 146), Bhatnagar,[76] studied the conductivity of emulsions by measuring the current flowing between two fixed platinum electrodes immersed in the emulsion. He found that oil-in-water emulsions passed currents of 10–13 milliamperes, while water-in-oil emulsions allowed currents of the order of 0.1 milliampere or less. Clayton[77] has recorded the usefulness of conductivity measurements in studies of the dilution of cream.

Recent studies of emulsion conductivity have largely concerned themselves with petroleum emulsions. This is to be expected since electrical methods are widely employed for the separation of such emulsions (cf. p. 288).

Ben'kovskii[78] has reported on the conductivity of W/O emulsions of petroleum which itself had a conductivity of $1-2 \times 10^{-6}$ ohm cm. It was found that the conductivity of emulsions containing one or more parts of water for one part of oil was determined by the conductivity of the water used. The greater the water-oil ratio, the greater the conductivity. On the other hand, the conductivity of aged emulsions (200 days old) was independent of the ratio, for emulsions containing between 5 and 70 per cent water. The conductivity of the emulsions increased with temperature.

Similar results were obtained by Lifshits and Teodorovich,[79] who found that a petroleum emulsion containing 50 per cent water had a conductivity two to three times higher than the dry petroleum, while a 25° to 90° C temperature rise increased the conductivity by 10 to 20 times. Microscopic examination of emulsions in an electric field of 1000–2000 v/cm showed alignment of water droplets in strings and merging of droplets into larger drops. Similar observation of string format on was made by Dixon and Bennet-Clark[80] in olive-oil emulsions, and by Muth[81] in milk and cream. It is the oil globules which form the strings in the latter two cases.

A more complete discussion of the use of conductivity measurements to determine emulsion type will be found in a later section (p. 327).

Electrophoresis

As will be elaborated in the following chapter, the droplets of a stable emulsion most often possess a definite electrical charge. While the origin and theory of this charge will be considered later (pp. 103–125), an important consequence of the charge is properly considered in the present chapter.

If an emulsion (or, indeed, any lyophobic colloid) is subjected to an electric field, the droplets, in consequence of their charge, will migrate to one or another of the electrodes. In most of the early work, this phenomenon was termed *cataphoresis*; the designation *electrophoresis* is now generally used.

This effect can be observed, and the rate of migration measured, by direct microscopic observation of the droplets. However, a more usual technique is by means of the moving-boundary apparatus. In this apparatus the rate of motion of a boundary between the emulsion or colloidal sol being studied and the dispersion medium is observed under the influence of an electric field. The rate of motion of the boundary is equal to the average rate of motion of the particles (cf. p. 333 for a more detailed description of this apparatus).

Smoluchowski[82] derived a theoretical equation (see infra Eq. 4.2) for the velocity of electrophoresis as a function of the applied voltage. It was pointed out by Debye and Hückel[83] that, while the Smoluchowski equation was valid in form, the constant term varied with the shape of the particle. Henry[84] was able to demonstrate that, for spherical particles, the equation takes the form

$$v = \zeta \epsilon E / 6\pi\eta, \tag{3.26}$$

where v is the particle velocity, ζ is the potential on the particle arising from the charge mentioned above (the so-called zeta-potential, ϵ is the dielectric constant of the dispersion medium, E the applied voltage per unit distance between the electrodes (volts/cm), and η is the viscosity of the dispersion medium. Since v can be determined experimentally as a function of E, as indicated above, while ϵ and η are known properties of the dispersion medium, Eq. 3.26 may be used to calculate ζ, the zeta-potential. Since the value and properties of the zeta-potential are significant in the theory of emulsion stability, being able to solve Eq. 3.26 for ζ is extremely useful.

Recently, Booth[85] has pointed out that Eq. 3.26 properly applies only to *solid* particles, and thus is not applicable to emulsions. In an elaborate calculation, he develops a theory of the electrophoresis of spherical fluid droplets in an electrolyte. A general formula is given for the electrophoretic velocity in terms of the applied field, the potential distribution in and near the drops, and the dielectric constants and coefficients of viscosity of the liquid and electrolyte. It is assumed that the liquid drop retains its spherical shape in electrophoresis and also that relaxation effects in the electrolyte may be neglected.

Detailed formulae are worked out for the case of a Debye-Hückel potential in the electrolyte, and for the following charge distributions in the field sphere:

1. No charge within the sphere; the whole charge is concentrated on the interface.

2. Uniform volume distribution of charge.

3. Ionic double layer potential in the sphere.

Booth is able to show that the derived equations agree with previous computations in special cases, e.g., the equation derived under assumed charge distribution (1) reduces to the Smoluchowski equation for nonconducting solid particles.

With suitable minor assumptions Booth's rather complex equations (for which the reader is referred to the original paper) can be used for the calculation of the zeta-potential from electrophoretic data, but the error involved in using the simpler Eq. 3.26 is rather small in view of other uncertainties.

Electrophoretic data is usually reported in terms of the *mobility* $u = v/E$, with usual units of $\mu/\sec/\text{volt}/\text{cm}$, i.e., the velocity of the droplet under a potential of one volt per centimeter.

Data on Electrophoresis of Emulsions. Early work on the electrophoresis of emulsions dealt with the effects of pH and of various ions on the mobility. Roberts[86] and Dickinson[87] made elaborate studies of these phenomena. The latter studied the effect of the pH in O/W systems in which the oil phase was *n*-octadecane, paraffin wax, cetyl and octyl iodide, cetyl bromide, and cetyl, octadecyl, and lauryl chloride. Measurements were carried out in the presence of 0.01 N Na^+ and at $25°$ C over the pH range 2.0 to 11.8.

All the emulsion particles were found to carry a small positive charge at the lowest pH values, but as the pH was increased the charge reversed and finally attained an almost constant negative value above pH 9. Dickinson discussed the source of the negative charge on the particle in relation to the effect of pH. The negative charge is ascribed to the adsorption of the hydroxyl ions, in agreement with Roberts.[86] The halide molecules at the interface orient so that their halide groups project toward the aqueous phase.

Moving boundary studies were carried out by Growney,[88] in order to measure the mobilities of droplets of ethyl laurate (liquid) and ethyl stearate (solid) containing various amounts of stearic acid. The measurements were carried out under conditions similar to those studied by Dickinson.[87] Growney observed that the mobilities for these compound droplets do not have an intermediate value between the mobilities of the two constituents by themselves. This indicates that the orientation of the surface layer of the stearic acid in the composite droplet differs from that which occurs in the case of particles of stearic acid by itself. In other words, the pattern of the tightly-bound surface charges is different, and hence the zeta-potential is different.

As will be discussed more elaborately in the case of stabilized emulsions, the surface charge on the droplet arises from the ions of the surface-active materials which are adsorbed at the interface. Thus, it would appear that there should be a clear-cut relation between the effect of surface-active agents on interfacial tension and upon electrophoretic mobility. That this is not the case, in some instances at least, is indicated by the measurements of Powney and Wood.[89]

Measurements of electrophoretic mobilities of oil drops in solutions of paraffin chain salts were compared with interfacial-tension data for the corresponding oil-paraffin chain salt solution interfaces. The systems examined were: *Nujol* with 1-dodecylpyridinium iodide, sodium tetradecyl sulfate, sodium dodecyl sulfate, and sodium laurate. Powney and Wood report that the mobility of *Nujol* (a highly-refined paraffin oil) in water alone is -4.35 μ/sec/volt/cm at 25° C. However, the addition of only 0.000075 per cent of the pyridinium salt reduced the mobility to zero. On further increasing the concentration to 0.0004 per cent the mobility attains a maximum value of $+5.8$. On the other hand, for this salt, lowering of the interfacial tensions (and hence strong surface adsorption) is inappreciable at concentrations below 0.01 per cent. It is only at the C.M.C. (approximately 0.4 per cent) that any correlation is found to exist between mobility and interfacial tension. The high positive mobility is, without doubt, due to the presence of positively charged dodecylpyridinium ions. In the absence of low interfacial tensions, however, they cannot be in the interface as part of the strongly-bound portion of the double layer; rather they are in the diffuse portion. It then becomes something of a problem to explain why these ions should have a stronger effect on the mobility than those of a simple uni-univalent electrolyte such as sodium chloride. This point will be returned to in the subsequent discussion.

More recently, Powell and Alexander[90] have examined the electrophoretic mobility of *Nujol* droplets in aqueous solutions of the sodium salts of dioctyl, dihexyl, and diamyl esters of sulfosuccinic acids. Measurements were carried out at 25° C over concentration ranges of 0–3 per cent in both water and in 0.05 M buffer of pH 6. In all cases the mobility versus detergent-concentration curve rises steeply to a plateau. The maximum mobility in water is about 8μ/sec/volt/cm, and 5.5 μ/sec/volt/cm in the pH 6 buffer. The buffer solution curve rises more steeply in the low concentration range, reaching its plateau at about 0.05 per cent; whereas in water the plateau occurred above 0.2 to 0.5 per cent, since in buffer solution the added ions tend to reduce the repulsive forces between adsorbed long-chain anions (thus increasing their tendency to be adsorbed). Powell and Alexander also made an estimate of the ratio of the number of gegen-ions (or counter-ions, i.e., the ions bound loosely in the outer layer of the double layer) adsorbed on the *Nujol* droplets to the number of long-chain anions. This was found to be of the order of 0.9, the value decreasing as the detergent concentration increased. In buffer solutions, the ratio was smaller than in water.

Jordan and Taylor[91] have discussed the technique of measuring the electrophoretic mobility of freshly formed hydrocarbon droplets. They also studied the influence of added ethyl alcohol on the mobility of decalin droplets.

A very complete study of the electrophoresis of resin emulsions was carried out by Munro and Sexsmith.[92] A glass cell was used for the direct measurement of the mobility of over 2500 particles of polyvinyl acetate emulsified in water with over twenty emulsifiers. Absolute mobilities at 20° C varied from 0.49 to 5.28 and from −0.49 to −3.22 μ/sec/volt/cm. No zero charges on the particles were observed. The nature of the emulsifying agent and electrolyte concentration, rather than concentration of the polymer, were the most significant variables affecting the mobility of these latices. The effect of pH changes from about 4 to 10, and of solids concentration changes from about 0.005 to 0.07 per cent were found to be unimportant.

Very little work has been on the subject of electrophoresis in nonaqueous media, although this is obviously of importance to our understanding of W/O emulsions. Recently, Hayek[93] has developed a technique which would apparently be suitable for such studies. In connection with studies on detergents for lubricating oils, a micro-electrophoresis cell was constructed and a technique developed for the measurement of mobilities of suspended particles in nonconducting media.

Bibliography

1. Sutheim, G. M. in Matiello, J. J., "Protective and Decorative Coatings," **4**, p. 282, New York, John Wiley & Sons, Inc., 1944.
2. Gabriel, L. G. in "Emulsion Technology," 2nd ed., p. 259, New York, Chemical Publishing Co., 1946.
3. Fischer, E. K. and Harkins, W. D., *J. Phys. Chem.* **36**, 98 (1932); Harkins, W. D., "The Physical Chemistry of Surface Films," pp. 86–90, New York, Reinhold Publishing Corp., 1952.
4. Berkman, S. and Egloff, G., "Emulsions and Foams," p. 55, New York, Reinhold Publishing Corp., 1941.
5. King, A. and Mukherjee, L. N., *J. Soc. Chem. Ind. (London)* **59**, 185 (1940).
6. Wiley, R. M., *J. Colloid Sci.* **9**, 427 (1954).
7. Lewis, W. C. McC., *Kolloid-Z.* **4**, 211 (1909); **5**, 91 (1909).
8. Bowcutt, J. E. and Schulman, J. H., *Z. Elektrochem.* **59**, 283 (1955).
9. Jürgen-Lohmann, L., *Kolloid-Z.* **124**, 41 (1951).
10. Kiyama, R., Kinoshita, H., and Suzuki, K., *Rev. Phys. Chem. Japan* **21**, 82 (1951).
11. Isakovich, M. A., *Zhur. Eksptl. Teoret. Fiz.* **18**, 907 (1948); *C.A.* **46**, 10783c.
12. Jellinek, H. H. G., *J. Soc. Chem. Ind. (London)* **69**, 225 (1950).
13. Harkins, W. D., *op. cit.*, p. 87.
14. Cooper, F. A., *J. Soc. Chem. Ind. (London)* **56**, 447T (1937).
15. Russ, A., *Seifen-Öle-Fette-Wachse* **76**, 537 (1950).
16. Griffin, W. C. in Kirk-Othmer "Encyclopedia of Chemical Technology," **5**, p. 695, New York, Interscience Encyclopedia, Inc., 1950.
17. Langlois, G. E., Gullberg, J. E., and Vermeulen, T., *Rev. Sci. Instruments*, **25**, 360 (1954).
18. Exner, S., *Ann. Physik* **2**, 843 (1900).
19. Clayton, W., "Theory of Emulsions," 4th. Ed. pp. 150–157, Philadelphia, The Blakiston Co., 1943.
20. Bailey, E. D., Nichols, J. B., and Kraemer, E. O., *J. Phys. Chem.* **40**, 1149 (1936).
21. Lothian, G. F. and Chappel, F. P., *J. Appl. Chem. (London)* **1**, 475 (1951).
22. Schulman, J. H. and Riley, D. P., *J. Colloid Sci.* **3**, 383 (1948); Oster, G. and Riley, D. P., *Acta Cryst.* **5**, 1 (1952); *Disc. Faraday Soc.* **11**, 107 (1951).
23. Wood, R. W., "Physical Optics," 3rd ed., pp. 115–117, New York, Macmillan Co., 1934.
23a. Francis, A. W., *J. Phys. Chem.* **56**, 510 (1952).
24. Andrade, E. N. da C., "Viscosity and Plasticity," p. 9, New York, Chemical Publishing Co., 1951.
25. Andrade, E. N. da C., *op. cit.*, p. 15.
26. Sherman, P., *Research (London)* **8**, 396 (1955).
27. Taylor, G. I., *Proc. Roy. Soc. (London)*, **A138**, 41 (1932).
28. Leviton, A. and Leighton, A., *J. Phys. Chem.* **40**, 71 (1936).
29. Oldroyd, J. G., *Proc. Roy. Soc. (London)*, **A218**, 122 (1953).
30. Bond, W. N. and Newton, D. A., *Phil. Mag.* **5**, 794 (1928).
31. Richardson, E. G., *Kolloid-Z.* **65**, 32 (1933).
32. Broughton, G. and Squires, L., *J. Phys. Chem.* **42**, 253 (1938).

33. Toms, B. A., *J. Chem. Soc.* **1941**, 542.
34. Sherman, P., *Mfg. Chemist* **26**, 306 (1955); *Kolloid-Z.* **6**, 141 (1955).
35. Einstein, A., *Ann. Physik* (4) **19**, 289 (1906); **34**, 591 (1911).
36. Guth, E. and Simha, R., *Kolloid-Z.* **74**, 266 (1936); Gold, O., *Dissertation*, Univ. of Vienna (1937).
37. Eilers, H., *Kolloid-Z.* **97**, 313 (1941); **102**, 154 (1943).
38. Sherman, P., *J. Soc. Chem. Ind. (London)* **69**, Suppl. No. 2, S70 (1950).
39. Oliver, D. R. and Ward, S. G., *Nature* **171**, 396 (1953).
40. Eirich, F., Bunzl, M., and Margaretha, H., *Kolloid-Z.* **74**, 276 (1936).
41. Broughton, G. and Windebank, C. S., *Ind. Eng. Chem.* **30**, 407 (1938).
42. Whitmore, R. L., *Dissertation*, Univ. of Birmingham (1949).
43. Eveson, G. F., *Dissertation*, Univ. of Birmingham (1950).
44. Nandi, H. N., *Dissertation*, Univ. of Birmingham (1949).
45. Higginbotham, G., *Dissertation*, Univ. of Birmingham (1951).
46. Williams, P. S., *Disc. Faraday Soc.* **11**, 47 (1951).
47. Simpson, G. K., *J. Oil & Colour Chemists' Assoc.* **32**, 60 (1949).
48. Hatschek, E., *Kolloid-Z.* **8**, 34 (1911).
49. Sibree, J. O., *Trans. Faraday Soc.* **26**, 26 (1930); **27**, 161 (1931).
50. Gabriel, L. G., *loc. cit.*, pp. 265–278.
51. Bredée, H. L. and de Booys, J., *Kolloid-Z.* **79**, 31, 43 (1937).
52. Maron, S. H. and Madow, B. P., *J. Colloid Sci.* **6**, 590 (1951); **8**, 130 (1953).
53. Arrhenius, S., *Z. phys. Chem.* **1**, 285 (1887); *Medd. Vetenskapakad. Nobel-Inst.* **4**, 13 (1916).
54. Fikentscher, H., *Cellulosechem.* **13**, 58 (1938).
55. Papkov, S., *Kunststoffe* **25**, 253 (1935).
56. Houwink, R., *Kolloid-Z.* **79**, 138 (1937).
57. Fikentscher, H. and Mark, H., *Kolloid-Z.* **49**, 135 (1930).
58. Sakurada, I., *Z. phys. Chem.* (B)**38**, 407 (1938).
59. Baker, F. and Mardles, E. W. J., *Trans. Faraday Soc.* **18**, 3 (1923).
60. Neogy, R. K. and Ghosh, B. N., *J. Indian Chem. Soc.* **29**, 573 (1952).
61. Neogy, R. K. and Ghosh, B. N., *ibid.* **30**, 113 (1953).
62. Sherman, P., *J. Soc. Chem. Ind. (London)* **69**, Suppl. No. 2, S74 (1950).
63. Mardles, E. W. J. and De Waele, A., *J. Colloid Sci.* **6**, 42 (1949).
64. Oldroyd, J. G., *Proc. Roy. Soc. (London)* **A232**, 567 (1955).
65. Criddle, D. W. and Meader, A. L., Jr., *J. Appl. Phys.* **26**, 838 (1955).
66. Serallach, J. A. and Jones, G., *Ind. Eng. Chem.* **23**, 1016 (1931); Serallach, J. A., Jones, G., and Owen, R. J., *ibid.* **25**, 816 (1933).
67. Smoluchowski, M. v., *Bull. acad. sci. Cracovie* **1903**, 182.
68. Van der Waarden, M., *J. Colloid Sci.* **9**, 215 (1954).
69. Lyttelton, D. U. and Traxler, R. N., *Ind. Eng. Chem.* **40**, 2115 (1948).
70. Richardson, E. G., *J. Colloid Sci.* **5**, 404 (1950); **8**, 367 (1953).
71. Ward, S. G. and Whitmore, R. L., *Brit. J. Appl. Phys.* **1**, 286 (1950).
72. Roscoe, R., *Brit. J. Appl. Phys.* **3**, 267 (1952).
73. Clayton, W., *op. cit.*, pp. 172–176.
74. Fricke, H. and Curtis, H. J., *J. Phys. Chem.* **41**, 729 (1937).
75. Kubo, M. and Nakamura, S., *Bull. Chem. Soc. Japan* **26**, 318 (1953).
75a. Frenkel, Ya. I., *Kolloid. Zhur.* **10**, 148 (1948); *C.A.* **43**, 7776f.
75b. Fradkina, E. M., *Zhur. Eksptl. Teoret. Fiz.* **20**, 1011 (1950); *C.A.* **45**, 2743c; cf. also Frenkel, Ya. I. and Fradkina, E. M., *Kolloid. Zhur.* **10**, 241 (1948); *C.A* **43**, 7776g.

75c. Wagner, K., *Arch. Elektrotech.* **2,** 378 (1914).
76. Bhatnagar, S. S., *J. Chem. Soc.* **117,** 542 (1920).
77. Clayton, W., *op. cit.*, p. 178.
78. Ben'kovskii, V. G., *Kolloid. Zhur.* **14,** 10 (1952); *Colloid J.* (U.S.S.R.) **14,** 11 (1952) (Engl. trans.); *C.A.* **46,** 4774*h.*
79. Lifshits, S. G. and Teodorovich, V. P., *Energet. Byull.* **1947,** No. 8, 16; *C.A.* **42,** 1411*f.*
80. Dixon, H. H. and Bennet-Clark, T. A., *Nature* **124,** 650 (1929).
81. Muth, E., *Kolloid-Z.* **41,** 97 (1927).
82. Smoluchowski, M. v., *Z. physik. Chem.* **93,** 129 (1918).
83. Debye, P. and Hückel, E., *Phys. Z.* **25,** 49 (1924).
84. Henry, D. C., *Proc. Roy. Soc.* (*London*) **A133,** 106 (1931).
85. Booth, F., *J. Chem. Phys.* **19,** 1331 (1951).
86. Roberts, A. L., *Trans. Faraday Soc.* **32,** 1705 (1936); **33,** 643 (1937).
87. Dickinson, W., *Trans. Faraday Soc.* **37,** 140 (1941).
88. Growney, G., *Trans. Faraday Soc.* **37,** 148 (1941).
89. Powney, J. and Wood, L. J., *Trans. Faraday Soc.* **37,** 152 (1941).
90. Powell, B. D. and Alexander, A. E., *Can. J. Chem.* **30,** 1044 (1952).
91. Jordan, D. O. and Taylor, A. J., *Trans. Faraday Soc.* **48,** 346 (1952).
92. Munro, L. A. and Sexsmith, F. H., *Can. J. Chem.* **31,** 287 (1953).
93. Hayek, M., *J. Phys. and Colloid Chem.* **55,** 1527 (1951).

CHAPTER 4

Theory of Emulsions: Stability

It is an unfortunate fact that there exists no single coherent theory of emulsion formation and stability. By and large, the various workers in the field, even in theoretical discussions, have limited themselves to specialized systems; and what has been found to be true for certain systems under certain conditions is not necessarily applicable to other systems. The author has pointed out, in an earlier work,[1] that emulsion theory has progressed to a point where some sort of theoretical interpretation of emulsion behavior is possible; the prediction of emulsion behavior is still largely a matter of art rather than science.

Schwartz and Perry[2] have indicated that a proper theory of emulsions should be capable of explaining the following important behavior characteristics:

1. Formation
2. Stability
3. Breaking and inversion
4. The role of emulsifying agents and other chemical factors such as pH and non-surface-active ions
5. The influence of physical factors

It is not possible to organize the following discussion strictly on the basis of this listing, but it can serve as a useful guide. In the following, the earlier theories of emulsions will be briefly mentioned, and the more recent work will be reviewed in detail.

Simple Theories of Emulsion Stability

It has been pointed out earlier that emulsions, in various forms, have been known for thousands of years. An indication of the antiquity of emulsions is the fact that the Greek physician Galen (131-c.201) was apparently the first person to record the emulsifying power of beeswax.

The very antiquity of emulsion formulation is a hinderance to postulating a scientific theory of emulsions. A great many accumulated "facts" must be jettisonned before any comprehensive theory can be constructed. For example, Bancroft[3] cites the following: an emulsion must be mixed with a wooden paddle; once an emulsion has been stirred in a given direction, the direction of mixing must not be changed; and (completely disqualifying this author!) a left-handed man cannot make a stable emulsion.

Early Work on the Effect of Emulsifying Agents. Only as recently as 1910 did Wo. Ostwald make the all-important distinction between the common form of emulsion, oil-in-water, and the then extremely uncommon water-in-oil form. It was immediately recognized that the type of emulsion which eventuated in a given system depended to a large extent on the choice of emulsifying agent. Thus Clowes[4] found quite early that sodium, potassium, and lithium soaps will give O/W emulsions; while magnesium, strontium, barium, iron, and aluminum soaps give W/O emulsions. In addition to these, Holmes[5] listed gelatin, saponin, albumin, lecithin, and casein (either acid or alkaline) as O/W emulsifiers, and gum dammar, lanolin, rosin, rubber, and cellulose nitrate as W/O emulsifiers. However, the mechanism involved, and the basic origin of stability or instability was not clearly understood.

It was recognized quite early that many of the most valuable emulsifying agents were surface-active materials, and Quincke[6] and Donnan and Potts[7] attached considerable importance to the effect of the emulsifier on the interfacial tension. That this is a significant factor can be readily shown.[8] For example, referring to Table 2–3, it is seen that the interfacial tension of olive oil against water at 20° C is 22.9 dynes/cm. Now, if ten cubic centimeters of this oil is emulsified into droplets having a radius of 0.1 μ, the total interfacial area created is 300 sq m.

Substituting these data into Eq. 2.2, the work required to do this would be 1.64 g-cal. On a more practical scale, the work necessary to emulsify one hundred pounds of olive oil would be 8.09 *kilogram-colaries*. Since this energy is contained in the system as potential energy, it represents a considerable degree of thermodynamic instability. On the other hand, as little as 2 per cent of a suitable soap could reduce the interfacial tension to as low as 2 dynes/cm. This would reduce the interfacial energy from 8.09 to 0.75 kg-cal.

Although this is a significant effect, the emulsion is still thermodynamically unstable; and the lowering of interfacial tension, while important, cannot begin to represent the totality of the effect of the emulsifying agent. In addition, this cannot explain the effectiveness of emulsions stabilized with non-surface-active materials, e.g., gums, fine solids, etc.

Bancroft[9] apparently was the first to point out the significance of the Gibbs equation (Eq. 2.24) in this connection. The lowering of the interfacial tension carries with it the correlary that the emulsifying agent is concentrated in the interface, and forms some sort of interfacial film which exerts a stabilizing influence. Although Bancroft apparently regarded the film as having an almost purely mechanical effect, and although interfacial films having a definate structure have been observed (cf. Chapter 3, ref. 66), it is doubtful that the interfacial films formed by surface-active materials owe much of their stabilizing effect to mechanical effects. Lewis[10] evidently recognized the importance of surface charge on the emulsion droplets (arising from the double layer) at an early date.

It was pointed out above that the type of emulsion was recognized as being derived, in part at least, from the nature of the emulsifying agent. In early work, no theoretical basis for this was given, but Bancroft[9] stated a general rule which has considerable general validity. The original statement is somewhat complicated, but it may be simply expressed by saying that the phase in which the emulsifying agent is more soluble will be the external one.

A more elaborate form of this theory was stated by Bancroft and Tucker.[11] Pointing out that the existence of an interfacial film requires the presence of *two* interfacial tensions (i.e., one on each side of the interfacial film), the film would curve in the direction of the higher interfacial tension. Thus the disperse phase would be on the side of the film with the higher interfacial tension.

Clowes[4] elaborated this theory, but his work properly relates to the phenomenon of inversion, which will be considered below.

Adsorption Theories of Emulsion Stability. While the adsorption of the surface-active emulsifying agent into the interface was recognized at an early date, a more sophisticated approach had to await the theoretical approach of Langmuir and Harkins.[12] Based on the results of orientation studies on surface-active materials by means of the hydrophil balance (pp. 15–21), these workers discarded the concept of a strong multimolecular film in favor of an oriented monomolecular one, with the polar groups oriented towards the water phase and the nonpolar hydrocarbon chains orientated toward the oil. Thus, the condition existing, for example, in an O/W emulsion stabilized by, say, a sodium soap, could be represented schematically as in Figure 4-1.

Experimental justification for this picture was found in a set of painstaking experiments by Fischer and Harkins.[13] In this work a large number of mineral oil-in-water emulsions were prepared at various emulsifier concentrations and the droplet size distribution determined as a function of time. It was found that emulsions stabilized with 0.1 M soap solutions showed scarcely any change in distribution, while those stabilized with lower soap concentrations, e.g., 0.005 M, changed radically as a function of time (cf. Figure 3-2, which gives some of their data). For example, calculations based

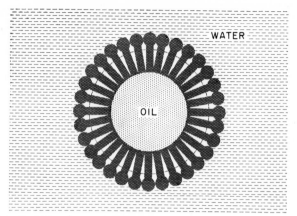

FIGURE 4-1. Stabilization of an O/W emulsion by the soap of a monovalent metal ("oriented wedge").

on the known concentration of emulsifier, and the calculated interfacial areas, showed that in an emulsion stabilized with 0.02 M sodium oleate the mean area per molecule of soap in the surface was reduced from 44.6 sq Å to 19.5 sq Å in 648 hours. This occurs as a result of the coagulation of smaller drops.

Additional studies, on ten different emulsions stabilized with sodium oleate, led to figures for the mean interfacial area per soap molecule of 27 to 45 sq Å. These mean areas are not sufficiently small to allow for more than one molecule as the thickness of the interfacial skin.

Table 4-1 reproduces a portion of Fischer and Harkins data, and gives a clear idea of the wide range of emulsions considered.

The areas given in the fourth column of Table 4-1 were calculated on the basis of analysis of the emulsion for oleic acid; values obtained for these areas by sodium analysis are reported in the original paper, and are in excellent agreement.

Oriented Wedge and Emulsion Type. The orientation concept leads to a simple explanation of the effect of emulsifying agent on emulsion type. As pointed out above, while the soaps of the univalent metals (e.g., sodium) give rise to O/W emulsions, those of the multivalent metals favor the formation of W/O.[4] Since the oriented adsorption concept requires a fairly close-packed monomolecular film at the interface, it should be recognized that geometrical considerations must play a part. Reference to Figure 4-2, portraying the situation in an emulsion stabilized by a bivalent metal soap, gives a clear idea of how the geometry virtually requires the formation of the water-in-oil type of emulsion.

TABLE 4-1. MOLECULAR AREAS FOR THE SOAP AT THE INTERFACIAL FILM IN
OIL-IN-WATER EMULSIONS[13]

Age of Emulsion (hrs.)	Specific Interfacial Area (cm² per cc × 10⁻³)	Molarity of Equilibrium Soap	Area per Sodium Oleate Molecule (sq Å)
18	11.2	0.0025	44.5
138	6.08	.0043	26.1
78	11.9	.0047	18.2
720	5.58	.0053	30.2
168	6.00	.0059	24.2
3	5.43	.00718	27.8
48	6.64	.00634	27.6
144	11.2	.00920	38.2
190	6.17	.0199	37.4
72	9.96	.0341	33.6
36	10.8	.0655	32.4
3	12.9	.112	27.0

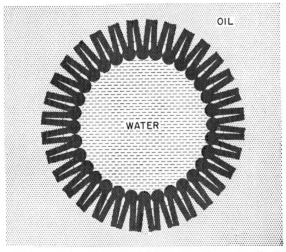

FIGURE 4-2. Stabilization of a W/O emulsion by the soap of a bivalent metal ("oriented wedge").

Examination of Figures 4-1 and 4-2 makes perfectly clear the designation, "oriented wedge," which Harkins[14] applied to this theory. Unfortunately, exceptions to this theory are known. For example, silver soaps which should, by the oriented wedge concept, yield O/W emulsions actually stabilize the opposite type. The simple solubility rule of Bancroft obviously has more validity in this case. Wellman and Tartar[15] have reported a number of other exceptions to the rule.

On the other hand, the oriented wedge theory has a certain conceptual value. The well known phenomenon of the inversion of soap-stabilized emulsions on the addition of salts of multivalent metals is clearly related to this explanation. Clowes[4] has shown that in mixtures of uni- and divalent ions the type of emulsion depends strongly on the ratio of the ionic concentrations; he has termed this phenomenon *ion antagonism.*

Phase Volume and Emulsion Type. It has been pointed out earlier, in the discussion of the effect of volume concentration on emulsion viscosity (pp. 59–64), that a relation also exists between emulsion type and the relative phase volumes. This *phase volume* theory was proposed by Wo. Ostwald[16] on purely stereometric grounds.

It is a fact of solid geometry that an assembly of spheres of equal radius can be placed in a position of densest packing in *two* ways. In either case, however, the spheres are found to occupy 74.02 per cent of the total assembly volume, the remaining 25.98 per cent being empty space. Ostwald's theory supposed that at a phase volume $\phi > 0.74$, the emulsion would, so to speak, be packed more densely than was possible. This means that any attempt to exceed a phase volume of 0.74 for the internal phase must result in either inversion or breaking.

In the previous chapter, reference was made to the effect of inversion on the viscosity of the emulsion (cf. Figure 3-5). It was indicated that inversion occurred sharply at the volume concentration $\phi = 0.74$. According to the Ostwald phase volume theory inversion must always take place at this concentration. That is, for a given system, both O/W and W/O emulsions are possible between the phase volume concentrations 0.26 and 0.74, below the one and above the other only one form can exist.

There exists a reasonable amount of experimental justification for this view. Bhatnagar,[17] for example, in a careful investigation of emulsions of olive oil in KOH by conductometric methods, found that the phase volume rule held for systems in which the alkali concentration was quite low (0.001 N); but at higher concentrations the values diverged from the theoretical in the direction of higher phase ratios than the theory would allow. A well known exception to the rule is found in emulsions of Pickering,[18] who prepared paraffin oil-in-soap-solution emulsions containing up to 99 volume per cent of oil. It has, however, been argued that these last were not strictly emulsions, but dispersions of the oil in soap gels.[19] The author has prepared stable, undoubted O/W emulsions containing only 4 per cent water, using nonionic emulsifiers.

However this may be, Sherman[20] has shown that the volume concentration at which inversion occurs can be a function of the emulsifier concentration (cf. Table 3-9). An interesting, if not especially important, result in this connection is that reported by Kremann, Griengl, and Schreiner.[21]

These investigators, working with olive oil, water, and aqueous solutions of sodium hydroxide and hydrochloric acid, measured the change of viscosity with olive oil content. With up to 23 per cent of olive oil, the emulsion (O/W) viscosity was little greater than that for pure water, but there was a sudden rise to a maximum at 25 per cent. Further increases in the oil concentration led to a fall in viscosity, and inversion to unstable W/O emulsions. In effect, they obtained the results of Figure 3-5 by starting in the opposite direction.

In subsequent discussion more elaborate theories of inversion will be discussed, and the simple theory of Ostwald will be seen in its proper perspective. It is an excellent example of a simple and attractive theory which is vitiated by some readily observable physical facts. The theory requires that the droplets of the disperse phase be perfect, non-deformable, uniform-diameter spheres. The droplets are, in fact, none of these. In dilute emulsions, to be sure, the droplets are spherical or nearly so; but they are certainly not resistant to deformation, nor are they uniform in diameter. The various distribution curves shown in the earlier discussion (cf. Figures 3-1, 3-2, and 3-6) bear witness to the last of these statements.

Now while the maximum volume which can be filled by uniform spheres is 74.02 per cent, it is obvious that a less homogeneous assembly of spheres can be packed more densely (indeed, the less homogeneous the spheres the more densely they can be packed) by virtue of the fact that the smaller spheres can be fitted into the interstices between the larger ones. If we add to this the possibility that the droplets may be deformed into polyhedra, even denser packing is possible.

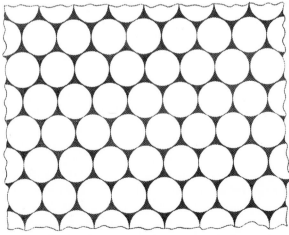

FIGURE 4-3. Schematic representation of a closely-packed emulsion of uniform spherical droplets. The droplets occupy 74.02 per cent of the volume.

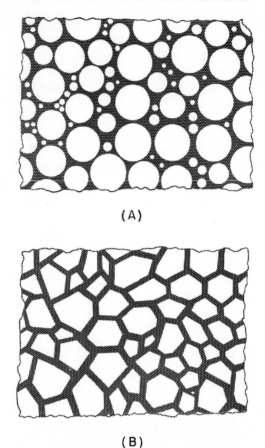

(A)

(B)

FIGURE 4-4. (A) Schematic representation of a closely-packed emulsion of non-uniform droplets. (B) A closely-packed emulsion of polyhedral droplets (extremely unstable).

Figure 4-3 indicates the idealized case of a homogeneous emulsion in which the droplets occupy 74.02 per cent of the volume. Figure 4-4(A) shows the more usual situation existing in a non-homogeneous emulsion, and Figure 4-4(B) represents the extreme case of an emulsion of closely packed nonuniform polyhedral droplets. Manegold[22] has called attention to the morphologic similarity between emulsions of this type and drained polyhedral foams. Figure 4-5, from Manegold, indicates the relationship between phase volume and the droplet form required for stability.

It should be recognized, however, that the polyhedral form (Figure 4-4B) represents an extreme case. An emulsion in which the concentration of disperse phase is so high that the droplets suffer deformation to polyhedra

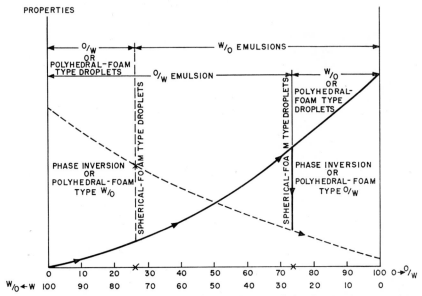

FIGURE 4-5. The relation between phase volume and the droplet form required for emulsion stability, according to Manegold.[22]

possesses an additional source of instability, inasmuch as the polyhedra have a greater surface area than the undistorted spheres.

Other Early Viewpoints. Hildebrand,[23] in 1941, reviewed the emulsion theories advanced to that date, considering particularly the question of which type of emulsion resulted. He pointed out that the factors contributing to the stability of one type of emulsion rather than the inverse type are supplemental rather than rival. The concept of the importance of the direction of film curvature for minimum energy[11] had probably been overemphasized, Hildebrand felt, at the expense of the more mechanical forces operating during the emulsification process. That the conditions of emulsification may very often control the emulsion type has been shown by more modern work, discussed in a subsequent section (p. 213).

Hildebrand[23] also pointed out that the rupture of a film separating two droplets can be resisted by the larger rise in the interfacial tension at the threatened point, which results if the reserve emulsifying agent is dissolved in the liquid forming the film, i.e., in the external phase, owing to the lower rate of adsorption in the case considered. This is, of course, a more sophisticated justification for Bancroft's rule, relating the solubility of the emulsifying agent with the nature of the external phase.

In considering these early theoretical approaches to a theory of emulsions, it will be apparent that most of the workers took the problem of emulsion type as being the basic one, evidently considering the question of the basic stability of emulsions in general as being a self-evident situation, completely covered by the theory of Quincke.[6]

An exception to this is the elaborate, if qualitative, theory of Roberts,[24] which represents in many respects a synthesis of the best elements of some of the previously quoted work. Roberts apparently anticipated the more modern concept of the hydrophile-lipophile balance* as regulating the emulsifying power of a particular emulsifying agent. He pointed out that the adsorption of the ions of the surface-active molecule in an interface increased as the ratio of charge to mass decreased (in the sense that the charge of the ion of, for example, a soap is localized in the hydrophilic carboxyl group). Thus, the emulsifying agent will be adsorbed in the interface in preference to smaller ions where the ratio charge/mass is not so favorable to the ions.

Roberts also pictured the molecule (or ion) of emulsifying agent as extending into both phases of the emulsion (in agreement with the Harkins-Langmuir concept). He recognized that this arrangement would lead to the formation of a Helmholtz double layer at the interface. In the case of a W/O emulsion this arrangement could lead to *two* double layers, one on each side sides of the interface, and that this double layer had an important stabilizing influence.

Modern Developments in the Theory of Emulsion Stability

Thus the early work in the field of emulsion stability concerned itself principally with emulsion type and with the concomitant phenomenon of inversion. Any theory of emulsion stability, as indicated at the beginning of the present chapter, must explain these phenomena; and these phenomena must arise as logical results of the theory's assumptions. It must be admitted at the outset that modern investigations (since about 1941) have fallen short of developing such a logical theory.

A number of important and somewhat related investigations will now be discussed, together with their theoretical implications. The effect of electric charge and the double layer will be covered in another section of this chapter.

Nature of the Interfacial Film. King[25] has pointed out that there are three main types of emulsions:

1. *Oil hydrosols*. These are unstabilized emulsions, mostly O/W, and contain less than one per cent internal phase.

---

* Abbreviated HLB (cf. pp. 189–196).

2. *Emulsions stabilized by electrolytes.* Cheesman and King[26] have shown that electrolytes in low concentrations are capable of stabilizing W/O emulsions.

3. *Emulsions stabilized by emulsifying agents.* This class includes what is commonly thought of as being an emulsion, the emulsifying agents being either colloidal materials (including surface-active agents) or finely divided solids.

King[25] supplied an interesting sidelight on the last two types by pointing out that sodium oleate at a concentration of one per cent normally gives an oil-in-water emulsion; at much lower concentrations, however, a water-in-oil emulsion may result, i.e., the sodium oleate is behaving as a normal electrolyte rather than as a colloidal electrolyte.

Considering only stabilized emulsions, King felt that the strength and compactness of the interfacial film are the most important factors favoring stability, and that other factors are subsidiary except insofar as they affect the properties of the films. Among these subsidiary factors the concentration of the emulsifying agent is significant, for the stability of an emulsion of a moderately high oil concentration is seriously affected by, for example, electrolytes only if the amount of emulsifying agent is insufficient to form a coherent film around the oil globule. King felt that interfacial tension plays a doubtful role in the stability of emulsions, but that the interfacial adsorption of the emulsifying agent which accompanied it was vital. Similarly, he felt that viscosity itself cannot explain emulsion stability, although it is generally true that viscous emulsions are more stable than mobile ones. Thus, as in the case of the interfacial tension, the high viscosity would be the symptom of the stability rather than the cause. This, of course, is in agreement with the more recent conclusions of Neogy and Ghosh[27] (p. 68); on the other hand, inhibition of creaming can be shown to be definitely related to emulsion viscosity (p. 135).

Rather similar conclusions as to the importance of the physical properties of the interfacial layer were reached by Coutinho[28] and Lachampt and Dervichian.[29] On the whole, however, these studies still represent a qualitative approach. For example, King[25] can suggest little more than the actual concentration of emulsifying agent as a quantitative source of film strength.

A more quantitative approach arises from observations by Schulman and co-workers.[30-32] In studies on monomolecular films carried out with the aid of the Langmuir-Harkins hydrophil balance, it was noted that, in certain cases, when a water-insoluble film was spread on a solution of a soluble material (e.g., a surface-active agent) a certain amount of the solute in the substrate penetrated into the insoluble monolayer. Schulman and Friend[32] point out that some substances which merely penetrate the unimolecular

. film on water are displaced when the film is compressed. In other cases, however, a definite molecular complex is formed between the insoluble material of the monolayer and the penetrating molecules. When this occurs, the complex film withstands greater pressures than does either component alone.

Films of Mixed Emulsifiers. The significance of this phenomenon for emulsions was clarified by Schulman and Cockbain.[33] It was reasoned that what was true for the air-water interface should also be valid for the oil-water interface. That is, if complex formation should take place at the interface, the resulting interfacial film should possess greater strength and resistance to rupture; hence, the emulsion droplets would be less liable to coalescence and the emulsion more stable.

These workers carried out an extensive investigation of oil-in-water emulsions using mineral oil (*Nujol*) as the disperse phase, and a large number of emulsifying agents, including sodium cetyl sulfate, cholesterol, oleyl alcohol, elaidyl alcohol, cetyl alcohol, sodium stearate, sodium elaidate, sodium oleate, cetyl sulfate, cetyl sulfonate, sodium palmitate, cetyl trimethylammonium bromide, sodium taurocholate, sodium glycocholate, and sodium desoxycholate. Emulsion stability was determined by measuring the time required for the first visible signs of separation of the two phases.

It was concluded that the conditions for maximum stability of O/W emulsions are satisfied when the interfacial film is charged, is completely covered with the charged molecules and is stable. This last serves to keep the interfacial tension at a minimum.[34]

The denseness of packing is apparently best realized in those cases where a distinct complex film is formed. Further, it was found that this was best realized for the emulsion system by having, in addition to the water-soluble emulsifying agent, an oil-soluble material capable of interacting with the detergent at the interface. The surface charges are produced by the use of the ionizable water-soluble emulsifying agent.

In general, it was also found that the same substances which form the stable complexes at the air-water interface also stabilize emulsions best; thus sodium cetyl sulfate in the water, with cholesterol or elaidyl alcohol in the oil, give both complex films at the air-water interface and stable emulsions. On the other hand, cholesteryl esters and oleyl alcohol (stereoisomeric to elaidyl alcohol) do neither. Figure 4-6, adapted from Schulman and Cockbain,[33] gives an idea of how the stereometrical packing of the molecules leads to the stable emulsion type. As implied by Figure 4-6(A), the proportion between the amounts of water-soluble and oil-soluble agents should be that required to give a condensed interfacial film containing equal numbers of molecules of each agent. It was also found the stability was

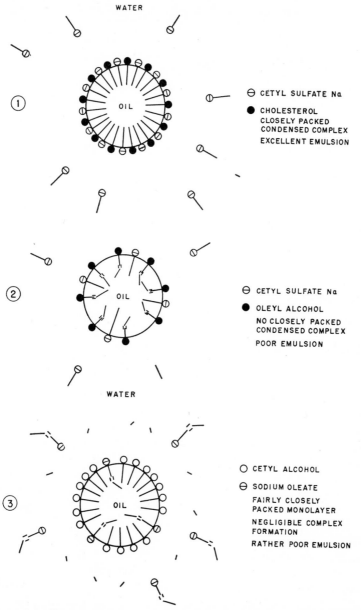

WATER

1. ⊖ CETYL SULFATE Na
   ● CHOLESTEROL
   CLOSELY PACKED
   CONDENSED COMPLEX
   EXCELLENT EMULSION

2. ⊖ CETYL SULFATE Na
   ● OLEYL ALCOHOL
   NO CLOSELY PACKED
   CONDENSED COMPLEX
   POOR EMULSION

WATER

3. ○ CETYL ALCOHOL
   ⊖ SODIUM OLEATE
   FAIRLY CLOSELY
   PACKED MONOLAYER
   NEGLIGIBLE COMPLEX
   FORMATION
   RATHER POOR EMULSION

FIGURE 4-6. Complex formation at the oil-water interface in emulsions, according to Schulman and Cockbain.[33] (1) Sodium cetyl sulfate and cholesterol form a closely-packed condensed complex, giving a good emulsion. (2) Sodium sulfate and oleyl alcohol form a poorly packed complex (because of the double bond in the alcohol), giving a poor emulsion. (3) Cetyl alcohol and sodium oleate form a fairly close-packed complex monolayer, hence form a fair emulsion.

best when the droplets were no larger than 3 μ in diameter, and smaller diameters were preferable.

The conclusion that can be drawn from these data is that considerable information may be obtained on the nature of the interfacial film in emulsions by the study of the air-water interface. This conclusion should be evaluated cautiously, even though Alexander[35] has shown that most of the properties of monolayers at the oil-water interface are obtainable from data at the air-water interface from the equation of state and the interfacial tension. He has established that, for small lowerings of the interfacial tension, the areas occupied per molecule at the oil-water interface are very much greater than at the air-water interface; but for increased lowerings of the interfacial tension, the two areas approach until an identical packing is obtained. At the oil-water interface all simple compounds were found to obey the following relation

$$(\gamma - \gamma_0)(A - A_0) = C, \qquad (4.1)$$

where γ is the interfacial tension, A is the area per molecule, and γ_0, A_0, and C are constants.

It is also interesting that complex formation of the type proposed by Schulman and Cockbain[33] has been found to have an important effect on the drainage properties of foams, and thus to contribute basically to the strength of the thin film making up the walls of the foam bubble.[36]

It must be conceded that the general picture which has been presented has considerable validity. For example, it has long been recognized that mixed emulsifying agents very often lead to more stable emulsions than do single ones, and, in many cases, this may be a result of complex formation at the interface. Aickin[37] has shown that the removal of mineral oil from wool is facilitated by the addition of simple polar compounds, e.g., oleyl alcohol, to the detergent solution; and, in agreement with the results quoted above, it is oil-soluble additives that have the desired effect. The end result of the wool-scouring procedure is, of course, a stable oil-in-water emulsion. Aickin points out that the addition of the oleyl alcohol reduces the surface tension, and apparently is inclined to regard this as the major operative effect. Earlier discussion[33, 34] will make it clear that this is a symptom of the improved packing rather than a cause of better emulsification.

Reservations to the Schulman-Cockbain picture must, however, be expressed. For example, the data of Fischer and Harkins[13] (cited earlier, Table 4-1) indicates that maximum stability for the emulsions studied was obtained over a large range of areas per molecule of stabilizer at the interface. In fact, if the soap molecules are oriented in the manner supposed, data obtained on the Harkins-Langmuir film balance indicates that the tightest packing of molecules (and hence, presumably, the most rigid interfacial film) would occur at an area per molecule of 20.5 sq Å (Table 2-5). However, Fischer and Harkins[13] found that stable emulsions can be formed when the surface area of the droplet per molecule of stabilizer varies between 24.2 and 44.5 sq Å (Table 4-1).

Further, the observations made by Schulman and Cockbain[38] on the subject of water-in-oil emulsions, while undoubtedly correct for the systems studied, are probably not generally correct. For example, in this type of system, stability is ascribed entirely to the physical nature of the interfacial film, and the presence of stabilizing charge is denied. Although the considerations embodied in this particular paper[38] lead to valuable conclusions regarding the process of inversion (discussed below, p. 138), evidence exists for the presence of charge in W/O emulsions.[39]

An interesting sidelight on the effect of mixed emulsifying agents (where the possibility of complex formation exists) is shown by the recent work of Dickinson and Iball.[40] Two emulsifying agents that together would be expected to give rise to complex formation, and separately to give rise to opposite emulsion types were employed in experiments on emulsions of cyclohexane-in-water and of mineral oil-in-water. Thus, sodium laurate was chosen for its ability to promote O/W emulsions and monoölein for W/O emulsions.

In the concentrations studied, sodium laurate is a poor emulsifier for the cyclohexane-water system, but the emulsion is reversed to a quite stable one by the addition of a small quantity of the monoölein. In mixtures of mineral oil, cyclohexane, and water emulsified with sodium laurate and monoölein, the emulsion type reversed at 50 per cent mineral oil. There appear to be two main factors contributing to this phenomenon: the purely hydrodynamic effect of viscosity on the emulsification *process*, and the effect of increased viscosity of the oil phase in retarding the diffusion of the monoölein to the interface during emulsification. Thus it appears that the rate and means of mixing and the phase volume ratios affect the emulsion type obtained. The precise physical structure of the interfacial film, while important, is not governing. This point will be returned to in the discussion of the physics of emulsification (p. 216 et seq.).

Kremnev[41] has recently studied the stabilization of concentrated emulsions by the use of mixed emulsifiers of the types studied by Schulman and Cockbain.[33] It was found that highly stable concentrated emulsions of benzene in water could be obtained by the use of thixotropic mixtures of aliphatic alcohols and sodium oleate, owing to their capacity of rapid restoration of the thixotropic structures, perturbed in the process of emulsification. With the alcohol introduced either in the benzene or in the 1 per cent sodium oleate solution, the stability of the emulsion increases with the chain length of the alcohol and with the concentration of the alcohol. Butyl, amyl, octyl, and cetyl alcohols were studied. Addition of 0.1 per cent of cetyl alcohol to the 1 per cent sodium oleate is enough to insure a very stable 60 volume per cent benzene emulsion. This is in contradiction to the observations of Schulman and Cockbain. Kremnev, in the same paper, reports on the thixotropic behavior of the sodium oleate-cetyl alcohol mixtures.

An interesting, if eclectic, paper by Mahler[42] combines, in a sense, the solubility concept of Bancroft, the phase-volume ratio concept of Ostwald, and the Schulman-Cockbain viewpoint. Mahler concludes that the preferred continuous phase of a stable emulsion is that which constitutes the preferential solvent of the contaminant* of the interfaces. The relative proportions of the two phases tend to favor one as a continuous phase, the tendency being for the phase present in greater proportion to become the continuous phase. An emulsion is, however, destroyed or reversed by any compound that breaks down the film of contaminant or substitutes therefore a film of contaminant having a reverse preferential solubility.

Dimensions of the Interfacial Film. In recent years a considerable body of work has been done, especially in the Soviet Union, in which attention has been directed more to the question of the *dimensions* of the interfacial film than to its precise structure.

Kremnev and Soskin[43] have studied the emulsification of benzene in 5 per cent aqueous sodium oleate solutions. Successive portions of the benzene were shaken with the aqueous phase until a new addition of the oil would have made the emulsion unstable; such limiting emulsions have been termed *gelatinized* emulsions by Kremnev. These emulsions are characterized by the quantity v_m, i.e., the maximum volume of benzene dispersed by 1 ml of the sodium oleate solution. On the basis of microscopic measurements of the volume and surface area of the dispersed droplets, it is possible to calculate the thickness of the interfacial film, from the relation

$$\delta = \theta v_1 / A,$$

where δ is the thickness of the film, θ is the ratio of the volume of the sodium oleate solution to v_2 (= volume of the disperse phase), and v_1 and A are the microscopically determined values of the volume and surface area, respectively, of the dispersed droplets in the emulsion.

---

* It is a little difficult to understand precisely what is meant by "contaminant" in this context, but it presumably has reference to the interfacial film, in much the same sense as used by Schulman and Cockbain.

Kremnev and Soskin apparently regard this quantity as a measure of the thickness of a *hydrated* film. It will be recalled that the existence of a hydated film was proposed to explain the value of the constant in the Sibree equation (Eq. 3.18), relating emulsion viscosity to phase volume.

In all the emulsions studied, the value of δ for the limiting emulsion was found to have approximately the same value, i.e., 0.01 μ. On the other hand, the limiting v_m depends on the method of preparing the emulsion, as does the droplet size distribution curves (cf. p. 216). The total surface area of the droplets, however, is independent of the mode of preparation. When emulsions were prepared by three different modes of shaking (i.e., sharp violent shaking after each addition of benzene; regular, more uniform shaking; and, perfectly uniform shaking) to the same constant $v_m = 100$, the corresponding values of δ were 0.010, 0.018, and 0.024 μ, respectively. The latter two emulsions could be dispersed still further by supplementary shaking, whereas the first, having a minimum value of δ, could not. On standing, a limiting emulsion changes spontaneously, in the direction of increased film thickness and decreased interfacial area. Within twenty-four hours a $v_m = 100$ emulsion changed from $\delta = 0.01$ μ to 0.025 μ and from a total interfacial area of 1×10^6 to 0.4×10^6 sq cm. This corresponds to an area per sodium oleate molecule of 100 sq Å, representing a rather dilute interfacial film.

In a later paper, Kremnev and Soskin[44] reported on the effect of emulsifier concentration on the dimensions of the interfacial film. They make the instructive observation that the emulsifying power of an emulsifying agent cannot be measured in terms of the maximum volume v_m of benzene which can be emulsified by 1 ml of the emulsifier solution, since, as indicated above, this is a function of the mode of preparation of the emulsion. However, the interfacial area in the limiting emulsion or the minimum thickness δ of the stabilizing layer can be considered as a true measure of the emulsifying power. When the concentration of sodium oleate in the aqueous phase was varied from 1 per cent to 5, 10, and 15 per cent, the limiting interfacial area was found to vary from 33 to 100, 100, and 110×10^4 sq cm, respectively. Thus, the interfacial area, and hence δ, are independent of the emulsifier concentration at concentrations higher than 5 per cent. The area per molecule of sodium oleate in the 1 per cent system was 160 sq Å. It would thus appear that, under certain conditions, extremely dilute monolayers can be efficient emulsion stabilizers. This is not in complete agreement with the conclusions of Schulman and Cockbain.[33]

It was found[44] that the volume of benzene which could be emulsified by one *gram* of sodium oleate was 3.3, 2.8, and 1.4 l when the concentration of the emulsifier was 1, 5, and 15 per cent, respectively. Thus, the dilute solu-

TABLE 4-2. CHARACTERIZATION OF THE EMULSIFYING ABILITY OF VARIOUS SOAPS[45]

Soap	v_m (0.32 N soln.)	$\delta(\mu)$	Area per molecule (sq. Å)
Sodium Oleate	178	0.01	50
Potassium Oleate	190	0.005	103
Rubidium Oleate	223	0.0035	147
Cesium Oleate	247	0.0020	258

tions of sodium oleate were better utilized; this is a measure of emulsifier *efficiency*, as opposed to emulsifying ability, as defined above. Measurements on emulsions stabilized with photographic gelatin indicated that its emulsifying ability was one-eighth that of sodium oleate.

Using the definition of emulsifying ability of Kremnev and Soskin,[44] a number of oleate soaps were characterized by Kremnev and Kagan.[45] These included the oleates of sodium, potassium, rubidium, and cesium. In agreement with earlier work, it was found that the amount of benzene emulsified by 1 ml of the aqueous solution of the soap varied with the mode of preparation and with the emulsifier concentration. Data is given in Table 4-2, showing the values of v_m obtained for 0.32 N solutions, as well as the limiting interfacial film thicknesses and areas per molecule of soap in the interfacial film.

It will be noted that the film thicknesses and areas per molecule vary in the same sense as the known radii of the cations, although the proportionality is not a simple one. It thus appears that the size of the gegen-ions has an effect on the nature of the interfacial film, and that soaps of quite large ions may even stabilize emulsions with dilute interfacial films.

Kremnev and Soskin[46] have also investigated the effect of the dimensions of the interfacial film on the stability of quite concentrated emulsions of benzene in 5 per cent sodium oleate solutions. These emulsions were prepared by forcing the mixtures through a sintered glass filter. Assuming that the droplets were in virtual contact, the distance between them, 2δ, may be calculated from the droplet concentration. Then, if the rate $-dm/dt$ of the decrease of the amount emulsified is determined as a function of this distance, it is found to obey the law:

$$(dm/dt)(2\delta - 2\delta_0) = \text{constant}$$

where $2\delta_0 = 0.02$ μ, and 2δ varies between 0.03 μ and 0.12 μ. The stability increases with increasing δ because the free stabilizer solution can repair damaged stabilizing membranes and acts as a lubricant for the slipping of the droplets past one another. The lubricating effect was demonstrated in experiments in which a glass sphere partly filled with mercury was allowed to descend through the emulsions. The time of descent was found to be inversely proportional to the parameter $(2\delta - 2\delta_0)$.

The studies of Kremnev and his coworkers have been extended to include W/O emulsions as well. Kremnev and Kuibina,[47] for example, made an elaborate study of emulsions of both types, stabilized by triethanolamine oleate. Maximally concentrated benzene emulsions of the O/W type, prepared in solutions of this emulsifying agent with emulsifier concentrations in the range 0.23 to 0.31 M, all showed a limiting interfacial film thickness $\delta = 100$ Å, corresponding to an interfacial area per molecule of 50 to 70 sq Å. A similar emulsion was prepared in 0.22 M triethanolamine stearate, with an interfacial film thickness of 110 Å, indicating that the double bond of the oleate molecule has no effect on the emulsifying action. This, of course, can only be true in cases such as those studied here, where the interfacial film is dilute. In concentrated films, as shown by Schulman and Cockbain,[33] the oleate double bond, because of its effect on the geometry of the molecule, may have a significant effect on emulsion stability.

In the case of W/O emulsions, Kremnev and Kuibina[47] found that aqueous ammonium carbonate solutions could be emulsified in solutions of triethanolamin oleate in benzene or toluene. In these emulsions, δ was as low as 20 Å, corresponding, for example, to a film concentration of 130 sq Å per molecule. Such emulsions were quite stiff, and consisted mostly of bubbles of 1 μ diameter. Triethanolamine oleate can stabilize emulsions of aqueous salt solutions (but not of water) in oil, because the distribution of the emulsifier between the water and benzene is shifted toward the benzene by the presence of the salts. Thus, the amount of triethanolamine oleate extracted by benzene from an aqueous solution is increased from 0.3 to 0.7 by 1.5 M ammonium carbonate and potassium chloride, to 0.8 by sodium sulfate, to 0.86 by potassium iodide, and to 0.9 by sodium iodide.

Interfacial Film and Droplet Diameter. Fihurovskii and Futran[48] observed the rate of rise of fine droplets of oil in emulsions under the microscope, in the presence and absence of emulsifying agent. In the absence of emulsifier, the rate of rise is that predicted by Stokes' law (cf. p. 135); in the presence of an interfacial layer however, the droplet radius r is replaced by an effective radius $\rho = r + \Delta r$. If V is the rate of rise in the absence of emulsifier, and W in the presence of emulsifier, then $W\rho = Vr$. Thus Δr can be calculated. Measurements on droplets ranging in diameter from 2 to 20 μ give the following approximate values for Δr: in the presence of 0.1 per cent sodium oleate, 0.25–0.3 μ; 0.82 per cent sodium oleate, 0.5–0.55 μ; 0.1 per cent saponin, 0.2–0.25 μ; 0.82 per cent saponin, 0.8–0.85 μ. It will be recognized that these values are somewhat larger than those quoted by Kremnev and coworkers as the thickness of the interfacial film. However, the authors[48] point out that their figures are effective, that is, no allowance was made for the possibility that the density of the films and the water may have been different. This could have a significant effect.

Emulsions of the O/W type stabilized by dilute interfacial films of sodium oleate have been found by Bromberg[49] to change in droplet distribution. On the other hand, when the interfacial film is very concentrated, i.e., 11 to 15 sq Å per molecule, the original size distribution is maintained for weeks at temperatures as high as 150° C. However, the amount of disperse phase (benzene) in the emulsions decreased. This decrease in disperse phase (apparently as a result of a definite separation) was a function of the shape of the containing vessel. It was accelerated in narrow vessels and decelerated in wide ones. Bromberg apparently contradicted this statement in writing that the average rate of decrease in disperse phase was greater the greater the emulsion-air or emulsion-benzene interfaces. Apparently, in this case, "surface coalescence" takes place.

Bromberg found that highly concentrated emulsions, protected by condensed interfacial films of sodium oleate, swelled in water until the mean thickness of the water interstices (2δ) was about 4 μ. Centrifuging had the effect of decreasing this to about 1 μ. As might be expected, more intense centrifuging led to breakdown of the emulsion. When water was frozen out, separation of benzene occurred when 2δ was about 0.9 μ. The presence of glycerine protected the emulsions during freezing.

Interfacial Film and Interfacial Viscosity. While considerable attention has been paid to the effect of the emulsifying agent on the total viscosity of the emulsion (pp. 54–75), as S. Ross[50] has recently pointed out, very little is apparently known about the viscous properties of the interfacial film. The possible effect of this property on the macroscopic viscosity of the emulsion has been touched upon (p. 64). More importantly, this property could also have a significant effect on the stability of the emulsion itself.

Sherman[51] has studied the properties of the interfacial films of sorbitan sesquioleate, a nonionic emulsifier, in water-in-oil emulsions. The films formed at the mineral oil-water interface were viscous and solid in appearance. Adsorptions was characterized by an overall rise in interfacial viscosity to a maximum value in about 200 hours, followed by a decrease to a constant value in 400 to 600 hours. Criddle and Meader,[52] in a series of similar measurements, also found an ageing effect.

In this connection it should be pointed out that, while techniques for the measurement of surface viscosity at the air-liquid interface are rather advanced (see, for example, Harkins[53]), interfacial measurements are practically limited to the papers cited above. Criddle and Meader[52] describe some very suitable equipment for the measurement of both the viscous and elastic properties of the interfacial films. The fact that elastic properties are measured may be of considerable importance, since these workers found that all the films studied were non-Newtonian.

General Observations on Emulsion Stability. Martynov[54] has applied surface-energy considerations to transition layers between two liquids. Such considerations lead to the conclusion that, in a given pair of liquids, emulsification probably takes place in such a direction as to result in an emulsion in which the liquid possessing the higher value of the ratio β/α (compressibility/thermal expansion) forms the disperse phase. Known effects of the addition of a third phase, e.g., emulsifying agent, can, in many cases, be explained by the effects of the solutes on the ratio. In particular, the increase of β/α of water, owing to dissolved substances, explains the fact that aqueous solutions may be emulsifiable in organic liquids which will not emulsify pure water. The theory also explains the inversion of emulsions with variation of concentration of emulsifier (cf. p. 67). Thermodynamic analysis points to the possibility of the existence of a critical pressure of mixing of two liquids, which, if correct, should have an effect on the mechanical factors involved in emulsion preparation.

Münzel[55] has pointed out that the term "emulsion stability" is somewhat inexact. There is a difference between the stability of the state of the emulsion which may be completely or partially instable, the stability of the degree of dispersion, and of the globules in all the volume-parts of the emulsion. The conditions under which these three different stabilities may be dependent upon each other or independent of one another may depend on such things as temperature, change of the physical properties of oil phase with time, nature and concentration of the emulsifying agent, and (in pharmaceutical or cosmetic emulsions) the addition of chemical agents or drugs to either the aqueous or oil phase. Münzel states that the stability of the emulsion itself has no relation to the stability of the degree of dispersion. While this is very possible, the further statement that the stability of the degree of dispersion is not affected by the degree of dispersion itself is certainly in disagreement with, e.g., the results of Fischer and Harkins.[13]

Spontaneous Emulsification. In certain cases, emulsification has been observed to take place in the absence of any mechanical agitation, and even downwards against the direction of gravity, e.g., a light oil emulsifying in the heavier water. The phenomenon was apparently discovered in 1878 by Gad.[56] McBain[57] has cited three possible explanations of this phenomenon.

The first, due to Quincke, is that the lowering of the interfacial tension by a solute is localized, and causes such violent spreading that the resulting turbulence entraps globules of the upper liquid.

A second explanation of the phenomenon, advanced by Gurwitsch to explain the spontaneous emulsification of petroleum oils containing naphthenic acids in aqueous alkali, proposes that diffusion across the interface, carrying solvent molecules along, is the cause of the emulsion formation. This supposition is supported by Raschevsky.[58]

A third theory is required to explain the cases in which a pure organic liquid spontaneously divides itself into many small globules when placed upon an aqueous solution of a surface-active agent. Kaminski and McBain[59] demonstrated that this phenomenon occurred when xylene was placed upon moderately dilute solutions of dodecylamine hydrochloride. In many cases violent disruption of the pure liquid may be observed. The emulsified droplets consist of pure liquid stabilized by a coating of the surface-active material. It is proposed that the source of the required energy is to be found in the energy of adsorption of the surface-active agent in the interface, as well as in the energy of solubilization of the hydrocarbon in the aqueous surface-active material.

In the experiments reported, benzene, toluene, mesitylene, and cyclohexane also produced spontaneous emulsification on the surface of 0.2 N dodecylamine hydrochloride solution. The rate of emulsification and the extent of the emulsion layer formed were found to decrease in the following order: cyclohexane, benzene, toluene, xylene, mesitylene. This is, of course, also the order of decreasing solubility and solubilization. On the other hand, the stability of the emulsions formed was greatest for mesitylene and least for cyclohexane. Kaminski and McBain thus believe that there is a strong parellelism between spontaneous emulsification and solubility and solubilization.

Recently, however, these particular experiments were repeated by Hartung and Rice.[60] It was found that xylene would not spontaneously emulsify when placed on dodecylamine hydrochloride solutions if the surface-active agent had been purified with extreme care. Apparently, however, stable emulsions can be prepared by shaking. It was also found that a solution of *unsaturated* amine hydrochlorides would not produce emulsions even on shaking; but, when mixed with the purified dodecylamine salt, the resulting solution was again capable of emulsifying xylene spontaneously. The solutions of the purified salt were tested with a large number of organic liquids, and, in no case, did spontaneous emulsification occur. Hartung and Rice conclude that some impurity, presumably an unsaturated one, seems essential in order to induce spontaneous emulsification of hydrocarbons by dodecylamine hydrochloride. It is also possible, then, as indicated by Kaminski and McBain,[59] that the simpler mechanism of Quincke is a sufficient explanation of the phenomenon.

In this connection, the work of Matalon[61] may be cited. He found that spontaneous emulsification of *Nujol* in water occurs at much lower concentrations of sodium dodecyl sulfate when cholesterol is present in the oil phase. This is consistent with the results quoted above[60] and with the more general observations of Schulman and Cockbain.[33]

An interesting study of spontaneous emulsification was carried out by Kling and Schwerdtner,[62] who suspended droplets of oleic acid about 0.3 mm in diameter from cover-glasses in aqueous solutions of eleven different surface-active agents. The behavior of the droplets was observed and photographed under the microscope. In some cases, spontaneous emulsification appeared as a smooth "corrosive" (Kling and Schwerdtner's term) action, proceeding from the periphery of the oil droplet inward, without any translational motion or agitation of the droplet as a whole. In other cases, the droplet underwent violent spontaneous agitation and appeared to shatter. It would appear that these phenomena are characteristic of, e.g., the Gurwitsch and Quincke mechanisms, respectively. The original paper is illustrated with seventy-eight photomicrographs.

An investigation by von Stackelberg, Klockner, and Mohrauer[63] leads to the conclusion that spontaneous emulsification may be connected with the existence of a negative interfacial tension. It was found that the interfacial tension between an oil layer containing 5 to 20 per cent of a long-chain fatty acid and alkaline aqueous phase decreases rapidly as the pH rises to 9, becomes negative in the range pH 9 to 12, and then rises again above pH 12. Spontaneous emulsification of a 10 per cent oleic acid solution in oil occurs in aqueous solutions of pH 9 to 12 at temperatures below 50° C, and takes place with concentrations as low as 2 per cent below 0° C. The size of the particle formed is ascribed to a balance of surface forces.

The concept of a negative surface tension, however, is a tenuous one. A careful study by Manfield[64] indicates that recourse to it is not necessary. Changes of interfacial tension accompanying the transfer of oleic acid from paraffin oil drops to surrounding alkaline solutions (pH 9.7 to 13.0) were followed by the sessile-drop method (Appendix A, p. 301). The interfacial tensions observed were positive and low, varying between 0.03 and 0.10 dynes/cm, during the transfer of acid, but rose to equilibrium values of about 2 to 3 dynes/cm when little acid remained in the oil. The phenomenon which determines the rate of transfer is apparently the slow transport of the oleic acid in the oil phase. On the other hand, high pH and high ionic strength of the aqueous phase inhibit both transfer and spontaneous emulsification. Oscillations occur in drop shape owing to differences of interfacial tension arising from acid depletion of thin sections of the oil drop. Under these conditions spontaneous emulsification occurs, since the oil flows away at regions of low tension. Thus, the transfer of the oleic acid and not negative interfacial tensions is the cause of spontaneous emulsification in this system.

Although many cases are known where the degree of dispersion of an emulsion decreases at a slow rate (p. 49), and a certain number are known where this takes place quite rapidly (p. 151), the opposite phenomenon does not often occur. Certainly, if an *increase* in the dispersity of the emulsion does occur, it may be regarded as a special case of spontaneous emulsification. Vándor[65] reports that when the ternary system containing 30 per cent phenol, 3 per cent sodium oleate, and 67 per cent xylene was diluted with water an emulsion of low dispersity was formed initially. On standing, however, a highly disperse emulsion was formed. This spontaneous increase in degree of dispersion was easily observed, and verified by measurements of the particle-size distribution. The dispersity of the emulsion was not affected whether the final dilution was reached in a single step or whether several steps were used in diluting the system.

A study of spontaneous emulsification of hydrocarbons in solutions of sodium alkylaryl sulfonates has been carried out by van der Waarden.[66] The distribution of the surface-active agents over the oil droplets and water phase in the emulsions was determined by measuring the surface tensions of the emulsions. The experimental results indicate that in spontaneous emulsification, part of the sulfonate originally present in the oil passes into the water phase. An adsorption equilibrium is established between the sulfonates in the interfacial layer and in the water phase at the interface of the oil droplets subsequently formed. A possible mechanism for the process involves the water penetrating between the plate-shaped layers of the sodium alkylaryl sulfonate micelles, splitting the micelles and forming oil-water interfaces.

Electrical Theories of Emulsion Stability

The previous sections of this chapter have dealt, in the main, with considerations of emulsion stability based on the geometry and physical properties of the interfacial layer. This approach will be returned to in a later section, when stabilization of emulsions by non-surface-active materials will be considered. The possibility that electrical effects may be of significance in stabilizing emulsions was also recognized at a fairly early date,[10] but elaboration of this theoretical approach has only been possible in recent years.

Origin and Sign of Charge on Emulsion Droplets. In Chapter 3 (p. 78–82) the existence of a charge on the droplets of an emulsion was referred to in connection with the phenomenon of electrophoresis, and the realization of this charge in the form of the so-called zeta-potential was indicated. A more elaborate discussion of this property will now be undertaken, and its consequences for emulsion stability discussed.

Alexander and Johnson[67] point out that the charge on particles in colloidal systems generally can arise in three different ways, i.e., by ionization, adsorption, or frictional contact. In the case of emulsion droplets, the difference between the first two mechanisms (completely real for, e.g., metal sols) tends to be somewhat blurred. As has been indicated, the stabilization of emulsions can, from one point of view, be regarded as arising purely from the presence of molecules of the emulsifying agent in the interface. Yet, when these adsorbed molecules are there, and in particular when oil-in-water emulsions are considered, the origin of a surface charge arising from the ionization of the water-soluble group at the interface is not hard to picture.

For example, in an O/W emulsion stabilized by a soap it is not at all unreasonable to expect that the carboxylic "heads" penetrating through the interface into the water phase will be, for the most part, ionized. That is, the group actually forming the surface of the droplet will be the carboxyl ion ($-COO^-$). The droplet will effectively be surrounded with a coating of negative charges.

On the other hand, in emulsions stabilized by non-ionic surface-active agents or other non-ionic materials or, for that matter, in the case of W/O emulsions, it is difficult to picture a surface charge arising by this mechanism. Nevertheless, adsorption in the more exact sense implied by Alexander and Johnson, i.e., adsorption of ions from the aqueous phase, is not improbable. Equally probable, is the existence of a charge arising from frictional contact between the droplets and the suspension medium, analogous to the frictional electricity generated when an amber rod is rubbed with a silk cloth. To be sure, there appears to be little experimental evidence which can be taken as a verification of this; Schulman and Cockbain[38] even deny that the droplets in W/O emulsions possess a charge. In systems other than those studied by these workers, however, there seems to be some evidence for such charges and the frictional mechanism may well be correct.

In the case of O/W emulsions stabilized by soaps, the surface charge will be negative, as indicated above. In emulsions stabilized by cationic agents, a positive charge is to be expected. In other words, the sign of the charge for emulsions stabilized by the combined ionization-adsorption mechanism is readily predicted. This is actually somewhat of an over-simplification, as subsequent data on the magnitude of the zeta-potential in emulsions will show. However, as a rule of thumb, it is reasonably accurate.

In connection with the charge arising from a possible frictional mechanism, the case is not so simple, and the empirical rule of Coehn[68] may apply. This rule states that *a substance having a high dielectric constant is positively charged when in contact with another substance having a lower dielectric constant.* Since water has a dielectric constant much higher than most of the substances which are likely to be the other phase of the emulsion, it appears that the droplets of an O/W emulsion will probably have a negative charge, whereas the water droplets of a W/O emulsion will probably be positively charged. It should be noted, however, that Coehn's rule was stated only for dielectrics and cannot be considered as precise for conducting materials. Supporting evidence for the rule was found by electro-ösmosis measurements of various liquids through glass and sulfur membranes.[69]

However these charges arise, their presence on the emulsion droplet, as on the particles of lyophobic colloids, contributes to the stability of the system, since the mutual repulsion of the charged particles prevents their close approach and coagulation, which could lead to breaking of the emulsion.

Helmholtz Double Layer. Quite early in the study of colloidal phenomena related to the existence of a surface charge, it was realized that the situation at the interface is not simple. In an attempt to explain these phenomena, Helmholtz[70] introduced the concept of the *electrical double layer*. Helmholtz assumed that the charge on the particles of a lyophobic colloid was due an unequal distribution of ions at the particle-water interface. Further, Helmholtz pointed out that if ions of one charge were closely bound to the particle, ions of opposite charge would line up parallel to them, forming a double layer of charges. This situation is presented in an idealized form for a spherical particle in Fig. 4-7. The corresponding potential distribution, as a function of distance from the surface of the particle, is represented schematically in Fig. 4-8(A). This distribution corresponds to a condenser formed by two concentric spherical shells; however,

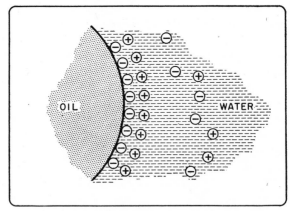

FIGURE 4-7. Idealized representation of the electrical double layer at an oil-water interface.

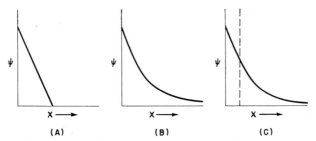

FIGURE 4-8. Potential at an oil-water interface according to (A) the simple Helmholtz theory, (B) Gouy diffuse double-layer theory,[74] (C) Stern diffuse double-layer theory.[78]

this can be shown to be mathematically equivalent to a simple parallel plate condenser,[71] with the plates less than a molecular distance apart.

If one now assumes that the algebraic sum of the charges on the particle and in the liquid is zero, then the electric center of gravity of the entire system cannot be moved by electrical attractive forces which arise from an applied E.M.F. (electromotive force). However, these forces *can* produce a displacement of, for example, positively charged water layer and the negatively charged particle, each being attracted by oppositely charged electrodes. Helmholtz applied hydrodynamic and electrodynamic methods to these assumptions, and derived a relation for the rate of electroösmosis[72]

$$v = \zeta\epsilon E/4\pi\eta \qquad (4.2)$$

where ζ is the zeta-potential, ϵ the dielectric constant, E the applied E.M.F., and η the viscosity of the dispersion medium.

It remained for Smoluchowski[73] to show that Eq. 4.2 was also applicable to the case of electrophoretic velocity, a result of more value in the study of emulsions. As has been pointed out earlier (p. 78), this equation does not apply exactly to spherical particles, and for this situation should be in the form

$$v = \zeta\epsilon E/6\pi\eta \qquad (4.3)$$

It should, however, be pointed out that many workers, using electrophoretic velocity measurements as a method of calculating the zeta-potential on emulsion droplets, have used the Smoluchowski result (Eq. 4.2). Indeed, considering the possible complications which can result from application of electrophoretic theory to emulsions, this can only introduce a relatively small error.

Gouy Diffuse Double Layer. Helmholtz' theory requires that the potential drop at the interface should be sharp (Fig. 4-8A). Considering the general mobility of the ions, however, it is doubtful if the regularly oriented ions required to form the true Helmholtz layer can have any real existence. In order to remedy this deficiency in the theory, Gouy[74] has proposed that the double layer is diffuse, with the outer ionic layer possessing an electric density falling off according to an exponential law (Fig. 4-8B). It should be noted that reduction of the radius of spherical particle to essentially point size reduces this problem to that considered in the Debye-Hückel theory of strong electrolytes. Introduction of a potential having the properties as sumed by Gouy into the Poisson equation leads to several interesting results. One of these is the existence of the parameter

$$\kappa = \sqrt{\frac{8\pi n z^2 e^2}{\epsilon k T}}, \qquad (4.4)$$

where z is the valence of the ions of opposite charge to the surface charge, n is the number of such ions per cm^3 in the solution at a distance from the double layer (i.e., the gross concentration), e is the elementary charge, ϵ the dielectric constant, k is Boltzmann's constant, T is the absolute temperature, and κ is the reciprocal of the distance from the particle of a plane containing most of the charge of the particle. Thus, $1/\kappa$ is the *effective* diameter of the particle, and the "thickness" of the double layer is seen to be proportional to $n^{-1/2}$, i.e., to the concentration of that ionic species which constitutes the counter-ions. This fact has important consequences for the theory of the coagulation of emulsions, as will be shown subsequently (p. 114).

On the basis of the exponential charge distribution proposed by Gouy, it may also be shown that the electrophoretic mobility, u, is given by

$$u = \frac{\sigma}{\eta} \frac{r}{(1 + \kappa r)} \qquad (4.5)$$

where σ is the surface charge (as calculated from the potential distribution), η is the viscosity, r is the radius of the particle, and κ is defined by Eq. 4.4 This indicates that the electrophoretic mobility depends on the particle *size* as well as on its shape, pointed out earlier.

A more elaborate examination of the effect of shape, due to Henry,[75] shows that Eq. 4.5 is itself not quite accurate, but must be multiplied by a function of κr

$$u = \frac{\sigma}{\eta \kappa} \frac{\kappa r}{(1 + \kappa r)} f(\kappa r) \qquad (4.6)$$

(The top and bottom of the equation has been multiplied by κ in order to put everything in terms of κr). Henry has calculated $f(\kappa r)$, and its behavior

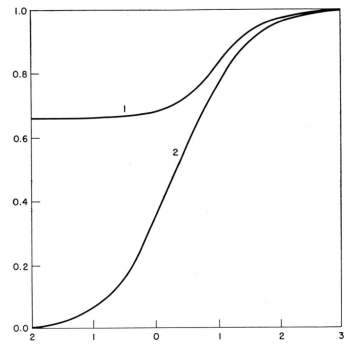

FIGURE 4-9. The effect of droplet size on the electrophoretic mobility is given by the function $f(\kappa r)$. In Curve 1, $f(\kappa r)$ is plotted against $\log(\kappa r)$; in Curve 2, $\dfrac{\kappa r}{(1+\kappa r)} f(\kappa r)$ is plotted against $\log(\kappa r)$.[75]

is shown graphically in Fig. 4-9. In Curve 1, $f(\kappa r)$ is plotted against $\log(\kappa r)$; in Curve 2, $(\kappa r/(1+\kappa r))f(\kappa r)$ is plotted against $\log(\kappa r)$.

A large series of elaborate measurements on the electrophoretic mobilities of emulsions in water and dilute electrolytes has been carried out by Mooney.[76] In this work emulsions of paraffin oil in water, in 0.0001 N hydrochloric acid, in 0.0008 N copper sulfate, and in 0.001 N sodium hydroxide, and emulsions of iodobenzene, tribromohydrin, and dimethyl aniline in water were studied as a function of droplet size. While a variation in mobility as a function of droplet radius was observed, it was not at all in accord with the predictions of Henry's theory. It has been suggested by Abramson[77] that the origin of the discrepancy may reside in a relationship between σ, the surface charge, and the droplet size, e.g., as a result of variation in vapor pressure (cf. p. 7) or solubility.

Stern Diffuse Double Layer. Although the Gouy concept of a diffuse double layer has definite validity, certain deficiencies in the theory exist. In order to correct these, Stern[78] has proposed a model which is, in effect, a compromise between the Helmholtz and Gouy theories (Fig. 4-8C). According to the theory of Stern the double layer is in two parts: one, which is approximately a single ion in thickness, remains essentially fixed to the interfacial surface. In this layer, therefore, there is a sharp drop in potential (Helmholtz layer). The second part (Gouy layer) extends some distance into the liquid dispersing phase and is diffuse, with a gradual fall in potential into the bulk of the liquid (Fig. 4-8B).

Repulsive Effect of Double Layer. It has been indicated above that the repulsive effect arising from the charged double layer is responsible for the stability of the system, since this prevents the close approach of the particles or droplets and subsequent coalescence. It is interesting now to get an estimate of this effect.

The calculation of this repulsive effect can only be done in a semi-quantitative way, since the electrostatic theory involved is extremely complicated. However, Verwey and Overbeek[79] have made a rather successful analysis of the problem of the interaction of two double layers. They show that the repulsive energy is a function of the quantity κH_0, where H_0 is the distance between the droplets or particles, and κ is the reciprocal of the effective radius of the double layer, as defined in Eq. 4.4. The repulsive potential between the two droplets is given to a good approximation by

$$V_R = 4.62 \times 10^{-6} \frac{r}{v^2} \gamma^2 e^{-\kappa H_0}, \tag{4.7}$$

where r is the radius of the particle, v is the valence of the counter-ions, and γ is given by

$$\gamma = \frac{e^{Z/2} - 1}{e^{Z/2} + 1};$$

and where $Z = v\epsilon\psi_0/kT$, ψ_0 being the double layer potential.

This function is plotted in Fig. 4-10A, for various values of the droplet radius. As can be seen, the repulsive energy rises quite sharply for values of κH_o smaller than about one. Of course, there is also a small attractive (van der Waals') force operating between the droplets, which can be given to a good approximation (for small distances) by

$$V_A = -Ar/12H_o, \tag{4.8}$$

where A is constant, depending on, among other things, the polarizability of the molecules of which the droplet is composed.

The total interaction between the droplets is thus given by the sum of these functions, i.e.,

$$V = V_A + V_R$$

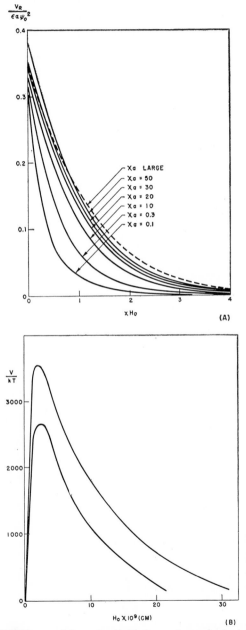

FIGURE 4-10. (A) The repulsive potential between two spherical droplets as a function of κr, for various values of Z, as calculated from Eq. 4.7. The solid lines are calculated from a more exact relation.[79] (B) Total interaction between droplets, obtained from summation of Eqs. 4.7 and 4.8, for two different values of κ and the surface charge.

The type of function which is obtained by this summation is shown in Fig. 4-10B, for reasonable values of the droplet radius r and of the van der Waals' constant A, and for two different values of κ and of the surface charge. It can be shown that if the plot of these functions is continued for quite large distances (of the order of about 100Å) a shallow secondary minimum occurs in the total interaction curve. This slight drop in potential energy is believed to permit the droplets to approach close enough for flocculation to take place under the appropriate conditions (cf. p. 157).

The Double Layer in Emulsions. The previous discussion has dealt with the phenomenon of the double layer in a general way. It will perhaps be noted that reference was made, for the most part, to "particles" rather than droplets. This is deliberate, since most theoretical work has concerned itself with the double layer existing at the solid-liquid interface. It is now necessary to consider what differences in theory will be required by the existence of a liquid-liquid interface.

The principal difference, according to Verwey,[80] resides in the possibility of a double layer existing on *both* sides of the interface. The exact form of the potential distribution at the interface, however, will depend on what ions, surface-active or otherwise, are to be found in the system.

This is illustrated, according to van den Tempel,[81] in Fig. 4-11. In the absence of any surface-active ions, the situation is as shown in Fig. 4-11(A). As can be seen, the potential on both sides of the interface is appreciable. In the case of O/W emulsions the smallest part of the double layer potential

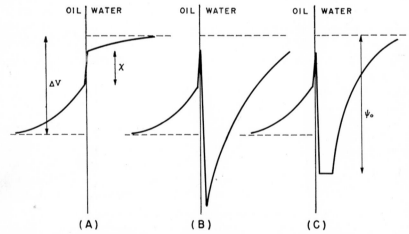

FIGURE 4-11. The double layer at the oil-water interface in emulsions, according to van den Tempel.[81] (A) In the absence of surface-active compounds. (B) In the presence of surface-active compounds. (C) In the presence of a large concentration of electrolyte in the water phase, in addition to the surface-active compound.

occurs in the outer aqueous phase (in the *absence* of emulsifying agents).[80] This is, of course, fatal to stability; and, as is well known, such systems have a strong tendency to coalescence.

The addition of surface-active materials, concentrated at the interface, changes the potential pattern markedly. The magnitude of the potential difference between the *interiors* of the two phases remains unchanged, as long as the ionic concentrations in the bulk phases is not affected by the adsorption process.[82] The initial change in the potential difference caused by the adsorption of the surface-active ions is compensated for by a rearrangement of the dissolved ions across the interface. This is shown in Fig. 4-11(B). As is evident, the major portion of the charge is now concentrated in the aqueous phase, and the zeta-potential is sufficiently large so that a stable O/W emulsion can result. Verwey[80] has estimated that the zeta-potential required for primary stability is of the order of ±100 mv.

Finally, Fig. 4-11(C) shows the changes in the potential pattern caused by the addition of a large amount of electrolyte in the water phase in addition to the surface-active agent. As would be expected on the basis of Eq. 4.4, the effective radius of the diffuse double layer is decreased by this addition. The counter-ions move in among the surface-active ions, producing a thin layer of uniform potential.

Van den Tempel[81] points out that in technically important emulsions, in which the oil phase has a very low conductivity and which are stabilized by ionic soaps, the potential drop at the aqueous side of the interface depends only on the amount of surface-active ions adsorbed, and on the electrolyte concentration in the aqueous phase. This follows from the electric neutrality of the entire interface,[80] which implies that the charge of the adsorbed surface-active ions (forming, as it were, the Helmholtz layer) is compensated by the charges in the diffuse double layer at *both* sides of the interface. The contribution of each of these double layers is proportional to

$$(n_i \epsilon_i)^{1/2} \sinh (e\psi_i/2kT),$$

in which n_i is the ionic concentration, ϵ_i the dielectric constant and ψ_i the potential drop in phase i. As van den Tempel points out, the value of $(n_i \epsilon_i)^{1/2}$ in the aqueous phase will be at least 10^3 times the value in the oil phase, while the hyperbolic sine terms will be of the same order of magnitude. This permits a theoretical treatment of the double layer in emulsions in which the capacity of the double layer in the oil phase can be neglected, even though the actual potential drop in the oil phase may have a fairly high value.

Verwey[80] expresses this in another way, by indicating that the presence of the surface-active ions renders the interface rather more like a solid-liquid interface than a liquid-liquid one. Indeed, the stability of the emulsion is related to the "solidity" of the interface.

Theoretical Calculation of Surface Charge. It is now possible to construct a model of an O/W emulsion, stabilized by soap molecules, from which a theoretical calculation of the surface charge is possible.[81] In this model, part of the counter-ions are considered to be situated *between* the ionic heads of the soap molecules, which will project some distance into the aqueous phase (cf. Fig. 4–1). The layer containing the soap ions and part of the counter-ions is considered to be situated in the immediate neighborhood of the interface and to have a depth of only a few Ångström units. The potential in this layer is assumed to have the value ψ_0 with respect to a point at a great distance from the interface in the aqueous phase (Fig. 4-11C). Van den Tempel refers to this as the Stern layer, although this is not strictly accurate, as he himself points out.

With the further assumption that the soap concentration in the bulk aqueous phase is always below the C.M.C., van den Tempel proceeds with the following calculation of the surface charge on the oil droplets, based on the fact that the only quantity which can be practically measured at the oil-water interface is the interfacial free energy, i.e., the interfacial tension.

From Eq. 2.22

$$-d\gamma = \Sigma_i \Gamma_i \, du_i, \tag{4.9}$$

in which Γ_i is the surface excess of component i, defined as (cf. p. 26)

$$\Gamma_i = \left(\frac{\partial n_i}{\partial \omega}\right) = \frac{1}{\omega}\left(n_i - m_i \frac{18}{1000} n_{H_2O}\right), \tag{4.10}$$

where μ_i is the thermodynamic potential of component i, n_i is the number of moles of i in the system, m_i is the molality of the homogeneous solution, and ω is the area of the interface. In applying Eq. 4.10, the assumption is made that the solubility of each of the components is negligible, as is the solubility of the oil in the aqueous phase. It is also assumed that the surface excesses of the oil and water are zero.

If the Gibbs equation is now applied to a system such as that under consideration, containing the salt BA and the anionic soap DS, both supposed to be completely dissociated in solution, there results the following

$$d\gamma = -\Gamma_{B^+}d\mu_{B^+} - \Gamma_{A^-}d\mu_{A^-} - \Gamma_{H^+}d\mu_{H^+}$$
$$- \Gamma_{OH^-}d\mu_{OH^-} - \Gamma_{D^+}d\mu_{D^+} - \Gamma_{S^-}d\mu_{S^-} \tag{4.11}$$

The number of independent variables is much lower than the number of terms in Eq. 4.11, as relations exist among them. However, this equation can be considerably simplified by making some further, reasonable assumptions. It is assumed that none of the ions present, except S^-, is specifically adsorbed at the interface. The term containing Γ_{H^+} may be omitted on the basis of experimental evidence showing no change in interfacial tension as a function of pH.[83] Similarly, the negative surface excess of OH^- at the negatively charged interface may be neglected, since the concentration of hydroxyl ions in the solution is only a very small fraction of that of the other anions. Together with the neutrality condition, this gives

$$-\Gamma_{A^-} + \Gamma_{B^+} + \Gamma_{D^+} = \Gamma_{S^-}$$

and

$$\Gamma_{H^+} = \Gamma_{OH^-} = 0 \tag{4.12}$$

Remembering also that

$$d\mu_{AB} = d\mu_{A^-} + d\mu_{B^+}, \tag{4.13}$$

and

$$d\mu_{DS} = d\mu_{D^+} + d\mu_{S^-},$$

Eq. 4.11 becomes

$$-d\gamma = \Gamma_{S^-}d\mu_{DS} + \Gamma_{A^-}d\mu_{AB} + (\Gamma_{A^-} - \Gamma_{B^+})(d\mu_{D^+} - d\mu_{B^+}) \tag{4.14}$$

Van den Tempel proceeds to apply Eq. 4.14 to his experimental data on the interfacial tensions existing between oil and aqueous solutions of surface-active agents at various concentrations. In his work, those solutions containing a non-surface-active electrolyte were maintained at a salt concentration which was high with respect to that of the surface-active material. This salt concentration was held constant as the soap concentration was varied.

For the case when no salt is present, Eq. 4.14 is modified to

$$-d\gamma = \Gamma_{S^-}d\mu_{DS} = 2\Gamma_{S^-}d\mu_{S^-} = 2\Gamma_{S^-}RT\ d\ln c_{S^-}f_{S^-}, \tag{4.15}$$

where c_{S^-} represents the concentration of surface-active anions, and f_{S^-} is the activity coefficient.

Since a large excess of salt is present it is possible to distinguish between experiments in which B^+ and D^+ are the same (e.g., Na^+) and those in which the cations are different. Obviously, in the first case the last term on the right-hand side of Eq. 4.14 vanishes. Furthermore, the thermodynamic potential of the sodium ions will remain unaffected by variation of the concentration of surface-active material (so long as the salt is present in excess), yielding, for this case

$$-d\gamma = \Gamma_{S^-} d\mu_{S^-} = \Gamma_{S^-} RT \, d \ln c_{S^-} f_{S^-} \qquad (4.16)$$

Turning now to the case where the cations are different, variation in the concentration of surface-active agent will result in:

$$d\mu_{AB} = 0$$
$$d\mu_{B^+} = 0$$
$$d\mu_{D^+} = d\mu_{S^-} \, ,$$

therefore

$$d\gamma = \Gamma_{S^-} d\mu_{S^-} + \Gamma_{D^+} d\mu_{D^+} \qquad (4.17)$$

Equation 4.17 may be further simplified by considering that the large excess of the cation B^+ insures that Γ_{B^+} will be of the same order as Γ_{S^-}, while Γ_{D^+} is only a small fraction of this quantity. Thus, only a small error is introduced by ignoring the second term on the right side of Eq. 4.17, making it identical with Eq. 4.16. It should be noted that if the cation B^+ has a valence greater than unity, a much smaller excess of electrolyte is required.

As pointed out above, van den Tempel assumes that the concentration of surface-activity agent is below the C.M.C. As a consequence, the theory of strong electrolytes of Debye and Hückel[84] may be employed. From this theory, the activity coefficient f_{S^-} may be readily calculated:

$$\ln f_{S^-} = -1.1 \sqrt{c_{S^-}} \qquad (4.18)$$

Substituting in Eq. 4.15 gives

$$-d\gamma = 2\Gamma_{S^-} RT(1 - 0.55 \sqrt{c_{S^-}}) d \ln c_{S^-} \qquad (4.19)$$

In Eq. 4.16 (and in Eq. 4.17) under the conditions stated above) the activity coefficient does not change appreciably when the concentration of surface-active agent is changed. As a consequence, Eq. 4.16 reduces to

$$-d\gamma = \Gamma_{S^-} RT \, d (\ln c_{S^-}) \qquad (4.20)$$

for the case considered.

Thus, Γ_{S^-} can be calculated from either Eq. 4.19 or 4.20, from the experimentally determined relation between the interfacial tension and the concentration of surface-active agent. If the surface charge, in the case of an O/W emulsion, is identified with the amount of surface-active ions adsorbed on unit interfacial area, the surface charge may finally be calculated rather readily from the relation

$$\sigma = ev\Gamma_{s-} , \qquad\qquad (4.21)$$

where e is the electronic charge, v is the valence of the ions, and Γ_{s-} is calculated from either Eq. 4.19 or 4.20.

Actually, if the electrolyte is not present in excess, Γ_{s-} is not exactly equal to σ/ev, because of a deficiency of anions of the surface-active material in the diffuse portion of the double layer. However, this represents only a small percentage of Γ_{s-}.

Potential Function of the Aqueous Phase. In this fruitful vein, van den Tempel[81] continues with a calculation of the nature of the potential in the aqueous phase of an O/W emulsion, based on the model described above.

In this model, part of the cations will be situated in the Stern layer, nestling among the ionic heads of the soap anions, and the remainder will be in the diffuse Gouy layer. The fraction of the counter-ions in the Stern layer is given by the Langmuir-Stern equation[78]

$$\sigma_1/ev = N_1\{1 + 0.6 \times 10^{24}/18n) \exp (ev\ \psi_0/kT\}^{-1}, \qquad (4.22)$$

where N_1 is the number of available sites in 1 cm^2 of the Stern layer, n is the number of counter-ions of valence v in 1 cc of the homogeneous solution, e is the electronic charge, and ψ_0 is the potential in the Stern layer with respect to a point distant from the interface in the aqueous phase.

In the Stern theory, the quantity $0.6 \times 10^{24}/18$ (which appears in Eq. 4.22) represents the number of positions available to the counter-ions in 1 cc of the homogeneous bulk solution. The corresponding quantity N_1 in the Stern layer may be estimated at 10^{15} for a Stern layer of depth 3 Å.

The charge of the diffuse layer σ_2 is given by the relationship

$$\sigma_2 = \{(\epsilon kT/\pi)\Sigma nv^2\}^{1/2} \sinh (ev\psi_0/2kT), \qquad (4.23)$$

where ψ_0 is the potential at the boundary between the Stern and the Gouy layers.

Taking into account the condition of electric neutrality for the entire double layer

$$\sigma_1 + \sigma_2 = \sigma \qquad\qquad (4.24)$$

The last three equations suffice to calculate ψ_0 as a function of n and σ,

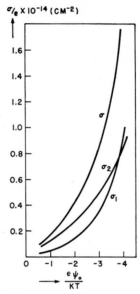

$\sigma/e \times 10^{-14}$ (CM^{-2})

FIGURE 4-12. The relation between the Stern layer potential and the amount of adsorbed soap σ/e for a 1-1 electrolyte for an aqueous phase which is $0.100M$ in electrolyte. The partition of the counter-ions between the Stern and Gouy layer is indicated.[81]

i.e., from a knowledge of surface-active agent and salt concentrations and experimentally determined interfacial tensions. Figure 4-12 shows the relation between the Stern layer potential and the amount of adsorbed soap σ/e for a 1-1 electrolyte (hence $v = 1$) for an aqueous phase which is $0.100 \ M$ in electrolyte. The partition of the counter-ions between the Stern and Gouy layer is also indicated.

Experimental Determination of the Layer Potential. Van den Tempel[81] next proceeded to an experimental determination of the potential of the Stern-layer, based on the theoretical considerations of the previous section. Interfacial tensions were measured, using an oil phase consisting of equal volumes of monochlorobenzene and paraffin oil. Such a mixture has a density of 0.988 at 20° C. Measurements were carried out against solutions of *Aerosol MA* in the presence of sodium and magnesium chloride, and with sodium laurate in the presence of sodium chloride only (salts of bivalent metals could not be used in this latter case because of the precipitation of the metal soap). Measurements were also carried out with *Aerosol OT*, but this material exhibits very low solubility in the presence of even monovalent electrolytes.

The surface excesses were calculated from the interfacial tension data, using Eq. 4.19 or 4.20 as appropriate. The results are shown graphically

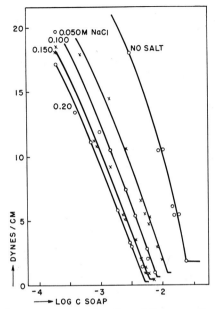

FIGURE 4-13. Surface excesses as a function of the concentration of surface-active material for systems containing varying amounts of electrolyte.[81]

in Figure 4-13, as a function of the concentration of surface-active material for systems containing varying amounts of electrolyte. Generally speaking, the addition of salt to a solution of surface-active material causes an increased adsorption at the interface. This is probably owing to the decreased electrostatic repulsion of the ionic heads in a more concentrated salt solution. On the other hand, at fairly high salt and soap concentrations (though still below the C.M.C.) a further increase of the salt concentration appears to decrease the amount of adsorption.

The interfacial-tension data indicate that this property decreases linearly with the logarithm of the soap concentration, which can be interpreted to mean that the amount of adsorbed soap is independent of the soap concentration. Presumably, this may be explained by the fact that there are many more ions of soap available than required for occupying sites in the interface.

Figure 4-13 also indicates that the laurate ion is, in all cases, more strongly adsorbed than the *Aerosol MA* anions. This may be ascribed to stereometric factors, since the *Aerosol MA* has a branched, and hence somewhat bulky, side-chain, while the laurate has, of course, a linear hydrocarbon chain.

Using the data presented in Figure 4-13 and graphs of the form of Figure 4-12, the surface potential was calculated as a function of the salt concentration. This is shown in Figure 4-14, in which the quantity $-e\psi_0/kT$ is

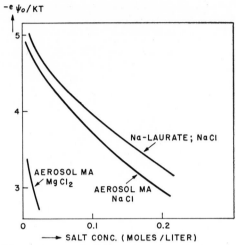

FIGURE 4-14. The surface potential as a function of salt concentration for sodium laurate in the presence of sodium chloride and for *Aerosol MA* in the presence of sodium and magnesium chlorides.[81]

plotted against the salt concentration for sodium laurate in the presence of sodium chloride and for *Aerosol MA* in the presence of sodium and magnesium chlorides. The potential decreases almost linearly with increasing salt concentration. However, the potential is practically independent of the concentration of surface-active agent in the range from 0.001 to 0.004 M for *Aerosol MA* and from 0.0014 to 0.006M for sodium laurate.

The influence of the bivalent cations is also indicated in Figure 4-14. The same decrease in potential obtained with a concentration of bivalent cations which is roughly one-tenth of the concentration of monovalent cations is required for the same effect. This has important consequences for the theory of demulsification (cf. p. 159).

TABLE 4-3. ZETA-POTENTIAL OF OIL DROPS* EMULSIFIED IN
AQUEOUS *AEROSOL MA* SOLUTION[81]

Soap Concentration mol/l	Electrolyte Concentration mol/l		ζ (mV)
0.00031	—	—	103
	NaCl	0.050	131
		0.100	124
		0.150	114
		0.200	117
0.00088	—	—	140
	NaCl	0.100	126
		0.150	105
	MgCl$_2$	0.005	66
		0.010	58

* The concentration of oil in the emulsions = 0.4%

Relation Between Calculated Potential and Zeta-Potential. Electrophoretic studies on emulsions containing 0.4 per cent of oil emulsified in *Aerosol MA* solutions containing various amounts of electrolyte are reported by van den Tempel.[81] Although the exact relation between ψ_0 and ζ (obtained from electrophoresis data) is not known, the general agreement in magnitude is what would be expected from Figure 4-14, although the values of the zeta-potential for the bivalent ions is somewhat lower than might be expected. Table 4-3 reproduces the data obtained for various electrolyte concentrations.

Stabilization by Solids

In previous sections, the emphasis has been on the stabilization of emulsions through the intermediary, principally, of surface-active compounds. However, it is well known that emulsions may be stabilized by materials which are surface-active only in a highly specialized sense, i.e., finely divided solids.

Experimental Observations on Solid-Stabilized Emulsions. The first extensive observations on emulsions stabilized by finely-divided solids were made by Pickering.[85] It was found that basic sulfates of iron, copper, nickel, zinc, and aluminum in a moist condition act as efficient dispersing agents for the formation of petroleum oil-in-water emulsions. Pickering also observed that materials such as calcium carbonate, lead arsenate, fine clays, etc., even in the dry state, can promote emulsification. However, such emulsions are relatively unstable, and Pickering termed them "quasi-emulsions."

Pickering's observations were solely on O/W emulsions. Briggs,[86] while noting that ferric hydroxide, arsenic sulfide, and silica led to the formation of O/W emulsions with kerosene and benzene, found that carbon black, rosin, and lanolin led to the formation of W/O emulsions in these systems. Weston[87] was able to produce both types of emulsions with colloidal clay. Cheesman and King[88] were able to demonstrate that, with a given solid, it is possible to form both types of emulsions merely by varying the technique of shaking.

As unexpected a material as sodium chloride has been found to stabilize emulsions of chloroform in brine-saturated solutions of Löffler methylene blue.[89]

An elaborate study of emulsions stabilized by hydroxides or hydrous oxides of metals has recently been carried out by Mukerjee and Srivastava.[90] Metals studied including aluminum, zinc, magnesium, cupric copper, beryllium, cadmium, mercury, tin, iron, manganese, cobalt, nickel, chromium, thorium, titanium, and lanthanum, the hydrous oxides or hydroxides being prepared in a variety of ways.

Kerosene-in-water systems were studied and periodic size-frequency

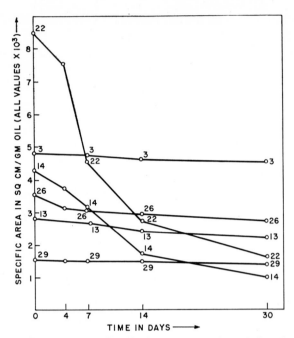

FIGURE 4-15. Size-frequency distribution as a function of time for kerosene-in-water emulsions stabilized by hydrous oxides. See text for identification of emulsions.[90]

analyses of the emulsions were made according to the method of King and Mukerjee.[91] Representative data obtained in this manner are shown graphically in Figure 4-15, for six of the emulsions studied. The numbers appended to the individual curves in the illustration refer to the emulsions studied.[91] Emulsion 3 was prepared with a smooth, highly-dispersed, aged alumina, emulsion 13 with a highly-dispersed gelatinous magnesium hydroxide, emulsion 14 with a highly-dispersed suspension formed by the addition of dilute ammonia at 5° C to magnesium chloride, emulsion 22 with a granular precipitate obtained by the addition of dilute ammonia to cadmium sulfate solution, emulsion 26 with gelatinous hydrous ferric oxide, and emulsion 29 with a gelatinous precipitate formed by the addition of sodium hydroxide to nickel sulfate.

The general conclusions drawn from this study are that moist precipitates are in general better emulsifiers than dry substances, and that the physical state of the precipitates appears to be a very important factor. Generally, highly gelatinous or highly-dispersed fine precipitates are more efficient than granular ones. The same hydroxide precipitated by different methods from different concentrations of the solutions at different temperatures gave emulsions with different stabilities.

Some agents improved on ageing, e.g., aluminum hydroxide. This was reflected, in some cases, by the stability of the emulsion improving with age, as measured by the particle size distributions. Apparently, there is also an effect related to the valence of the metal ion, since emulsions stabilized with hydroxides of metals of valence three or greater gave very stable emulsions, showing little or no separation of the disperse phase during a six month period.

Emulsion Type Produced by Finely-Divided Solids. As indicated above, there appears to be a relation between the emulsion type obtained and the solid used. Pickering[85] suggested that the fundamental condition of the formation of oil-in-water emulsions is that the solid must be more easily wetted by the water than by the oil. Schlaepfer[92] was able to produce a highly concentrated W/O emulsion by the use of soot, which represents the converse situation.

The theoretical rule governing the wetting of solids by liquids is, of course, Young's equation (Eq. 2.11). In this connection, the work of von Reinders[93] must be cited. This worker reported a study of liquid-solid interfaces, using powders, and the distribution of such powders between two mutually insoluble liquids, e.g., water and oil. In such a system, the distribution of the solid will depend on the relation between three interfacial tensions: that between solid and water, γ_{sw} ; that between water and oil, γ_{wo} ; and that between solid and oil, γ_{so} . Von Reinders points out that three situations are realizable:

1. If $\gamma_{so} > \gamma_{wo} + \gamma_{sw}$, the solid will remain suspended in the aqueous phase.
2. If $\gamma_{sw} > \gamma_{wo} + \gamma_{so}$, the solid will remain suspended in the oil phase.
3. If $\gamma_{wo} > \gamma_{sw} + \gamma_{so}$, or if none of the three interfacial tensions is greater than the sum of the other two, the solid particles will concentrate in the boundary.

Clearly, if the solid remains entirely in one phase or another, it will not be particularly useful as an emulsion stabilizer. On the other hand, for the case where the solid collects in the liquid-liquid interface, we may apply Eq. 2.11 in the form

$$\gamma_{so} - \gamma_{sw} = \gamma_{wo} \cos \theta$$

If $\gamma_{sw} < \gamma_{so}$ then $\cos \theta$ is positive and $\theta < 90°$; this will result in the major portion of the solid particle being in the aqueous phase. Similarly, if $\gamma_{so} < \gamma_{sw}$, $\cos \theta$ is negative, $\theta > 90°$, and if the particle will be principally in the oil phase. There obviously exists the possibility of the unusual situation in which the solid particle is equally wetted by both the oil and aqueous phases. These three possibilities are illustrated in Figure 4-16.

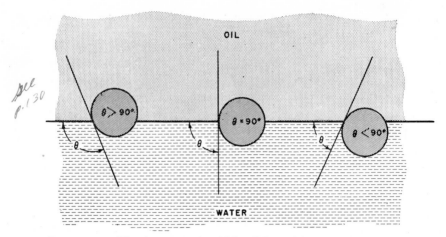

see
p. 130

FIGURE 4-16. The three ways in which solid particles may distribute themselves in an oil-water interface, according to von Reinders.[93] In the case shown on the right, the solid is wetted better by the water than the oil, and thus is principally in the aqueous phase. In the left-hand case, the reverse situation occurs; in the center case the particle is wetted equally well by both phases.

The significance of these effects are made evident by consideration of Figure 4-17, adapted from Thomas.[94] As can be seen, the fact that the solid particles are principally in the disperse phase, and only slightly wetted by the internal phase, leads to a minimum interfacial area. All other things being equal, it is clear that an emulsion system which has a low interfacial area will be more stable than one in which this parameter is larger. Thus the

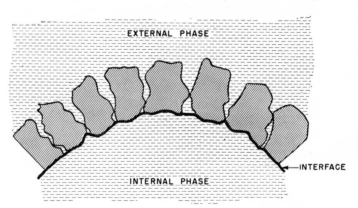

FIGURE 4-17. Stabilization of an emulsion by finely-divided solids, according to Thomas.[94] The fact that the solid particles are principally in the disperse phase leads to a minimum in the interfacial area.

observations of Pickering[85] and Schlaepfer[92] may be regarded as having theoretical justification.

Recently, Schulman and Leja[95] have reported an elaborate investigation illustrating this point. In this work, emulsions stabilized by solid barium sulfate were investigated. However, the surface of the solid was modified by the controlled adsorption of various amphiphilic compounds, thus affecting the contact angle. The effect of chain length of the hydrophobic groups, and the effect of pH were studied. Figure 4-18 shows the dependence of the contact angle at the oil-water and air-water interfaces on the chain length for alkyl sulfates and carboxylic soaps. Figure 4-19 illustrates the variation of the contact angle with pH.

From the data presented in Figure 4-18, it would be expected that barium sulfate powder coated in an 0.001 M solution of sodium laurate at pH 12 would give a O/W emulsion, whereas that coated with sodium dodecyl sulfate would be W/O. This is verified experimentally. Since the contact angles are close to 90° (on either side), quite stable emulsions are formed. On the other hand, with the salts of the higher alkyl sulfates, and with the fatty acid soaps, the contact angles are high, and there is a tendency for complete and preferential wetting by the oil phase (von Reinders' case 2); while W/O emulsions are formed, they tend to be unstable.

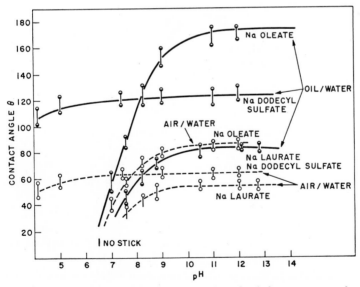

FIGURE 4-18. The dependence on the chain length of the contact angle at the oil-water and the air-water interface for barium sulfate coated with an adsorbed layer of amphiphilic compounds.[95]

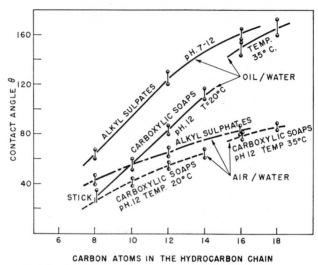

FIGURE 4-19. The dependence of the contact angle at the oil-water and air-water interface as a function of pH for barium sulfate coated with an adsorbed layer of amphiphilic compounds.[95]

From Figure 4-19, similar conclusions can be drawn as to the effect of pH.

Since adsorption of the surface-active agent to the solid presumably occurs via the anionic group, Schulman and Leja conclude that the preferential wetting arises as a result of the packing of the adsorbed material, and of the chain-length required for similar packing. Results obtained when the concentration of surface-active material is varied are consistent with this, as is the effect of the addition of a long-chain fatty alcohol to the oil phase. Figure 4-20 illustrates the situation which is believed to exist.

Stability of Solid-Stabilized Emulsions. As is very well illustrated by the work of Schulman and Leja,[95] cited above, the most stable emulsions are obtained when the contact angle with the solid at the interface is close to 90°, the type depending on whether the angle is greater or lesser than 90°. Obviously, a concentration of solids at the interface represents an interfacial "film" of considerable strength and stability, which will serve to stabilize such emulsions. Indeed, Verwey[80] has pointed out that emulsions stabilized by surface-active agents tend to become more stable as the interfacial properties of the dispersed droplet approach those of a solid.

It is also probable that, in many cases, such concentrations of powdered solid at the interface lead to high zeta-potentials, which themselves have a stabilizing effect.

DENSITY OF PACKING OF AMPHIPATHIC MOLECULES
ADSORBED ON SOLID SURFACES.

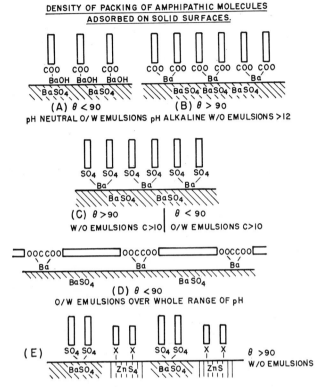

FIGURE 4-20. Schematic representation of the condition existing when amphiphilic compounds are adsorbed to a crystal surface. The effect of the addition of a long-chain alcohol on the packing is indicated.[95]

Bibliography

1. Becher, P., "Principles of Emulsion Technology," p. 2, New York, Reinhold Publishing Corp., 1955.
2. Schwartz, A. M. and Perry, J. W., "Surface Active Agents," p. 345, New York, Interscience Publishers, Inc., 1949.
3. Bancroft, W. D., "Applied Colloid Chemistry," 2nd Ed., p. 359, New York, McGraw-Hill Book Co., 1926.
4. Clowes, G. H. A., *J. Phys. Chem.* **20**, 407 (1916).
5. Holmes, H. H., in Bogue, R. H. (ed.), "Colloidal Behavior," **1**, p. 227, New York, McGraw-Hill Book Co., 1924.
6. Quincke, G., *Pogg. Ann.* **139**, 1 (1870); *Wied. Ann.* **35**, 571 (1888).
7. Donnan, F. G. and Potts, H. E., *Kolloid-Z.* **7**, 208 (1910).
8. Becher, P., *op. cit.*, pp. 9–10, 32–33.
9. Bancroft, W. D., *J. Phys. Chem.* **17**, 501 (1913); **19**, 275 (1915).

10. Lewis, W. C. McC., *Kolloid-Z.* **4,** 211 (1909).
11. Bancroft, W. D. and Tucker, C. W., *J. Phys. Chem.* **31,** 1681 (1927).
12. Harkins, W. D., "The Physical Chemistry of Surface Films," pp. 83–91, New York, Reinhold Publishing Corp., 1952.
13. Fischer, E. K. and Harkins, W. D., *J. Phys. Chem.* **36,** 98 (1932).
14. Harkins, W. D., Davies, E. C. H., and Clark, G. L., *J. Am. Chem. Soc.* **39,** 541 (1917).
15. Wellman, V. E. and Tartar, H. V., *J. Phys. Chem.* **34,** 370 (1930).
16. Ostwald, Wo., *Kolloid-Z.* **6,** 103 (1910); **7,** 64 (1910).
17. Bhatnagar, S. S., *J. Chem. Soc. (London)* **117,** 542 (1920).
18. Pickering, S. U., *J. Chem. Soc. (London)* **91,** 2002 (1907).
19. McBain, J. W., "Colloid Science," p. 36 (ref. 26), Boston, D. C. Heath & Co., 1950.
20. Sherman, P., *Research (London)* **8,** 396 (1955).
21. Kremann, R., Griengl, F., and Scheiner, H., *Kolloid-Z.* **62,** 61 (1933).
22. Manegold, E., "Emulsionen," p. 23, Heidelberg, Strassenbau, Chemie und Technik, 1952.
23. Hildebrand, J. H., *J. Phys. Chem.* **45,** 1303 (1941).
24. Roberts, C. H. M., *J. Phys. Chem.* **36,** 3087 (1932).
25. King, A., *Trans. Faraday Soc.* **37,** 168 (1941).
26. Cheesman, D. F. and King, A., *Trans. Faraday Soc.* **36,** 1241 (1940).
27. Neogy, R. K. and Ghosh, B. N., *J. Indian Chem. Soc.* **30,** 113 (1953).
28. Coutinho, H., *Seifensieder Ztg.* **68,** 349, 362, 371, 382, 392, 401 (1941).
29. Lachampt, F. and Dervichian, D., *Bull. soc. chim.* **1946,** 491.
30. Schulman, J. H. and Rideal, E. K., *Proc. Roy. Soc. (London)* **122B,** 29, 46 (1937).
31. Schulman, J. H. and Stenhagen, E., *Proc. Roy. Soc. (London)* **126B,** 356 (1938).
32. Schulman, J. H. and Friend, J. A., *Kolloid-Z.* **115,** 67 (1949).
33. Schulman, J. H. and Cockbain, E. G., *Trans. Faraday Soc.* **36,** 651 (1940).
34. Alexander, A. E. and Schulman, J. H., *Trans. Faraday Soc.* **36,** 960 (1940).
35. Alexander, A. E., *Trans. Faraday Soc.* **37,** 117 (1941).
36. Epstein, M. B., Wilson, A., Jakob, C. W., Conroy, L. E., and Ross, J., *J. Phys. Chem.* **58,** 860 (1954).
37. Aickin, R. G., *J. Soc. Dyers Colourists* **60,** 41 (1944).
38. Schulman, J. H. and Cockbain, E. G., *Trans. Faraday Soc.* **36,** 661 (1940).
39. Albers, W., private communication.
40. Dickinson, W. and Iball, J., *Research (London)* **1,** Suppl. 614 (1948).
41. Kremnev, L. Ya., *Kolloid. Zhur.* **10,** 18 (1948); *C.A.* **43,** 4923c.
42. Mahler, E., *Chimie & Industrie* **53,** 12 (1945).
43. Kremnev, L. Ya. and Soskin, S. A., *J. Gen. Chem. (U.S.S.R.)* **16,** 2000 (1946); *C.A.* **41,** 6791i.
44. Kremnev, L. Ya. and Soskin, S. A., *Kolloid. Zhur.* **9,** 269 (1947); *C.A.* **47,** 943f.
45. Kremnev, L. Ya. and Kagan, R. N., *Kolloid. Zhur.* **10,** 436 (1948); *C.A.* **43,** 7775i.
46. Kremnev, L. Ya. and Soskin, S. A., *Kolloid Zhur.* **11,** 24 (1949); *C.A.* **43,** 6884b.
47. Kremnev, L. Ya. and Kuibina, N. I., *Kolloid. Zhur.* **13,** 38 (1951).
48. Fihurovskii, N. A. and Futran, M. F., *Kolloid. Zhur.* **9,** 392 (1947); *C.A.* **43,** 5256h.
49. Bromberg, A. V., *Kolloid. Zhur.* **9,** 23 (1947).
50. Ross, S., *J. Soc. Cosmetic Chemists* **6,** 184 (1955).
51. Sherman, P., *J. Colloid Sci.* **8,** 35 (1953).
52. Criddle, D. W. and Meader, A. L., Jr., *J. Appl. Phys.* **26,** 838 (1955).
53. Harkins, W. D., *op. cit.*, pp. 140–149.

54. Martynov, V. M., *Kolloid. Zhur.* **10**, 33 (1948).
55. Münzel, K., *Pharm. Acta Helv.* **21**, 145 (1946).
56. Gad, J., *Arch. Anat. u. Physiol.* **1878**, 181.
57. McBain, J. W., *op. cit.*, pp. 19–21.
58. Raschevsky, N., *Z. Physik.* **48**, 513 (1947).
59. Kaminski, A. and McBain, J. W., *Proc. Roy. Soc. (London)* **A198**, 447 (1949).
60. Hartung, H. A. and Rice, O. K., *J. Colloid Sci.* **10**, 436 (1955).
61. Matalon, R., *Trans. Faraday Soc.* **46**, 674 (1950).
62. Kling, W. and Schwerdtner, H., *Melliand Textilber.* **22**, 21 (1941).
63. Van Stackelberg, M., Klockner, E., and Mohrauer, P., *Kolloid-Z.* **115**, 53 (1949).
64. Manfield, W. W., *Australian J. Sci. Research*, Ser. A., **5**, 331 (1952).
65. Vándor, J., *Magyar Kém. Lapja* **4**, 592 (1946); *C.A.* **45**, 926g.
66. Van der Waarden, M., *J. Colloid Sci.* **7**, 140 (1952).
67. Alexander, A. E. and Johnson, P., "Colloid Science," I, pp. 43–48, London, Oxford University Press, 1949.
68. Coehn, A., *Ann. Physik* **66**, 217 (1898).
69. Abramson, H. A., "Electrokinetic Phenomena," pp. 56–57, New York, Reinhold Publishing Corp., 1934.
70. Helmholtz, H., *Wied. Ann.* **7**, 537 (1879).
71. Abramson, H. A., *op. cit.*, pp. 40–41.
72. Abramson, H. A., *op. cit.*, pp. 42–48.
73. Smoluchowski, M. v., *Z. physik. Chem.* **93**, 129 (1918).
74. Gouy, G., *Compt. rend.* **149**, 654 (1909); *J. phys. radium* **9**, 457 (1910); cf. also Chapman, D., *Phil. Mag.* **25**, 475 (1913).
75. Henry, D. C., *Proc. Roy. Soc. (London)* **A133**, 106 (1931).
76. Mooney, M., *J. Phys. Chem.* **35**, 331 (1931).
77. Abramson, H. A., *J. Phys. Chem.* **35**, 299 (1931).
78. Stern, O., *Z. Elektrochem.* **30**, 508 (1924).
79. Verwey, E. J. W. and Overbeek, J. Th. G., "Theory of the Stability of Lyophobic Colloids," *passim*, New York, Elsevier Publishing Co., 1948.
80. Verwey, E. J. W., *Trans. Faraday Soc.* **36**, 192 (1940).
81. Van den Tempel, M., *Rec. trav. chim.* **72**, 419 (1953).
82. Dean, R. B., Gatty, O., and Rideal, E. K., *Trans. Faraday Soc.* **36**, 161 (1940).
83. Van den Tempel, M., *ibid.*, 423.
84. McBain, J. W. and Bolduan, O. E. A., *J. Phys. Chem.* **47**, 94 (1943).
85. Pickering, S. U., *loc. cit.; J. Soc. Chem. Ind.* **29**, 129 (1910).
86. Briggs, T. R., *Ind. Eng. Chem.* **13**, 1008 (1921).
87. Weston, ——, *Chem. Age (London)* **4**, 604, 638 (1921).
88. Cheesman, D. F., and King, A., *Trans. Faraday Soc.* **34**, 594 (1938).
89. Liesegang, R. E., *Kolloid-Z.* **45**, 370 (1928).
90. Mukerjee, L. N. and Srivastava, S. N., *Kolloid-Z.* **147**, 146 (1956).
91. King, A. and Mukerjee, L. N., *J. Soc. Chem. Ind.* **57**, 431 (1938).
92. Schlaepfer, A. U. M., *J. Chem. Soc. (London)* **113**, 522 (1918).
93. Von Reinders, W., *Kolloid-Z.* **13**, 235 (1913).
94. Thomas, A. W., *J. Am. Leather Chem. Assoc.* **22**, 171 (1927).
95. Schulman, J. H. and Leja, J., *Trans. Faraday Soc.* **50**, 598 (1954).

CHAPTER 5

Theory of Emulsions: Creaming, Inversion, and Demulsification

In the previous chapter, those properties of emulsions which contribute to their stability have been discussed. It is now time to consider the conditions which lead to emulsion instability.

It has been well said that an emulsion is stable only when it is broken. Thermodynamically speaking, this is quite correct; but from the point of view of the practical user of emulsions this is small comfort. Nonetheless, the fact is that even the most stable emulsion bears within itself the seeds of its own destruction. In certain cases, to be sure, stability may not be a desideratum, and a knowledge of the factors which lead to the breaking of emulsions may often be of value (cf. Chapter 9).

In the present chapter, the phenomena accompanying emulsion breakdown will be discussed, and the theory of this process, to the extent that it has been developed, will be introduced.

Emulsions can show instability in three ways: by creaming, by inversion, and by breaking (demulsification). Each of these processes represents a somewhat different situation. In specific cases, however, they may be related, e.g., creaming may precede complete emulsion breakdown, or creaming may be accompanied by inversion. However, the mechanisms are sufficiently different so that these three forms of emulsion instability can be discussed separately.

Creaming

The phenomenon of creaming receives its name from the most common instance: the separation of the cream of unhomogenized milk. What occurs in this case, and indeed in all cases of true creaming, is not so much a breaking of the emulsion as a *separation into two emulsions*, one of which is richer in the disperse phase, the other poorer, than the original emulsion. The emulsion which is more concentrated is the "cream." In the case of milk, the cream represents an emulsion much richer in the dispersed butter-fat phase than the depleted milk phase, i.e., approximately 35 per cent as opposed to about 8 per cent.

The most usual situation occurs when, as in the case of milk, the cream rises to the top. As the discussion of the pertinent theory will show, however, it is entirely possible for the richer phase to sink to the bottom. Such cases are encountered occasionally in practice; the phenomenon is called "downward creaming."

Sedimentation. In many emulsion systems, creaming, although undesirable, may occur to some extent. However, certain adjustments in manufacturing technique or formulation can reduce the rate of creaming to a point where it may be considered to have a negligible effect. This may be understood from a consideration of the sedimentation phenomenon.

According to Stokes[1] the sedimentation rate of a spherical particle in a viscous liquid is given by

$$u = \frac{2gr^2(d_1 - d_2)}{9\eta}, \qquad (5.1)$$

where u is the rate of sedimentation, g is the acceleration of gravity, r the droplet radius, d_1 is the density of the sphere, d_2 that of the liquid, and η the viscosity of the liquid. Clearly, the sign of u, and hence the direction in which the particle will move, depends on the relative values of the densities. In an O/W emulsion the oil density (d_1) is usually the smaller, hence upward sedimentation (creaming) will occur.

Examination of Eq. 5.1 leads to the conclusion that emulsion stability is favored by small droplet radius, small density differences between the disperse and continuous phases, and by high viscosity of the continuous phase. It should be pointed out, however, that small droplet radius implies a large interfacial area, which may be, in itself, a source of instability. It will also be recalled that Neogy and Ghosh,[2] on the basis of fairly extensive studies of emulsion viscosity, denied that emulsion stability was favored by high viscosities (cf. p. 70). In this case, however, it is possible that a confusion exists between emulsion viscosities per se and the viscosity of the continuous phase. Most workers in the field of practical emulsion formulation would agree with the qualitative conclusions drawn from Eq. 5.1.

Rate of Creaming. Numerous authors in the field of emulsions (including this writer) have implied that Eq. 5.1 described the rate of creaming of an emulsion. However, Greenwald[3] has correctly pointed out that it gives the rate of creaming of only a *single* drop. What one is actually interested in is, of course, the mass rate of creaming, i.e., the motion of the center of gravity of the disperse phase. This requires a consideration of the form of Stokes' law involving a distribution of droplet radii.

An emulsion is considered as consisting of a system of n_i droplets of radius r_i, each droplet thus possessing a mass $\frac{4}{3} \pi r_i^3 d_1$. The velocity of the center of gravity of the disperse phase of an emulsion in a long tube will then be

$$u = \sum_i \frac{4}{3V} \pi r_i^3 n_i u_i .$$ (5.2)

where V is the total volume of the disperse phase

$$\Sigma_i \tfrac{4}{3}\pi r_i^3 n_i$$

and where u_i is the sedimentation velocity of a droplet of radius r_i. Substituting for u_i, from Eq. 5.1, the mass creaming rate is found to be

$$\bar{u} = \sum_i \frac{8\pi}{27\eta V} g n_i r_i^5 (d_1 - d_2)$$ (5.3)

Consideration of Eq. 5.3 leads to the same general conclusions as Eq. 5.1. Obviously, for an emulsion of uniform droplet sizes, Eq. 5.3 reduces to Eq. 5.1.

The order of magnitude of the effect may be estimated from the fact that a droplet of density 0.8 and radius 0.1 μ will cream upward in an aqueous disperse phase at a rate of 5×10^{-7} cm/sec; increasing the droplet radius to 1 μ raises the rate to 0.2 cm/hr.

Greenwald[3] further points out that this discussion applies to a creaming process occurring in a long tube (or, properly, in one of infinite length). A more significant question is to determine the behavior of droplets in a bottle of finite size which has been stored for some time. In this case it will be found that the droplets of disperse phase distribute themselves through the length of the bottle according to the general Boltzmann law. That is, the concentration n_h of disperse phase droplets with a radius r_i at a height h in the container is

$$n_h = n_0 e^{-4\pi r_i^3 (d_1 - d_2) gh/3kT};$$ (5.4)

where k is the Boltzmann constant, T is the absolute temperature, and n_0 is the concentration at the bottom of the container. This will be recognized as similar to the distributions found by Perrin[4] in gamboge suspensions. Comparison of Eq. 5.4 with Eq. 5.1. makes it clear that the same considerations which govern emulsion stability in the case of more rapid creaming will result in a more uniform distribution for this equilibrium case.

Dombrowsky[5] has considered the time dependence of this sort of Boltzmann effect. His treatment, however, considers the case in which complete separation of the disperse phase occurs.

Practical Aspects of Creaming. Although creaming in general represents an undesirable situation, there are instances where it is useful. Thus, in the classic instance, the separation of cream from milk, it is usually desirable to accelerate the process.

This is done by the use of centrifugal separators, operating at high speeds, e.g., of the order of 6000 rpm. Stokes' law still applies, but the gravitational acceleration is now replaced by a term depending upon the geometry and speed of the centrifuge, *i.e.*,

$$ u = \frac{2\omega^2 R r^2 (d_1 - d_2)}{9\eta}, \tag{5.5} $$

where ω is the angular velocity of the centrifuge and R is the distance of the droplet from the center of rotation. By the use of speeds of several thousand revolutions per minute, forces equal to many thousands times that of gravity act on the emulsion droplets. Clayton[6] gives a brief discussion of the more usual types of cream separators.

The use of the centrifuge to accelerate creaming, as a convenient method of examining emulsion stability, will be considered subsequently (cf. p. 331).

Another instance of desirable creaming arises in the processing of natural rubber latex. The rubber latex, for example that of *Hevea braziliensis*, is a moderately fine O/W emulsion, in which the droplets of the disperse phase vary in diameter from a maximum of 2 μ downwards, the majority being in the neighborhood of 0.5 μ, although there are a large number of barely visible ones. Such fine division does not favor creaming. On the other hand, the specific gravity of the rubber hydrocarbon is about 0.9042, while that of the serum is 1.024, so that creaming will take place, albeit at a very slow rate.[7]

The separation of the rubber from the latex, however, is facilitated by concentrating the material in the form of cream. It has been found that the addition of various materials, termed "creaming agents" or "coagulating agents," serve to increase the speed of creaming. These operate by causing the individual droplets to cluster together, or coagulate,* in larger assemblies. These clusters, consisting in some cases of several hundred droplets, were apparently first observed by Baker.[8] Although each cluster is not a single droplet (nor is it spherical), the general conclusions of Stokes' law can be applied, i.e., the clusters act effectively like very large single droplets, and the speed of creaming is markedly increased.

Electrolytes are especially suitable as creaming agents, and their effectiveness varies with, among other things, the charge of the cation. This indicates

---

* A distinction must be made between this phenomenon, coagulation, or the formation of clusters, and coalescence, in which a single large droplet is formed. This distinction is explored further (cf. p. 149).

that double layer effects are involved. A recent discussion of these, and other effects, has been given by van Gils and Kraay.[9]

On the other hand, creaming is most undesirable in latices formed in the manufacture of synthetic rubbers of, for example, the *GR-S* variety. The factors governing creaming in these systems have been investigated by Howland and Nisonoff.[10] These investigators found that creaming was favored by the formation of large particles, a conclusion in accord with Stokes' law. It was found that the concentration of these giant particles could be prevented by the timely addition of sufficient stabilizing soap, or by the use of a soap which initiates a small enough number of particles in relation to its stabilizing capacity. From the point of view of emulsion theory generally, this is equivalent to requiring a sufficient number of surface-active molecules to stabilize an emulsion with a large interfacial area.

Creaming as a result of coagulation or aggregation has been used by Cockbain[11] to study the phenomenon of coagulation. This investigation is more properly discussed, however, in connection with the general problem of demulsification (cf. p. 157).

Alsop and Percy[12] showed how the rate of creaming varied with emulsifier concentration, in a study of mineral oil-in-water emulsions stabilized by a mixed emulsifying agent consisting of a coco monoglyceride together with a potassium coco soap and with a coco monoglyceride sulfate. Figure 5-1 illustrates the amount of oil (i.e., cream) and water (i.e., serum) separated after eighteen hours for the system containing soap and monoglyceride. Figure 5-2 gives the same data for the system stabilized by monoglyceride sulfate detergent and monoglyceride. It will be noted that the upper left portion of the diagram so plotted corresponds to systems in which the opposite type of emulsion was obtained. All points correspond to systems in which the phase volumes were equal; the determinations were carried out at 25° C. It is striking to note, in the latter case (Figure 5-2) in what a relatively small concentration range a stable emulsion (for at least the period of test) can exist.

Inversion

Another type of emulsion instability consists of instability with regard to type, i.e., the emulsion may suddenly change from O/W to W/O, and vice versa. Such an emulsion is said to have *inverted*. This phenomenon has already been briefly touched upon in connection with the earlier discussion of emulsion viscosity (p. 65), and many of the considerations relating to the type of emulsion which will form under given conditions are germane (pp. 213 et seq.). Inversion, although the subject of considerable investiga-

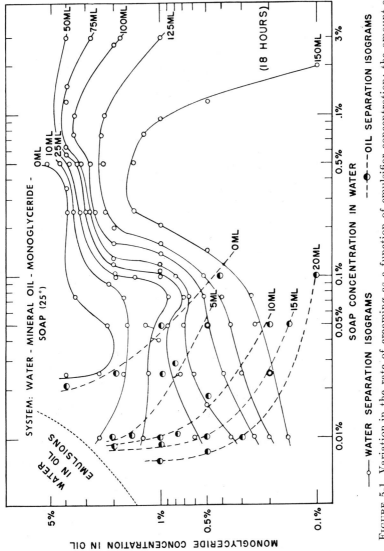

FIGURE 5-1. Variation in the rate of creaming as a function of emulsifier concentration; the amount of cream separated in eighteen hours for a system stabilized by soap and monoglyceride.[12]

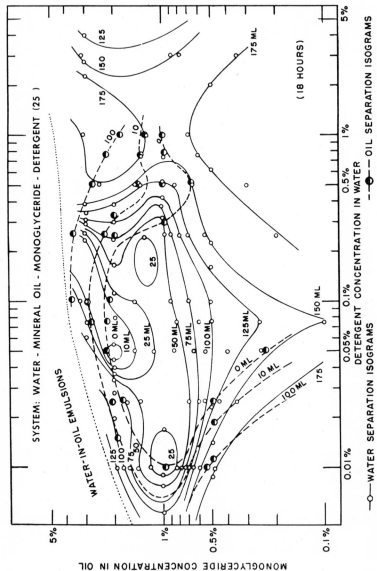

FIGURE 5-2. Variation in the rate of creaming as a function of emulsifier concentration; the amount of cream separated in eighteen hours for a system stabilized by monoglyceride sulfate and monoglyceride.

ion, is not well understood in many respects. A most important difficulty resides in the purely conceptual problem of conceiving of a physical mechanism whereby the actual process is carried out.

Simple Theories of Inversion. The simple stereometric theory of Ostwald has been discussed earlier (p. 90) in connection with the more general question of emulsion type, and its shortcomings indicated. If this theory were regarded as rigorously true, it is evident that inversion must occur in emulsions where the phase volume of the disperse phase is greater than 0.74. As was pointed out, systems are known in which this is actually the case, but the actual geometry of an emulsion often does not fit the ideal case envisioned by Ostwald. Inversion (although a concentration effect) may well occur at other values of the phase volume.

Indeed, as pointed out earlier, there is a considerable body of evidence that the concentration of disperse phase is only one consideration, and that not too important. For example, Sherman[13] has shown that for a given emulsion system the inversion concentration varies with the emulsifier concentration (cf. Table 3-9). An interesting observation in connection with these data is that, although the maximum emulsion viscosity obtained prior to inversion increases steadily with the concentration of emulsifying agent, the phase volume at inversion goes through a maximum at an emulsifier concentration of 3.5 per cent.

In the same connection, the oriented wedge theory of Harkins (pp. 88–90), whatever its limitations, serves to explain the inversion of sodium soap-stabilized O/W emulsions on the addition of bivalent cations. Generally speaking, of course, in the case when a specific emulsifier capable of stabilizing one type of emulsion is chemically changed into one of the opposite type, inversion may be expected to occur. This sort of change can be brought about by such things as change of pH or bacterial action.

Mechanism of Inversion. As indicated above, a considerable difficulty in the theoretical discussion of inversion resides in the physical mechanism involved. Thus, Clowes[14] and Sutheim[15] have presented diagrammatic representations of the inversion process. It must be admitted that this writer finds these unclear, but apparently the process involves the formation of filamentary droplets of the disperse phase, which, forming a species of network, then break the continuous phase into similar filaments. The initial step is then reversed, except that now the filaments of what was originally the continuous phase form the droplet phase.

Actually, this does not appear very satisfactory. A more realistic mechanism (at least for the case considered) is advanced by Schulman and Cockbain.[16] The process is illustrated in Figure 5-3, adapted from their paper.

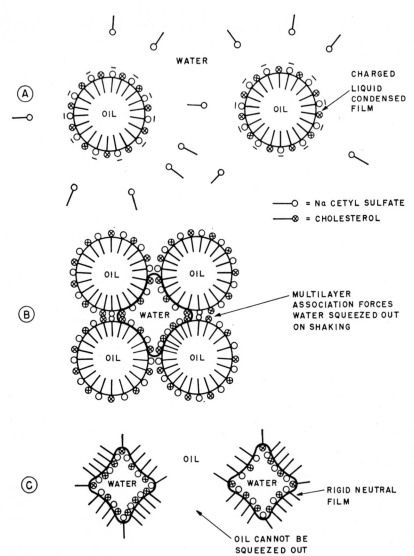

FIGURE 5-3. Mechanism of inversion of oil-in-water emulsion according to Schulman and Cockbain.[14] (A) Emulsion is stabilized by a condensed mixed film of cholesterol and sodium cetyl sulfate, and further stabilized by a negative surface charge. (B) Addition of multivalent ions discharges the surface charge, and the interfacial film realigns itself so that irregularly shaped droplets of water are formed. (C) Coalescence of oil droplets to form new continuous phase completes the inversion process.

An oil-in-water emulsion stabilized by a condensed mixed film of cholesterol and sodium cetyl sulfate is further stabilized by the existence of a negative charge (Figure 5-3A). If a multivalent cation (e.g., Ba^{++} or Ca^{++}) is now added to this system, it serves essentially to neutralize the surface charge, making coagulation of the droplets possible. This is recognizable as the same process involved in the accelerated creaming of latex.

In this case, however, it is assumed that small amounts of the aqueous phase are trapped among the clumped oil droplets. It is now possible for the molecules of the interfacial film to realign themselves in such a way that irregularly shaped droplets of water, stabilized by a rigid uncharged film, are formed (Figure 5-3B). Coalescence of the oil droplets to form the continuous phase now completes the inversion process (Figure 5-3C). In the case in question, Schulman and Cockbain indeed found that the inversion was accompanied by the formation of irregularly shaped, uncharged, rigid "sacks" of water dispersed in the oil.

It is interesting to note that the phenomenon of coalescence is of importance as an intial step in both creaming and inversion. As will be seen below, it is also significant in the question of total demulsification.

Hysteresis Effects on Inversion. Although it has apparently not been noted in the literature, the process of inversion, considered as a reversible phenomenon, must involve hysteresis. For example, consider an inversion phenomenon caused solely by change in phase volume. If viscosity measurements are taken, the curve represented in Figure 3-5 will be observed, i.e., at inversion the viscosity will drop sharply. However, if the direction of addition is now reversed, the system will not, in general, retrace the same path. This situation is indicated schematically in Figure 5-4. That this is indeed a true hysteresis phenomenon is made clearer if such a property as conductance is used as the criterion of inversion. Figure 5-5 is a schematic representation of the change in conductance occurring in the reversible inversion of an initially O/W emulsion.

An interesting exception to this was found by Kremann, Griengl, and Schreiner,[17] working with olive oil, water, and aqueous solutions of sodium hydroxide and hydrochloric acid. As the oil concentration was increased, low viscosity O/W emulsions were formed. At about 24 per cent olive oil, however, the emulsion viscosity *increased* sharply, and further addition of oil resulted in lower viscosities and inversion to W/O.

Temperature Effects. Inversion is apparently a function of the temperature, although data on the subject is meager. Probably the most elaborate study of this effect is due to Wellman and Tartar.[18] These investigators found that water-in-benzene emulsions, stabilized by sodium soaps, can

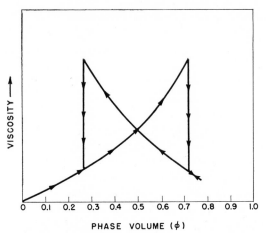

FIGURE 5-4. Inversion hysteresis as measured by viscosity. An initially O/W emulsion is inverted by addition of oil, with a consequent change in viscosity. When aqueous phase is now added, the emulsion does not retrace the original viscosity-concentration path.

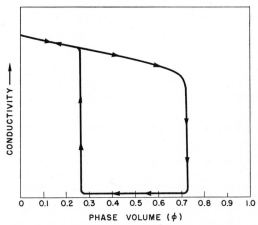

FIGURE 5-5. Inversion hysteresis as measured by conductance. In the same system as illustrated in Figure 5-4, the existence of hysteresis is better illustrated by conductance.

be inverted by increasing the temperature, together with gentle shaking. These O/W systems will reinvert if allowed to cool and stand about 30 minutes. It was also found that the temperature at which inversion occurs was sensitive to the concentration of emulsifying agent. Their results for emulsions stabilized by sodium stearate and sodium palmitate are shown in Figure 5-6.

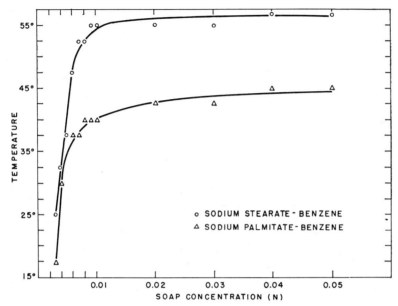

FIGURE 5-6. Temperature dependence of inversion, as shown by water-in-benzene emulsions stabilized by sodium stearate and sodium palmitate.[18]

As can be seen, when the concentration of emulsifier is very low, the inversion temperature is extremely sensitive to concentration. At higher concentrations, however, the curve tends to level out; the inversion temperature no longer changes with concentration. It is interesting that the temperatures corresponding to the flat portion of the curve are quite close to the so-called Krafft temperatures for the soaps, i.e., the temperatures at which the soap solutions are no longer turbid, but are transparent, isotropic systems. This suggests that the phenomenon is, to some extent, related to the solubility of the emulsifier. This suggests that it may, in some way, be connected with the phenomenon of micelle formation.

Rebinder, Gol'denberg and Ab[19] found that in emulsions stabilized by oleate soaps plus free oleic acid inversion is related to the acid-soap ratio (a point which will be discussed in more detail below, p. 148). What is of significance in the present connection is that this ratio is also temperature dependent. For example, it was found that at 0 to 1° C the required ratio was lower and the time required for inversion was longer than at 20°.

Isemura and Kimura[20] have noted that in benzene-water emulsions stabilized with sodium oleate or stearate formed *in situ* (cf. p. 210), the type of emulsion obtained varied with the type of shaking, the original sodium hydroxide concentration in the aqueous phase, and the temperature. Thus,

W/O emulsions were generally obtained with vertical shaking at 10–20° C with sodium hydroxide concentrations less that 0.002 N and acid concentrations of 0.0025–0.01 N. On the other hand, O/W emulsions tended to form at sodium hydroxide concentrations greater than 0.2 N, and at temperatures of 28–30°.

General Observations on Inversion. As indicated earlier, quite a large body of work has been done on inversion, much of it, however, quite old. Clowes[14] and Bhatnagar[21] carried out classic investigations of the effect of adding electrolytes to soap-stabilized emulsions. The general results obtained are explicable, in part at least, by the stereometric considerations of the oriented wedge hypothesis. However, the matter cannot be entirely as simple as this, since Bhatnagar found that the valency of the added ion had a significant effect on the concentration of added electrolyte required to cause inversion. The inverting capacity of the electrolytes fall in the order $Al^{+++} > Cr^{+++} > Ni^{++} > Pb^{++} > Ba^{++} > Sr^{++}$ (calcium, bivalent iron, and magnesium showing the same effectiveness as strontium). This suggests that discharge of double layer charges may be involved, in agreement with Schulman and Cockbain.[16]

Similar results have been found by Wellman and Tartar[18] and others. On the other hand, King and Wrzeszinski[22] have found that emulsions stabilized by synthetic surface-active agents (as distinguished from soaps) are less affected by added electrolyte.

An important phenomenon connected with inversion is the so-called "lecithin-cholesterol antagonism," which has been discussed by Corran.[23] This arises in connection with the production of mayonnaise, although it is probably significant in biological applications as well. Lecithin is an extremely good emulsifying agent, producing O/W emulsions; cholesterol, on the other hand, is an agent favoring the production of W/O emulsions. Unfortunately, however, the source of stabilizing agent used for mayonnaise is found in dried egg yolk, which contains *both* lecithin and cholesterol. Corran investigated the system olive oil-water, as stabilized by lecithin-cholesterol mixtures. Table 5-1 summarizes his results. Clearly, inversion

TABLE 5-1. THE EFFECT OF MIXTURES OF LECITHIN AND CHOLESTEROL ON
EMULSION SYSTEMS[23]

Ratio Lecithin/Cholesterol	Emulsion Type
19.4	O/W
10.0	O/W
8.0	Indefinite
6.0	W/O
4.1	W/O
2.0	W/O

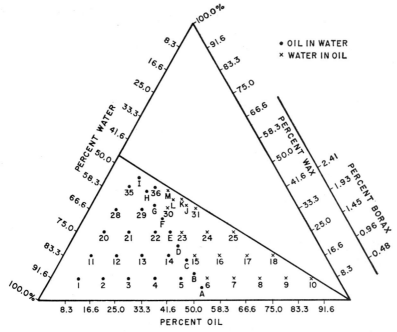

FIGURE 5-7. Inversion as a function of phase volume in cold cream formulations, according to Salisbury, Leuallen, and Chavkin.[24]

will occur when the lecithin-cholesterol ratio is 8.0; the emulsion described by Corran as "indefinite" is presumably a multiple emulsion (cf. below, p. 149).

Corran also discusses the effects of other additives to mayonnaise, the effect of mixing, pH, etc. These will be covered in more detail in the discussion of food emulsions (p. 261–264).

Salisbury, Leuallen and Chavkin[24] have reported an elaborate study of inversion as a function of phase volume in cold cream formulations. The results of a large number of experimental results are indicated in Figure 5-7. As can be seen, the triangular diagram is divided into two well-defined areas (with one unexplained exception), and inversion will result if the phase boundary is crossed. Although Pickthall[25] has expressed some reservations concerning the general validity of some of the results, the basic conclusions of this paper are sound. In this case the inversion point occurs at an aqueous concentration of 45 volume per cent, considerably distant from the theoretical value of 24 per cent (or 78 per cent) required by the simple Ostwald theory.

The work of Rebinder, Gol'denberg and Ab,[19] previously referred to in connection with temperature effects on inversion, should now be considered in more detail. This investigation dealt with phase inversion in emulsions prepared from purified kerosene and 5 per cent aqueous solutions of sodium, potassium, or lithium oleate. Each of the soap solutions contained a certain amount of free oleic acid.* The quantity of free oleic acid (m) was expressed as per cent of soap, i.e., of neutralized acid, in the solution. It was found that up to a certain value of m the emulsions first formed were O/W, and remained so. When this critical value of m was exceeded, the emulsions initially were of the inverse type (W/O), and remained so for a period of time proportional to m. For sodium oleate the critical value of m was 8 per cent. At this point the emulsion first formed was water-in-oil; the emulsion inverted at the end of one day. As m was increased from 8 to 17.5 per cent the life of the W/O emulsions increased from one to fifty days. For a m of 20 per cent inversion did not take place even after several months, at which time the original W/O emulsions broke into layers. Adding the free oleic acid to the hydrocarbon phase rather than the aqueous produced the same general results except that the critical value of m rose from 8 per cent to 14 per cent, and the life of the W/O emulsions was much shorter. Thus, at $m = 30$ per cent the inversion of W/O into O/W took place after only three days.

These researchers explain the formation of the W/O emulsions in the presence of an excess of free oleic acid by assuming that the free acid formed protective envelopes, preventing the coalescence of droplets of the aqueous phase. Actually, it is perhaps more reasonable to assume that complex Schulman-Cockbain formation between the soap anions and the undissociated acid (or what amounts to the same thing, formation of so-called "acid soap") is responsible for the difference in emulsion types. On the other hand, the researchers' contention that inversion occurs as the result of gradual diffusion of the free acid into the oil phase, is a reasonable explanation of the time dependence of the inversion on m. It also serves to explain the temperature dependence noted earlier (p. 143).

Phase inversion was also studied by Rebinder, Gol'denberg and Ab, by a titration method which also indicated the volume ratio of the two phases. The volume of the oil phase was kept constant at 3 ml. To it was added an aqueous soap solution in which m was varied from 8 to 26 per cent. The soap solution was added dropwise, with shaking after each addition. The volume of soap solution present when inversion took place was noted.

---

* Some free oleic acid would, in any case, be present in such solutions, as a result of hydrolysis. The data presented may be regarded as indicating the effect of the degree of hydrolysis on emulsion type.

The ratio of volume of aqueous phase to oil phase was greater the greater the value of m. Thus for $m = 8$ per cent, the ratio was 1.5, whereas for $m = 26$ per cent, the ratio was 6.7. A comparison of the results obtained with potassium, sodium, and lithium soaps showed that the critical value of m increases as the atomic value of the cation decreases.

Multiple Emulsions. Although significant exceptions to the general phase-volume theory of Ostwald have been shown to exist, the concept that certain emulsion types are favored, all other things being equal, in extreme concentration ranges is still valid. This is indicated schematically in Figure 5-5, where the ranges are given according to the close-packing view. However, it should be noted that, *pari passu*, the possibility exists that in the *intermediate* range of volume concentrations *no* particular form of emulsion is favored. From this fact arises the possibility of hysteresis effects, already considered, and of multiple emulsions.

As the name indicates, a multiple emulsion is one in which both types of emulsions exist simultaneously. What is meant by this is that an oil droplet may be suspended in an aqueous phase, which in turn encloses a water droplet, thus giving what might be described as a W/O/W emulsion. Actually, there is no reason why this process cannot be continued indefinately, and Seifriz[26] has published a photograph of a quinqui-multiple emulsion.

As indicated in the discussion of the work of Corran,[23] multiple emulsions may be expected in the neighborhood of the inversion point, a fact borne out by numerous investigations.

Demulsification

It is now necessary to consider the most important and the most complete example of emulsion stability: complete demulsification or breaking of the emulsion. As has been pointed out above, demulsification is often accompanied by creaming or inversion, and some of the considerations of the previous sections are of significance here.

It is important to note that the coagulation of the disperse phase occurs as a two-stage process.[27, 28] In the first stage, *flocculation*, the droplets of the disperse phase form aggregates in which the drops have not entirely lost their identity (such aggregation is often reversible). However, from the point of view of creaming, these aggregates behave as single drops, and acceleration of the rate of creaming will be observed in systems in which the aggregate density differences are sufficiently large. Also, in sufficiently concentrated emulsions, a perceptible increase in emulsion viscosity may be noted.

In the second stage, termed *coalescence*, each aggregate combines to form a single drop. This is an essentially irreversible process, leading to a decrease in the number of oil droplets and finally to complete demulsification.

Flocculation and Coalescence. The flocculation of emulsion droplets, as the first stage in demulsification, is governed by the identical theoretical considerations applied by von Smoluchowski[29] to the coagulation of hydrophobic sols. Indeed, Lawrence and Mills[30] make the point that it is somewhat remarkable that studies on emulsion systems have not been used to test the Smoluchowski equations.

Lawrence and Mills[30] and van den Tempel[28] have worked out theoretical interpretations of the flocculation process in terms of the Smoluchowski theory. Although the general equations derived are similar, only van den Tempel's treatment will be given *in extenso*.

A useful theory requires a consideration of how the process is to be experimentally followed. In this connection, it should be remembered that in a process consisting of two consecutive reactions, the overall reaction rate is determined by the slower of the two. In very dilute oil-in-water emulsions the rate of flocculation can be made much smaller than the rate of coalescence. As a consequence, the stability of the emulsion will be affected by those factors which affect the rate of flocculation. Increasing the concentration of oil phase in the emulsion will result in a *slowly* increasing rate of coalescence, and a much *faster* increasing rate of flocculation. Finally, in highly concentrated emulsions, the coalescence can be made rate-determining.

In a certain range of concentrations the two processes will be of the same order of magnitude. Consequently, the effect of particle concentration on the rate of coagulation will be most evident. Van den Tempel has pointed out that, even in very dilute emulsions, it is possible to make coalescence rate-determining by the addition of surface-active agents which may have little or no effect on the rate of flocculation, but which inhibit coalescence.

Coagulation can be measured by observations on the rate of creaming as, for example, has been done by Cockbain.[11] The rate of oil separation, however, depends not only on the interactions between the droplets, but also on those factors whose effect is described by the Stokes' equation (pp. 135 et seq.). Van den Tempel has indicated that a clearer insight into the mechanism of coagulation can be obtained by studies on systems in which creaming effects have been ruled out, e.g., by adjusting the densities of the two phases.

The most exact data on coagulation has been obtained by measurements of the specific interface, as in a most elaborate study by King and Mukherjee.[31] This is also the technique employed, in principle, by Lawrence and Mills,[30] although the actual parameter reported is the mean drop volume. It can be shown, however, that the determination of the number of

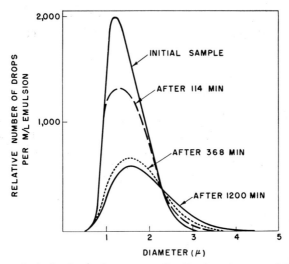

FIGURE 5-8. Variation in the droplet-size distribution of an unstabilized emulsion with time. The curves are unimodal leptokurtic, the kurtosis increasing with time.[30]

particles as a function of time is more sensitive than the specific interface method, since a 10 per cent decrease in interfacial area is accompanied by a 27 per cent decrease in the number of particles, provided that the general shape of the size-frequency distribution curve does not change appreciably during the coagulation.[28]

This last point is not insignificant, as indicated by Figure 5-8, showing the variation in size-frequency of an unstabilized emulsion with time.[30] These curves are unimodal and leptokurtic, the kurtosis increasing with time. These curves are probably sufficiently similar in shape for the assumption indicated above to be valid.

If one may use the particle concentration as a measure of flocculation, the following theoretical interpretation, due to van den Tempel,[28] may be used.

Decrease of Particle Concentration with Time. The number of particles, n, in unit volume, is defined in such a manner that no distinction is made between single droplets, or primary particles, and aggregates consisting of a number of primary particles.* The number n is given by Smoluchowski's theory as

$$n = n_0/(1 + an_0t), \tag{5.6}$$

---

* It should also be noted that no distinction is made between primary particles comprising the original emulsion droplets, and those arising from coalescence of an aggregate.

where n_0 is the number of particles at time $t = 0$, and a is a rate-determining constant which (in the case of rapid coagulation of a monodisperse sol) has the value

$$a = 8\pi DR \backsim 8\pi \frac{kT}{6\pi\eta R} R \backsim 10^{-11} \text{ cm}^3 \text{ sec}^{-1} \qquad (5.7)$$

In Smoluchowski's derivation of Eq. 5.6 it is tacitly assumed that the coagulation process has been going on for a length of time sufficient to insure the existence of "nearly steady-state" conditions in the environment of each particle. That is, the number of particles diffusing in unit time through a sphere surrounding a given particle equals the number of particles adhering to the central particle. This state will be reached in a time $t > R^2 D$.

In applying the results of Smoluchowski's theory to emulsions, it must be remembered that the flocculation of oil drops is often easily reversible. Thus, in many cases the aggregates can be easily redispersed by stirring. This action of shearing forces can be enhanced by dilution with a solution of a suitable surface-active agent.

Furthermore, the coagulation of an emulsion is a function not only of the rate of formation of these more or less reversible aggregates, but also of the rate at which the particles coalesce to form larger droplets. By recognizing that only those primary particles which have formed an aggregate are capable of coalescing, the number of particles existing after some time t can be found. The average number of primary particles in an aggregate, n_a, at time t can be obtained from the Smoluchowski theory. Thus, the number of primary particles which *have not* yet combined into aggregates is given by

$$n_1 = n_0(1 + an_0t)^{-2}, \qquad (5.8)$$

while the number of aggregates is given by

$$n_v = an_0^2 t(1 + an_0t)^{-2} \qquad (5.9)$$

Thus, the total number of primary particles in unit volume associated into aggregates is

$$n_0 - n_1 = n_0 \left\{ 1 - \frac{1}{(1 + an_0t)^2} \right\}, \qquad (5.10)$$

and therefore

$$n_a = (n_0 - n_1)/n_v = 2 + an_0t \qquad (5.11)$$

If one designates the average number of separate particles existing in an aggregate at time t as m, it is clear that this number must always be smaller than n_a, since a certain amount of coalescence will have occurred. The quantity m will be only slightly less than n_a if coalescence is quite slow; on the other hand, it will have a value close to unity if coalescence is rapid.

It may be assumed that the rate of coalescence will be proportional to the number of points of contact between the particles comprising an aggregate, a quantity which may be taken as equal to $(m - 1)$ if the aggregate is made up of a relatively small number of particles. Van den Tempel reports that observation has, in fact, shown that in sufficiently dilute emulsions small aggregates generally consist of one large particle together with one or two small ones, and are built up linearly. Thus, m will decrease in direct proportion to $(m - 1)$, while m increases by adherence of new particles. The overall rate of increase of m caused by flocculation is thus (from Eq. 5.11)

$$dm/dt = an_0 - K(m - 1),\qquad(5.12)$$

where K is a measure of the rate of coalescence. Equation 5.12 may be integrated using the boundary condition $m = 2$ when $t = 0$, with the result

$$m - 1 = an_0/K + (1 - an_0/K)\exp(-Kt)\qquad(5.13)$$

The *total* number of particles, whether flocculated or not, in a coagulating emulsion is found by adding the number of unreacted primary particles to the number of particles in the aggregates, i.e.,

$$n = n_1 + n_v \cdot m = \frac{n_0}{1 + an_0 t} + \frac{an_0^2 t}{(1 + an_0 t)^2}\left\{\frac{an_0}{K} + \left(1 - \frac{an_0}{K}\right)e^{-Kt}\right\}\qquad(5.14)$$

In Eq. 5.14, the first term on the right-hand side of the equation corresponds to the number of particles which would be found if each aggregate had been counted as a single particle. The second term gives the number of particles which enters in addition to that of the classical Smoluchowski treatment when the composition of the aggregates is taken into account. This is, essentially, unnecessary in the regular theory of the flocculation of hydrophobic sols.

When $K = \infty$ (i.e., immediate coalescence), Eq. 5.14 reduces to Eq. 5.6, in agreement with the above. When $K = 0$ (i.e., no coalescence whatever), Eq. 5.14 gives $n = n_0$ for all values of t. For intermediate values $(0 < K < \infty)$, the effect of a change in particle concentration on the rate of coagulation is given by Eq. 5.14. That this is the case is made clear by rewriting Eq. 5.6 in the form

$$1/n - 1/n_0 = at, \qquad (5.15)$$

which shows that the rate of increase of $1/n$ in a coagulating sol is inde pendent of the concentration. This is not the case for an emulsion, as ind cated in Figure 5-9, showing the rate of increase of $1/n$ (i.e., decrease i particle concentration) for various values of n_0, as calculated by means c Eq. 5.14. Reasonable values of a and K have been used.

A more striking illustration of the effect of the particle concentration i shown in Figure 5-10. From this it appears that, in both dilute and con centrated emulsions, the rate of coagulation after a predetermined time (a measured by $1/n$) is not appreciably changed by a change in particle con centration. On the other hand, in the region where an_0/K is of the order c unity, the rate of coagulation is strongly dependent on n_0.

Approximate Treatment. Van den Tempel proceeds to make severa useful approximations in the equations derived above. These approxima tions, given below, make the equation more useful in interpreting experi mental results.

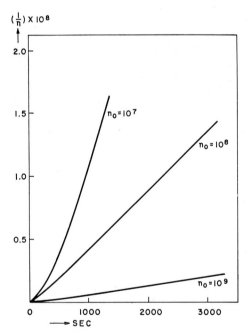

FIGURE 5-9. Rate of decrease in particle concentration as a function of the initia particle concentration for a flocculating emulsion, calculated from Eq. 5.14.[28]

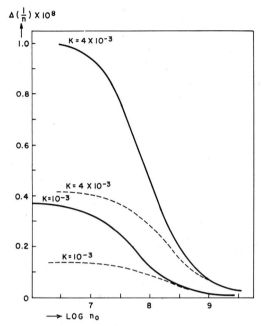

FIGURE 5-10. Rate of decrease in particle concentration as a function of time. The rate of coagulation after a predetermined time is not a strong function of the initial particle concentration for dilute and concentrated emulsions, but is strongly dependent for intermediate concentrations $(an_0/K \approx 1)$.[28]

1. In a flocculating concentrated emulsion it is possible to make the quantity $an_0/K \gg 1$. In actual systems, K is usually much smaller than one (of the order 10^{-3} sec^{-1}) and it is sufficient to make $an_0 \geqq 1$ to satisfy this condition. When this is true $an_0 t$ becomes much larger than unity very quickly, and the contribution of the primary particles may be neglected. Equation 5.14 then reduces to

$$n = an_0^2 t(1 + an_0 t)^{-2}\{an_0/K(1 - e^{-Kt})\} \qquad (5.16)$$

Since $an_0 t \gg 1$, a further approximation yields

$$n = \frac{n_0}{Kt}(1 - e^{-Kt}) \qquad (5.17)$$

From Eq. 5.17, it may be concluded that *in concentrated emulsions, the rate of coagulation no longer depends on the rate of flocculation*, i.e., coalescence is completely rate-determining. In Figure 5-10, the dashed line represents the rate of coagulation for values of $n_0 > 10^{10}$ as calculated either from Eq. 5.14, 5.16, or 5.17; there being little difference in the curves as determined from these different approximations. On the other hand, for $n_0 = 10^9$, Eq. 5.16 gives a serious divergence for values of t less than 2000 sec. Equation 5.17 would, of course, show even worse agreement.

There is, however, a limitation imposed on the use of any of these equations for concentrated emulsions. The assumption has been made, in deriving Eq. 5.12, that the number of points of contact between the m particles making up an aggregate was equal to $m - 1$. This is probably not the case with the larger aggregates which are encountered in a flocculating concentrated emulsion. In a closely packed assembly of uniform spheres, each sphere is in contact with twelve others. The number of points of contact is thus proportional to m rather than to $m - 1$. When the spheres are not uniform (the usual case), each particle may be in contact with more than twelve others. It is thus necessary to rewrite Eq. 5.12 as

$$dm/dt = an_0 - pKm, \qquad (5.18)$$

where p has a value between 1 and about 6. On integration, this equation yields:

$$m = an_0/pK + (2 - an_0/pK) \exp(-pKt) \qquad (5.19)$$

which should be used in place of Eq. 5.13 for the case of concentrated emulsions.

2. Approximations depending upon the existence of extremely dilute emulsions can also be made. Thus, in such a case an_0/K may be made very much smaller than unity, provided coalescence occurs at a sufficiently high rate. After coagulation has taken place for a sufficiently long time such that $Kt \gg 1$, the second term in the right-hand member of Eq. 5.14 may be neglected, and the equation reduces to Eq. 5.6. This equation does not contain K, and hence flocculation, rather than coalescence becomes rate-determining.

3. A third approximation may be made in the case where coalescence is very slow. In this instance, the exponential term may be expanded in a power-series, which may be cut off after the second term when $Kt \ll 1$. Then

$$n = n_0 \{1 - Kt(1 + an_0t)^{-1} + Kt(1 + an_0t)^{-2}\} \qquad (5.20)$$

In accordance with the assumption made, this equation predicts a very slow decrease in particle number.

4. After coagulation has gone on for a sufficiently long time, Kt may be very much greater than unity. In this case, the exponential term may be neglected, and since an_0/t (which is in the denominator) is also very much greater than unity, one obtains

$$n = n_0/Kt + 1/at \qquad (5.21)$$

It is interesting to compare the equations obtained by this treatment with that obtained by Lawrence and Mills[30] in terms of the average volume of the individual droplets:

$$\bar{v} = v_0 + \beta\phi t, \qquad (5.22)$$

where v_0 is the volume of the primary particles, ϕ is the phase volume of the disperse phase, i.e., $n_0 v_0$, and β has the value $2kT/3\eta$. If one takes into account the possibility that each collision between primary particles may not lead to flocculation into an aggregate, Eq. 5.22 is modified to

$$\bar{v} = v_0 + q\beta\phi t \qquad (5.23)$$

where $q = A \exp(-E/RT)$, i.e., represents a term containing an energy of activation.

Experimental Studies. A considerable number of studies have been reported on the coagulation of emulsions. An early and interesting paper is due to Kraemer and Stamm.[32] However, these workers were principally interested in using sedimentation results as a method for determining the droplet-size distribution. The more elaborate study of King and Mukherjee[31] has already been referred to, but these workers were principally concerned with coagulation as a measure of the efficiency of various emulsifying agents.

More recently, Cockbain[11] has studied the reversible aggregation and disaggregation of the oil droplets in emulsions stabilized by soaps. Rate of creaming was used as criterion of aggregation, but no attempt was made to separate the two steps (i.e., flocculation and coalescence) involved. In all the systems investigated in this study, aggregation of the oil particles commences only at a soap concentration equal to, or a little greater than, the critical micellar concentration of the soap. In many cases, states of maximum and minimum particle aggregation are found to occur at concentrations much higher than the C.M.C. Such maxima and minima are most pronounced in emulsions where benzene is the disperse phase and unbranched hydrocarbon chain soaps are employed as the stabilizers.

Cockbain has also demonstrated that a correlation exists between the effect of aliphatic alcohols (of chain-length less than six carbon atoms) in reducing the concentration of soap required for aggregation of the emulsion droplets and the effect of these alcohols on the C.M.C. of the surface-active materials. Addition of inorganic salts prevents disaggregation of the emulsion particles and reduces the soap concentration at which aggregation commences. The latter process is independent of the nature of the salt anion when an anionic soap is used as stabilizer.

Table 5-2 reproduces those data of Cockbain which show the relationship between the C.M.C. and the concentration of surface-active stabilizer at

TABLE 5-2. INFLUENCE OF SOAPS ON PARTICLE AGGREGATION IN 10%
PARAFFIN OR BENZENE EMULSIONS[11]

Soap	C.M.C. (%)	Conc. of Soap (%) for					
		Init. aggrgn.	Max. aggrgn.	Dis-aggrgn.	Init. aggrgn.	Max. aggrgn.	Dis-aggrgn.
		In benzene emulsion			In paraffin emulsion		
CTAB	0.03	ca. 0.07	0.14	0.26	ca. 0.08	0.27	0.67
NaLS	0.21	0.30	0.65	1.05	0.45	0.95	ca. 1.6
K laurate	0.57	0.80	1.4	3.0	0.95	—	—
Aerosol MA	1.6	1.9	—	—	2.2	—	—
Aerosol OT	0.18	0.20	0.55	—	0.30	—	—

which aggregation occurs. These results are perhaps a trifle surprising at first glance, since one might very well expect emulsion stability to increase with increasing stabilizer concentration. However, Cockbain points out that if one considers that multimolecular adsorption at the interface can very well occur, the results become not unreasonable; and are, indeed, consistent with observations on the sedimentation rate of solids stabilized by surface-active materials.

It is also possible that this is related to the phenomenon of "limited" coalescence, reported by Wiley (cf. below, p. 163).

Brady, Mandelcorn, and Winkler[33] have made a study of the coagulation of a GR-S latex with sulfuric acid. The dependence on the concentration of latex and of acid was measured, as was the effect of temperature. An attempt was also made to determine the rate of desorption of soap during the coagulation and to seek a corresponding relation between this phenomenon and the rate of coagulation. It was found that scarcely any coagulation occurred when the sulfuric acid concentration was as low as $N/90$, while coagulation was almost immediate at $N/10$. At intermediate acid concentrations, the time-dependence of the latex concentration followed the law

$$C = C_0/(1 + k[H^+]C_0 t) \qquad (5.24)$$

where $[H^+]$ stands for the concentration of hydrogen ion in excess of the minimum required to cause coagulation. This will be recognized as identical in form to Eq. 5.6, to which Eq. 5.14 reduces, under conditions of immediate coalescence (p. 153).

Experimental studies by Lawrence and Mills[30] on unstable systems of hydrocarbon oil and water indicated that, even in the absence of emulsifying agent, some electric stabilization must occur. Indeed, this would be implied by the well-known stability of oil hydrosols. These authors found, on the basis of their treatment, that only one collision in 10^9 leads to coagulation. This corresponds to an activation energy of 3.9 kcal/mole, if the non-exponential term in the expression for q is put equal to unity.

Lawrence and Mills also analyse the data of Jellinek and Anson[34] on soap-stabilized emulsions at a 50 per cent phase volume and a temperature of 70° C, as compared to their own data on emulsions at a 1 per cent phase volume at 25° C. For the dilute emulsion they find $q = 1.8 \times 10^{-5}$ and $E = 6.5$ kcal/mole. For the least stable emulsion of Jellinek and Anson it is found that $q = 8.8 \times 10^{-6}$ and $E = 7.9$ kcal/mole; for another emulsion $q = 1.5 \times 10^{-6}$, and hence $E = 9.1$ kcal/mole.

Effect of Electrolytes. It has been pointed out earlier (p. 146) that the addition of electrolyte often leads to emulsion inversion. However, it is also well known that complete demulsification can also accompany such addition; indeed, this is the basis of certain commercial demulsifying compositions (cf. p. 288).

Recently, van den Tempel[35] has investigated the stability of O/W emulsions stabilized by *Aerosol OT*, *Aerosol MA*, and sodium laurate under the influence of added electrolyte. These results have been treated in accordance with the theory of coagulation outlined above.

The oil phase in the emulsions studied consisted of a mixture of monochlorobenzene and paraffin oil, of a composition such that the density was equal or close to that of the aqueous salt solutions. In this way, purely kinetic effects due to creaming could, in most cases, be disregarded. In order to make measurements as a function of time, samples of the coagulating emulsion were diluted by the addition of a solution of a nonionic surface active agent (*Emulphor O*). This had the effect of effectively "freezing" the reaction, and permitting a count of the number of particles.

In accordance with van den Tempel's extension of the Smoluchowski theory to emulsions, the quantity $1/n$ should vary linearly with the time. That this is indeed the case is shown in Figure 5-11, after van den Tempel, which shows the influence of concentration of sodium chloride on the coagulation of emulsions stabilized with a 0.00068 M solution of *Aerosol OT*. Similarly, Figure 5-12 shows the effect of various polyvalent electrolytes on emulsions stabilized with *Aerosol MA*. In general, the results obtained are in agreement with the well-known Schulze-Hardy rule,[36] according to which the concentration of a cation required to obtain a certain rate of coagulation with a sol consisting of negatively charged particles is mainly determined by the charge of the cation, and to a much lesser extent by its specific nature.

In the case of emulsions, however, an exception must be noted for the case where the coagulating electrolyte has an effect on the rate of coales-

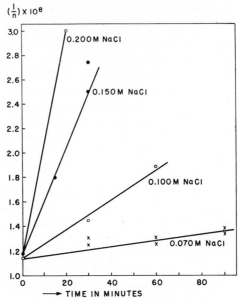

FIGURE 5-11. Effect of the concentration of sodium chloride on the rate of decrease of particle concentration in an emulsion stabilized by *Aerosol OT*.[28]

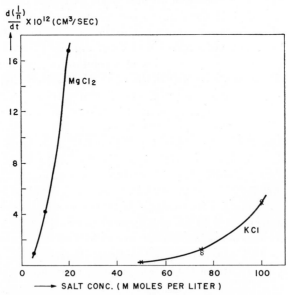

FIGURE 5-12. Effect of polyvalent electrolytes on the rate of decrease of particle concentration in an emulsion stabilized by *Aerosol MA*.[28] The predictions of the Schulze-Hardy rule are, in general, in agreement with the data.

cence, i.e., where, for example, insoluble salts of the surface active anion are formed. Such insoluble salts will, of course, tend to retard coalescence; soaps in the presence of the bivalent alkaline earth cations are an example of this effect.

The general conclusions drawn by van den Tempel[35] may be summarized. When all collisions between particles are effective in removing one separate particle from the system, the rate of flocculation has a value of about 10^{-11} cm^3 sec^{-1} in a monodisperse system of spherical particles. If higher values are obtained (as were found by van den Tempel), factors other than Brownian motion must be considered as operative in causing flocculation. (Among these are the van der Waals attraction which occurs as soon as the particles approach to within about one particle radius. This is possible since the addition of electrolyte has removed or decreased the stabilizing zeta-potential.)

Secondly, a non-uniform particle-size distribution will have the effect of increasing the rate of flocculation. In this connection, Müller[37] has shown that rate increases of up to 15 per cent can occur with symmetrical particle distributions, while increases of 50 per cent are possible with asymmetrical distributions.

There are also two factors inherent in the experimental method which may lead to apparent high flocculation rates. One of these is the possibility of some flocculation occuring from mechanical causes when the coagulant electrolyte solution is stirred in. The other is the possibility of orthokinetic flocculation, i.e., as a result of creaming. As has been indicated above, the density of the oil phase was adjusted to reduce this factor to a minimum. In some experiments, however, especially with monovalent cations, such high electrolyte concentrations were reached that the density difference became appreciable.

The order of magnitude of the results obtained are indicated in Table 5-3, showing the rates of flocculation and coalescence for a number of systems, as calculated from the dependence of the rate of coagulation on the initial particle concentration. These values are obtained by comparison with a series of graphs of the type of Figure 5-10; that is, Eq. 5.14 is solved graphically.

Some earlier work on the effect of electrolytes was carried out by Martin and Hermann.[38] In these experiments the rate of coagulation was not measured; rather, the total amount of the oil phase coagulated was observed as a function of the amount and type of electrolyte used. A standard emulsion consisting of 60 ml of xylene in 40 ml of $N/30$ sodium oleate was used, and the amount of oil separated as the result of the addition of varying quantities of 0.1 N electrolyte observed. The quantity of emulsion used was such that 4.5 ml was the maximum volume of oil which could separate. Some of

TABLE 5-3. REACTION RATES IN STABILIZED O/W EMULSIONS[35]

Stabilizer	Salt	Rate of Floccula-tion (cm³/sec × 10^{11}) = a	Rate of Coalescence sec⁻¹ × 10^3) = K
Aerosol OT:			
0.0035 M	0.070 M NaCl	1	0.5
0.0007 M	0.050 M KCl	(30)	(0.1)
0.0007 M	0.070 M KCl	30	0.6
Aerosol MA:			
—	0.075 M NaCl	5	0.1
—	0.100 M KCl	30	2
Sodium Laurate:			
0.0028 M*	0.175 M NaCl	25	1

* Adjusted to pH 11 by addition of sodium hydroxide.

TABLE 5-4. VOLUME OF OIL SEPARATED FROM STANDARD XYLENE EMULSION ON ADDITION OF ELECTROLYTE[38]

0.1 N Electrolyte Added (ml)	Volume of Oil Separated (ml)*					
	$CuCl_2$	$MgCl_2$	$BaCl_2$	$CuSO_4$	$ZnSO_4$	$CaCl_2$
1.0	4.23	3.54	3.48	4.00	2.91	2.84
1.0	4.38	4.20	4.35	4.30	4.42	3.39
0.8	2.63	2.25	1.97	2.05	1.49	1.38
0.8	3.24	2.77	2.28	2.80	1.69	1.66
0.6	1.68	1.50	1.38	1.40	0.38	1.01
0.6	1.74	1.75	1.56	1.40	0.72	1.22
0.4	1.02	1.05	0.83	0.94	0.24	0.70
0.4	1.13	1.15	0.98	0.94	0.37	0.88
0.2	0.56	0.55	0.43	0.47	0.16	0.35
0.2	0.57	0.55	0.54	0.47	0.23	0.46

* Total volume of oil present = 4.5 ml.

the data of Martin and Hermann is shown in Table 5-4, for a number of bivalent cations. Duplicate results are reported for each electrolyte to show the reproducibility. With some exceptions, the quantities separated for each salt concentration are reasonably comparable. Furthermore, the nature of the anion appears to be unimportant. This, of course, is in accordance with the Schulze-Hardy rule, and suggests, as does the work of van den Tempel,[35] that discharge of the double layer potential is involved.

Kremnev and Kuibina[39] found that O/W emulsions were broken by the addition of salts. On increasing the salt concentration, however, the inverted emulsion was obtained. This emulsion was stable, but could be caused to break on addition of water.

Limited Coalescence. Coalescence of emulsion droplets, at a rate which decreases to zero as the droplets approach a limiting size and relatively uniform size distribution, has been termed "limited coalescence" by Wiley.[40] This phenomenon, first described by Hardy,[41] is apparently not uncommon. It appears to occur quite frequently in emulsions stabilized by finely-divided solids, as indicated by data reported by Pickering[42] and Bennister, King, and Thomas.[43]

Wiley[40] investigated styrene-in-water emulsions stabilized by *Dowex 50* ion-exchange resin beads of known diameter. It was found that the observed limiting droplet diameter could be predicted with reasonable accuracy, on the basis of the Gibbs-Kelvin relation between the radius of the droplet curvature and the adsorption of the emulsifier particles

$$RT \ln (S'/S) = 2\gamma M/rd, \qquad (5.25)$$

in which S' is the escaping tendency of the emulsifier particle from a surface of radius r, S the escaping tendency from a flat surface, γ the interfacial tension, M the molecular weight of the particle, and d the density of the particle at the surface. R and T have their usual meanings. This equation will be recognized as a more general form of Eq. 2.3.

The basis of Wiley's treatment lies in consideration of the fact that the stability of the individual droplet may well be a function of the amount of stabilizer adsorbed to the interface. Hence, the droplet radius corresponding to maximum stability will be the one at which the tendency of the stabilizer to escape from the interface into the aqueous phase is a minimum. This is calculated to be

$$D = \pi \, dVh/kw \qquad (5.26)$$

where D is the limiting diameter, V the volume of the oil phase, h the diameter of the stabilizer particle, w and d the dry weight and density of the stabilizer particles, respectively, and k the ratio of the wet volume to the dry volume of the particle.

Although it is perhaps more difficult to state so explicitly, similar considerations would apply to emulsion systems stabilized by surface-active agents in which phenomena similar to those reported by Wiley occur. Indeed, it should be pointed out that the observations of Cockbain[11] (cf. pp. 157–158) may be better explained by such a hypothesis than the one advanced. Cockbain's explanation of his data requires the initial assumption of an adsorbed layer of surface-active agent which, in effect, shields, if it does not discharge, the double layer. Thus the observed coagulation may be expected to take place. The subsequent disaggregation, however, requires the adsorption of micelles to the droplet. For stereometric reasons, if for no other, this seems improbable.

On the other hand, if one considers that the surface-active agents, at concentrations lower than the C.M.C., are relatively insoluble, a high concentration of the surface-active species present may be expected to be found at the interface. However, when the critical micelle concentration is reached the solubility increases sharply (e.g., Figure 2-23). When this happens, the escaping tendency of the surface-active molecules becomes quite high, and the stabilizer molecules are, in effect, "bled off" the interface to make up the micelles. After micelle formation is complete, however, the additional surface-active molecules will again find their way to interface and disaggregation will occur.

A related, and not exclusive, consideration is the possibility that impurities in the surface-active agent which form stabilizing complexes at the interface (cf. p. 96), may be preferentially solubilized in the micelles, thus causing instability.[44]

Other Causes of Demulsification. In the previous sections, the implication has been made that a prime cause of demulsification may be found in treatments which lead to a change in the properties of the electrical double layer. Emulsions may also be broken by other treatments, however, and some of these are of industrial significance.

For example, the electrical double layer may be completely discharged by some sort of electrolysis treatment.[45] Agitation, under certain conditions,[46] may be of value. Extremes of temperature may lead to breakdown,[47-49] as may the addition of surface-active agents of a type different from that employed to stabilize the emulsion.[27] Many of these techniques will be discussed in Chapter 9, in connection with the practical problems of demulsification.

Bibliography

1. Stokes, G. G., *Trans. Cambridge Phil. Soc.* **9**, (1851).
2. Neogy, R. K. and Ghosh, B. N., *J. Indian Chem. Soc.* **30**, 113 (1953).
3. Greenwald, H. L., *J. Soc. Cosmetic Chem.* **6**, 164 (1955).
4. Cf. Weiser, H. B., "Colloid Chemistry," p. 202, New York, John Wiley & Sons, 1953.
5. Dombrowsky, A., *Kolloid-Z.* **95**, 286 (1941).
6. Clayton, W., "Theory of Emulsions," 4th Ed., pp. 435–439, Philadelphia, The Blakiston Co., 1943.
7. Stevens, H. P. and Stevens, W. H., in "Emulsion Technology," 2nd Ed., p. 242, Brooklyn, Chemical Publishing Co., 1946.
8. Baker, H. C., *Trans. Inst. Rubber Ind.* **13**, 70 (1937).
9. Van Gils, G. E. and Kraay, G. M. in Kramer, E. O. (ed.), "Advances in Colloid Science," I, pp. 247–268, New York, Interscience Publishers Inc., 1942.
10. Howland, L. H. and Nisonoff, A., *Ind. Eng. Chem.* **46**, 2580 (1954).
11. Cockbain, E. G., *Trans. Faraday Soc.* **48**, 185 (1952).

12. Alsop, W. G. and Percy, J. H., *Proc. Sci. Sect. Toilet Goods Assoc.*, No. 4, p. 24 (1945).
13. Sherman, P., *Research* (London) **8**, 396 (1955).
14. Clowes, G. H. A., *J. Phys. Chem.* **20**, 407 (1916).
15. Sutheim, G. M., "Introduction to Emulsions," p. 132, Brooklyn, Chemical Publishing Co., 1946.
16. Schulman, J. H. and Cockbain, E. G., *Trans. Faraday Soc.* **36**, 661 (1940).
17. Kremann, R., Griengl, F., and Scheiner, H., *Kolloid-Z.* **62**, 61 (1933).
18. Wellman, V. E. and Tartar, H. V., *J. Phys. Chem.* **34**, 379 (1930).
19. Rebinder, P. A., Gol'denberg, N. L., and Ab, G. A., *Kolloid. Zhur.* **9**, 67 (1947); *C.A.* **41**, 4352*f*.
20. Isemura, T. and Kimura, Y., *Mem. Inst. Sci. Ind. Research Osaka Univ.* **6**, 104 (1948); *C.A.* **45**, 7508*c*.
21. Bhatnagar, S. S., *J. Chem. Soc. (London)* **117**, 542 (1920).
22. King, A. and Wrzeszinski, G. W., *Trans. Faraday Soc.* **35**, 741 (1939).
23. Corran, J. W., in "Emulsion Technology," 2nd Ed., pp. 176–192, Brooklyn, Chemical Publishing Co., 1946.
24. Salisbury, R., Leuallen, E. E., and Chavkin, L. T., *J. Am. Pharm. Assoc. Sci. Ed.* **43**, 117 (1954).
25. Pickthall, J., *Soap, Perfumery, Cosmetics* **27**, 1270 (1954).
26. Seifriz, W., *J. Phys. Chem.* **29**, 745 (1925).
27. Lawrence, A. S. C., *Chem. and Ind.* **1948**, 615.
28. Van den Tempel, M., *Rec. trav. chim.* **72**, 433 (1953).
29. Smoluchowski, M. v., *Physik. Z.* **17**, 557, 583 (1916).
30. Lawrence, A. S. C. and Mills, O. S., *Disc. Faraday Soc.* **18**, 98 (1954).
30. King, A. and Mukherjee, L., *J. Soc. Chem. Ind. (London)* **58**, 243 (1939); **59**, 185 (1940).
32. Kraemer, E. O. and Stamm, A. J., *J. Am. Chem. Soc.* **46**, 2716 (1924).
33. Brady, G., Mandelcorn, L., and Winkler, C. A., *Can. J. Chem.* **31**, 55 (1953).
34. Jellinek, H. H. A. and Anson, H. A., *J. Soc. Chem. Ind.* **69**, 229 (1950).
35. Van den Tempel, M., *Rec. trav. chim.* **72**, 442 (1953).
36. McBain, J. W., "Colloid Science," p. 177, Boston, D. C. Heath & Co., 1950.
37. Müller, H., *Kolloidchem. Beih.* **26**, 257 (1928).
38. Martin, A. R. and Hermann, R. N., *Trans. Faraday Soc.* **37**, 30 (1941).
39. Kremnev, L. Ya. and Kuibina, N. I., *Kolloid. Zhur.* **17**, 31 (1955).
40. Wiley, R. M., *J. Colloid Sci.* **9**, 427 (1954).
41. Hardy, W. B., *Colloid Symposium Monograph* **VI**, 8 (1928).
42. Pickering, S. U., *J. Chem. Soc.* **1907**, 91, 2001.
43. Bennister, H. L., King, A., and Thomas, R. K., *J. Soc. Chem. Ind.* **59**, 185 (1940)
44. I am indebted to L. Shedlovsky for pointing this out.
45. Beaver, C. E., *Trans. Am. Inst. Chem. Engrs.* **42**, 251 (1946).
46. Ben'kovskii, V. G., *Neftyanoe Khoz.* **24**, 42 (1946); *C.A.* **41**, 6391*d*.
47. Levius, H. P. and Drommond, F. G., *J. Pharm. Pharmacol.* **5**, 743, 755 (1953).
48. Walker, H. W., *J. Phys. and Colloid Chem.* **51**, 451 (1947).
49. Bromberg, A. V., *Kolloid. Zhur.* **9**, 231 (1947); *C.A.* **47**, 944*c*.

CHAPTER 6

The Chemistry of Emulsifying Agents

From much of the foregoing discussion, it is evident that the emulsifying agent plays an important part in any consideration of emulsion stability. It is thus important to devote some space to the descriptive chemistry of such materials. Only a few years ago, comparatively few emulsifying agents were commercially available. Today, an inclusive list and description would almost require a separate monograph; one company alone lists more than sixty separate emulsifying agents in its catalogue.*

Following the necessarily brief discussion of the chemistry of emulsifying agents, the present chapter will review the literature on the subject of emulsifier efficiency.

Classification of Emulsifying Agents

For purposes of discussion, it is useful to have some method of classifying emulsifying agents. Although such classifications are rather arbitrary and may even tend to obscure certain relationships, their usefulness outweighs their disadvantages. The following discussion is based on a rather broad system of classification.

A rather simple classification into three major classes of emulsifying agents may be made:

1. Surface-active materials
2. Naturally-occurring materials
3. Finely-divided solids

These divisions are arbitrary. For example, many of the naturally-occurring materials are, by any definition, surface-active. Nevertheless, the division has merit if it is recognized that the first group comprises essentially the so-called synthetic detergent materials, while the second group contains such materials as alginates, cellulose derivatives, gums, lipids and sterols, etc. The third group requires no particular elaboration.

Classification of Surface-Active Materials. The surface-active emulsifying agents, in the sense indicated above, probably represent the prin-

---

* In Appendix B an attempt has been made to list most of the commercially available materials at the time of writing, together with some indication of their compositions and uses. It should be recognized, however, that such a list is inevitably incomplete, even as it is finished.

166

cipal type used in industry, on a weight basis at least. This class will be given a moderately elaborate classification, to indicate the wide variety of materials available to the formulator of emulsions. The classification given below is based on one due to Schwartz and Perry.[1] The agents are classified according to the hydrophilic group in the molecule.

I. Anionic

A. *Carboxylic Acids*
  1. Carboxyl joined directly to hydrophobic group.
  2. Carboxyl joined through an intermediate linkage.
B. *Sulfuric Esters (Sulfates)*
  1. Sulfate joined directly to hydrophobic group.
  2. Sulfate group joined through intermediate linkage.
C. *Alkane Sulfonic Acids*
  1. Sulfonic group directly linked to hydrophobic group.
  2. Sulfonic group joined through intermediate linkage.
D. *Alkyl Aromatic Sulfonic Acids*
  1. Hydrophobic group joined directly to sulfonated aromatic nucleus.
  2. Hydrophobic group joined to sulfonated aromatic nucleus through intermediate linkage.
E. *Miscellaneous Anionic Hydrophilic Groups*
  1. Phosphates and phosphonic acids.
  2. Persulfates, thiosulfates, etc.
  3. Sulfonamides.
  4. Sulfamic acids, etc.

II. Cationic

A. *Amine Salts (Primary, Secondary, and Tertiary)*
  1. Amino group joined directly to hydrophobic group.
  2. Amino group joined through intermediate link.
B. *Quaternary Ammonium Compounds*
  1. Nitrogen joined directly to hydrophilic group.
  2. Nitrogen joined through an intermediate group.
C. *Other Nitrogenous Bases*
  1. Nonquaternary bases (e.g., guanidine, thiuronium salts, etc.)
  2. Quaternary bases.
D. *Nonnitrogenous Bases*
  1. Phosphonium compounds.
  2. Sulfonium compounds, etc.

III. Nonionic

A. *Ether Linkage to Solubilizing Groups*
B. *Ester Linkage*

C. *Amide Linkage*
D. *Miscellaneous Linkages*
E. *Multiple Linkages*

IV. Ampholytic

A. *Amino and Carboxy*
  1. Nonquaternary
  2. Quaternary
B. *Amino and Sulfuric Ester*
  1. Nonquaternary
  2. Quaternary
C. *Amino and Alkane Sulfonic Acid*
D. *Amino and Aromatic Sulfonic Acid*
E. *Miscellaneous Combinations of Basic and Acidic Groups*

V. Water-Insoluble Emulsifying Agents

A. *Ionic Hydrophilic Group*
B. *Nonionic Hydrophilic Group*

It is perhaps unnecessary to point out that the division into, for example, anionic and cationic materials depends on the nature of the surface-active *ion* species in the detergent.

Many of the above classifications are capable of further subdivision; the interested reader is referred to Schwartz and Perry.[1] Another, more elaborate, classification scheme is found in the compendium of Sisley and Wood.[2]

In order for the classification scheme outlined above to be completely useful, it is necessary to know corresponding hydrophobic groups of at least major importance. These are also outlined by Schwartz and Perry, and may be summarized as follows:

1. The straight-chain alkyl groups of eight to eighteen carbon atoms derived from the natural fatty acids.

2. The lower alkyl groups of three to eight carbon atoms, frequently attached to aromatic nuclei such as benzene or naphthalene, the combination forming the hydrophobic group. (The lower alkyl groups are also occasionally used themselves.)

3. Propene, isobutene, and some of the isomers of pentene and hexene can be readily polymerized to a low degree, yielding branched-chain mono-ölefins of eight to twenty carbon atoms. These may be condensed with benzene, to form an alkyl aromatic hydrophobe.

4. Petroleum hydrocarbons in the range of eight to twenty or more carbon atoms (e.g., the kerosene, light oil, and paraffin wax fractions) may be modified by a number of reactions to form suitable hydrophobic groups.

5. Naphthenic acids from certain types of petroleum are surface-active in themselves. In addition, they may be used as a source of hydrophobic groups.

6. The higher alcohols and hydrocarbons obtained from the Fischer-Tropsch and related syntheses.

7. Rosin acids may be used in the same way as the natural fatty acids.

8. Terpenes and terpene alcohols have been used, largely as alkylating agents for aromatic nuclei.

Classification of Other Types of Emulsifying Agents. Unfortunately, the members of the other two classes of emulsifying agents do not lend themselves to as systematic a classification as do the surface-active materials.

The major types of compounds included in the second class (i.e., naturally-occurring materials) have already been indicated as alginates, cellulose derivatives, water-soluble gums, lipids and sterols.*

As for the finely-divided solids, it will be recognized that any material which is sufficiently insoluble and sufficiently finely divided will suffice. Only a limited number of such substances are used as emulsifying agents.

Anionic Surface-Active Emulsifying Agents

As is apparent from the moderately elaborate classification given in the previous section, a large number of compounds can be comprised under the heading of surface-active emulsifying agents. Actually, of course, only a limited number of the total possible surface-active agents find any employment as emulsifying agents. This arises from considerations of cost, availability and efficiency. In other words, not every surface-active compound is a satisfactory emulsifying agent, in spite of the fact that it may be effective in reducing interfacial tension.

On the other hand, it cannot be denied that, under appropriate conditions, virtually any surface-active agent will suffice to stabilize an emulsion. The following discussion will primarily concern itself with those surface-active agents which are most used as emulsifying agents (some unimportant, or even relatively useless, compounds need be discussed in order to best illustrate certain concepts).

Soaps (Class IA1). Probably the oldest, and certainly the best-known, of the anionic-active emulsifying agents are the soaps. Soaps are the salts of the long-chain fatty acids, derived from naturally occurring fats and oils, in which the acids are found as the triglycerides. The soaps used in emulsi-

---

* The lipids and sterols might quite legitimately be classed in Group VB, among the surface-active materials.

fication may be obtained from the natural oils, in which case they will consist of a mixture of fatty acids, the precise nature of the mixture depending on the fat or oil employed.[4]

The mixed fatty acids of tallow, coconut oil, palm oil, etc., are those most commonly employed. The acids derived from tallow may be partially separated by filtration or pressing into "red oil" (principally oleic acid) and the so-called "stearic acid" of commerce, which is sold as single-, double-, or triple-pressed, depending on the extent to which the oleic acid is separated. Such "stearic acid" is actually a mixture of stearic and palmitic acids.

Other, more or less pure, single acids are also commonly employed. Perhaps lauric acid, derived from coconut oil, is the most common of these.

For the preparation of O/W emulsions, it is primarily the sodium soaps of these fatty acids which are employed. Potassium soaps are rarely used. As has been pointed out earlier, the soaps of the multivalent metals (calcium, magnesium, zinc, aluminum, etc.) favor the formation of W/O emulsions, and are used to some extent for that purpose. They are, however, somewhat inefficient emulsifiers by themselves, but are often used in conjunction with other agents.

The sodium soaps, in solution, are distinctly alkaline, with a pH in the neighborhood of 10. As a result, they are effective only in emulsions in which a relatively high pH can be tolerated. For systems which must be maintained at a lower pH, the amine soaps have won considerable acceptance. These materials have the advantage of being usable at lower pH (around 8.0) and, in some cases, have certain formulation advantages (cf. p. 253).

The amine soaps are the salts of various substituted amines of the general formula

$$N \diagup \overset{\displaystyle R}{\underset{\displaystyle R''}{\diagdown}} \!\!\! -R'$$

Here R, R', and R'' may be hydrogen or various organic groupings. Thus, one of the most widely-used amines for this purpose is triethanolamine:

$$N \diagup \overset{\displaystyle C_2H_4OH}{\underset{\displaystyle C_2H_4OH}{\diagdown}} \!\!\! -C_2H_4OH$$

Similarly, cyclic amines may be used. An extremely important one is morpholine

$$\begin{array}{c} \text{CH}_2\text{—CH}_2 \\ \diagup \qquad \diagdown \\ \text{HN} \qquad\qquad \text{O} \\ \diagdown \qquad \diagup \\ \text{CH}_2\text{—CH}_2 \end{array}$$

Amines of a more complex structure may also be used, e.g., 2-amino-2-methyl-1,3-propanediol:

$$\begin{array}{c} \text{NH}_2 \\ | \\ \text{CH}_2\text{OH—C—CH}_2\text{OH} \\ | \\ \text{CH}_3 \end{array}$$

The amine soaps are particularly useful in cases where the emulsion is to break after application, and re-emulsification is not desired, e.g., floor-waxes. In cases of this sort, the particular amine chosen will depend on, among other things, its volatility under the conditions of use.

All of the amino-compounds form substituted ammonium salts of the fatty acids, e.g., triethanolammonium stearate. However, the amine salts are rarely used as such. They are usually formed *in situ*, by the direct reaction of the appropriate amounts of amine and fatty acid (cf. p. 210).

The alkali and amine soaps are widely used as stabilizers for O/W emulsions, since they have the advantage of being relatively inexpensive as well as quite effective. They possess the disadvantage of being sensitive to hard water, however, as a result of the formation of lime soaps. This leads to inversion or, frequently, to breaking of the emulsion. Such emulsifying agents are said to be *calcium-sensitive.* They cannot be used in applications where hard water will be used, for example, in insecticidal emulsions which are often diluted just prior to use.

Needless to say, in the preparation of emulsions with these stabilizers it is necessary to operate with a source of soft water, or to soften the water by distillation, demineralization, or by the addition of water-softening agents such as the complex phosphates or chelating agents (e.g. the derivatives of ethylenediamine-tetra-acetic acid). The efficacy of these last compounds in cosmetics has been discussed by Goodyear and Hathorne.[5] While such methods solve the water-hardness problem, they are somewhat expensive. Therefore, the number of synthetic calcium-insensitive emulsifying agents is steadily on the increase.

Other Fatty Acid Soaps. In addition to the soaps derived from the straight-chain fatty acids derived from natural fats and oils, carboxylic acids derived from other sources are used. For example, the principal acidic component of rosin is abietic acid:

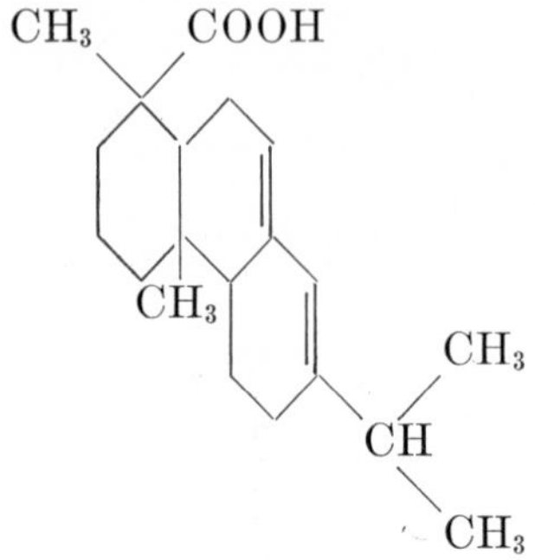

Other acids are also present, but these are apparently isomers of abietic acid. When used as the sodium salt, these soaps have many of the disadvantages of the fat-derived soaps. However, the rosin may be modified by hydrogenation, dehydrogenation, or polymerization, and the resulting derivatives of abietic acid are often significantly better in performance.[6]

Another source of carboxylic acids is tall oil, a by-product of the sulfate process for making wood pulp. This consists largely of a mixture of fatty acid and resin-type acids. While the raw tall oil is very dark in color, and of an extremely unpleasant odor, the material may be considerably improved by refining.

Certain types of petroleum contain the so-called naphthenic acids, which can be separated by means of alkali refining. Like tall oil, the crude material is dark and odoriferous, but can be refined. A principal use of these acids, as the heavy metal salts, is as driers in the paint industry. However, Levinson and Minich[7] have found that the potassium, sodium, and ethanolamine soaps of a highly purified naphthenic acid have good emulsifying properties.

The bile acids are also of some value in this connection; however, these may be more profitably considered as naturally-occurring emulsifiers.

Sulfated Oils (Class IA1, IB1). What is perhaps the first attempt to produce a more calcium-insensitive agent dates back to about 1875, with the introduction of *turkey-red oil*, so named because of its importance in "Turkey red" dyeing with ground-madder root. Turkey-red oil is obtained by treating castor oil with sulfuric acid (which reacts principally with the free hydroxyl group of the ricinoleic acid residue in the triglyceride), and by subsequently neutralizing to the sodium salt.

Reaction does not stop with the sulfation of the hydroxyl group, however, but may continue to sulfation of the double bond, and by hydrolysis to diglycerides, monoglycerides, and free acids. Thus there is present, for example, an appreciable amount of the salt of the sulfated ricinoleic acid:

$$CH_3(CH_2)_5CHCH_2CH\!\!=\!\!CH(CH_2)_7COONa$$
$$\underset{OSO_3Na}{|}$$

The reactions which occur have been studied in detail by Burton and Robertshaw,[8] and more recently by Rueggeberg and Sauls.[9] Glicher,[10] in a discussion of the manufacture of turkey-red oil, lists among the possible products of reaction:

1. Unchanged castor oil.
2. Free ricinoleic acid.
3. Dihydroxystearic acid.
4. Sulfate esters of the above.
5. Polymerized ricinoleic acid of
   the type $C_{17}H_{32}CO \cdot C_{17}H_{32}COOH$.

Although turkey-red oil is somewhat less sensitive to hard water than regular soaps, it still presents a problem in this connection.

Sulfation of the fatty acid residues can also be carried out at the double bonds of the unsaturated acids, e.g., sulfated olive oil, sulfated neatsfoot oil, etc. Although the sulfated oils are not any longer of primary importance as emulsifying agents, considerable quantities are still used in special applications. Trask[11] has recently reviewed the manufacture of these compounds.

Compounds of Class IA2. The hard-water serviceability of a carboxylic surface-active agent can be increased by increasing the total number of hydrophilic residues in the molecule. This may be accomplished by linking the hydrophobic chain to the carboxy-containing radical through an intermediate linkage of polar or hydrophilic character.

A large number of compounds of this type are known, in which the linkage may be acid amides, sulfonamides, or ester linkages. Most of these compounds are of limited value as emulsifiers. However, an important emulsifier base is the group which may be listed under the generic name *Lamepons*, in which the intermediate group is of the acid amide type.

Lamepons are prepared by hydrolyzing waste protein (scrap leather, etc.), obtaining a mixture of amino acids and lower peptides. This mixture is then condensed with the desired fatty acyl chloride to produce the agent.

Sulfated Alcohols (Class IB1). This extremely important group of surface-active agents was introduced about 1930. They probably were the first important source of anionic emulsifying agents which were not seriously calcium-sensitive.* A typical material of this group is the sodium salt of lauryl alcohol sulfate:

$$C_{11}H_{22}CH_2OSO_3Na$$

The alcohols are produced by the reduction of the corresponding fatty acids derived from natural fats, e.g., coconut or palm-kernel oil. Recently, the sulfates of the alcohols derived from hydrogenated tallow have been introduced. These are considerably less soluble than those derived from the shorter chain-length vegetable oils. Potentially, however, they are quite inexpensive, and will probably find wide application.

The hydrophilic grouping of the alcohol may be modified, prior to sulfation, by condensing the alcohol with one or more moles of ethylene oxide. This produces alkyl polyoxyethylene sulfates of the type:

$$R(OC_2H_4)_nOSO_3Na,$$

where n is usually in the range 1 to 5. In the case of the tallow-derived materials, such treatment with ethylene oxide will increase solubility without an appreciable increase in cost, and thus widen their field of application.

Alcohols from other sources have been sulfated and find some application, e.g., abietyl and hydroabietyl alcohols from rosin acid esters, naphthenyl alcohols, and synthetic alcohols derived from the Fischer-Tropsch or Oxo processes.

Similar compounds are also prepared by the sulfation of olefins, although in this case the sulfate group is not terminal. For example, upon sulfation, cetene-1 (produced by the dry distillation of spermaceti) yields cetyl-2-sulfuric acid. Olefins useful in this process may be obtained by the high-temperature cracking of paraffin wax followed by fractionally distilling to obtain suitable cuts of olefins.

Sulfated Alcohols (Class IB2). Properly speaking, the ethylene-oxide modified alcohol sulfates, described in the previous section, belong in this class. Similar compounds will be described below. However, the most important member of this class is probably the sulfated monoglyceride, usually of coco fatty acids:

---

* These compounds, and indeed all of the so-called anionic "synthetic detergents," are calcium-insensitive *only* in the sense that no insoluble calcium or magnesium salts precipitate in hard water. The calcium and magnesium salts *do* form, and in many cases the properties of these salts are different from those of the sodium salts, from the point of view of their applicability as emulsifying agents. It is therefore still true that, in cases where hard water is or may become a part of the emulsion system, particular care in selecting the emulsifying agent is necessary.

$$CH_2OOCR$$
$$|$$
$$CH_2OH$$
$$|$$
$$CH_2OSO_3Na$$

These may be prepared by a number of methods. At present, the most usual method is a continuous operation in which the coconut oil triglyceride, glycerine, and sulfuric acid are caused to react in one stage in the proper molecular proportions.

Sulfated fatty monoethanolamides,

$$RCONHC_2H_4OSO_3Na,$$

although more useful as detergents, have some applicability as emulsifying agents. The sulfated ethanolamides of blown or bodied castor oil fatty acids have been described as good emulsifying agents.

The possibility of modifying fatty alcohol sulfates by the introduction of ethylene oxide has already been mentioned. The preparation of similar compounds based on a different hydrophobic grouping leads to useful types of compounds. For example,

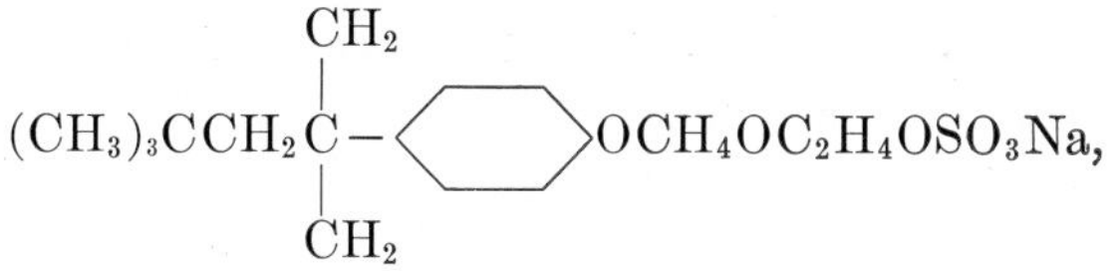

$$CH_2$$
$$|$$
$$(CH_3)_3CCH_2C - \langle \rangle OCH_4OC_2H_4OSO_3Na,$$
$$|$$
$$CH_2$$

which is a member of the *Triton* class of trade-named surface-active agents.

Alkane Sulfonates (Class IC1). The sulfates, discussed in the previous sections, are characterized by the fact that the sulfur is linked to the carbon through an oxygen atom. That is, the group characteristic of this class of emulsifying agents is

$$-\overset{|}{\underset{|}{C}}-$$
$$OSO_3^-$$

There exists another class of sulfate derivatives, however, in which the sulfur is directly linked to the carbon atom, i.e.,

$$-\overset{|}{\underset{|}{C}}-$$
$$SO_3^-$$

This group is known as the *sulfonates*, and includes an extremely important class of emulsifiers.

The properties of the purified straight-chain sulfonates of the paraffins, in the range from octane to octadecane, have been described by Reed and Tartar.[12] Although these sulfonates are producible by a number of reactions, the principal source of these agents is via the so-called *Reed reaction*.[13] This reaction takes place between an aliphatic hydrocarbon, sulfur dioxide, and chlorine to produce an alkane sulfonyl chloride:

$$RH + SO_2 + Cl_2 \rightarrow RSO_2Cl + HCl$$

The reaction is activated by visible light of the shorter wavelengths. The sulfonyl chloride is then hydrolyzed to the sulfonic acids and their salts.

A similar type of emulsifying agent is the once worthless *petroleum sulfonates*, obtained as a by-product in petroleum refining.

Alkane Sulfonates (Class IC2). A class of alkane sulfonates of much wider applicability is that in which there is an intermediate linkage between the sulfonate group and the hydrophobic group. Such intermediate linkages may be ester, amide, or ether groupings. A prominent example of the first are those members of the so-called *Igepon* series, which may be looked on as derivatives of isethionic acid, $HOC_2H_4SO_3H$. In this case the alcoholic group is esterified by an appropriate fatty acid, leading to a compound of the type

$$RCOOC_2H_4SO_3Na$$

The most common fatty acid employed is perhaps oleic. The intermediate linkage may be turned around, as it were, in which case the compounds may be considered as derivatives of sulfoacetic acid, i.e., $HOOCCH_2SO_3H$. In this case fatty alcohols are esterified with the carboxyl group of this compound. This type of compound is, however, rather expensive, and probably does not find wide application as an emulsifying agent.

Probably the most important compounds where an amide group forms the intermediate linkage are those which are derivatives of N-methyl taurine, e.g.

$$\begin{array}{c} RCONCH_2CH_2SO_3Na \\ | \\ CH_3 \end{array}$$

The most common source of the fatty acid residue is again oleic acid. Similar compounds are known in which the methyl group on the nitrogen is replaced by hydrogen or by some other short-chain group.

The amides of sulfosuccinic acid are also compounds of this type, e.g.,

$$C_{18}H_{37}NHCOCH_2CHCOONa$$
$$| $$
$$SO_3Na$$

Under the trade-name of *Aerosol 18*, this has considerable application as an emulsifier in emulsion polymerization.

Compounds analogous to the oxyethylated alkaryl sulfates also exist, e.g.,

$$C_8H_{17}\langle \rangle OC_2H_4OC_2H_4SO_3Na$$

Compounds of this type have the advantage over other members of the series under discussion in that they contain no hydrolyzable linkages, as, for example, the ester and amide linkages. This, of course, permits them to be used under conditions where other materials of the same general type would be inadmissable.[14]

Alkyl Aromatic Sulfonates (Classes ID1 and ID2). These compounds are somewhat similar to these described at the end of the preceding section, except that the intermediate ether-grouping is not present. As a consequence, the sulfonate group is attached directly to the aromatic nucleus.

When the alkyl side-chain consists principally of highly branched alkyl groups (usually formed from a low degree of polymerization of propylene and attached to the benzene ring via Friedel-Crafts reaction), the compounds thus formed are the well-known alkyl aryl sulfonates. On a tonnage basis, these materials are probably the leading surface-active compounds. They are mainly used as the basis for the so-called "synthetic detergents" of common household applications. Otherwise, they have only very limited applicability as emulsifying agents.

Alkylated naphthalene sulfonates, however, are quite effective as emulsifying agents, and are quite widely used.

A large number of variations on the basic alkyl aromatic sulfonates are described in the patent literature.[15] Nevertheless, very few of these are commercially available, or, for that matter, of value as emulsifiers.

Cationic Surface-Active Emulsifying Agents

The cationic surface-active agents are characterized by the fact that the hydrophobic grouping is found in the cation. Principally, these compounds are amines or quaternary ammonium salts. Although they have been known as a class for a long time, and their effectiveness as emulsifying agents recognized, they have heretofore represented only a small fraction of the emulsifying agents employed. However, in recent years, their relative insensitivity to hardness, and their effectiveness at low pH, have commended them to formulators of cosmetic emulsions, especially of those which remain in prolonged contact with the skin.

Amine Salts (Class IIA) and Quaternary Ammonium Compounds (Class IIB). Although the amine salts, e.g., octadecylamine hydrochloride (or, more correctly octadecylammonium chloride), $C_{18}H_{37} \cdot NH_3Cl$, have several interesting applications, their value in the emulsion field is rather limited. Indeed, the primary, secondary, and tertiary amines are, in general, too insoluble to be of use. In systems with a pH lower than 7, however, they find some application, especially as the acetate.

On the other hand, the so-called quaternary amine salts are appreciably soluble in both acidic and basic solutions. These are prepared by the addition of alkyl halides or sulfates to tertiary amines. Thus, a typical quaternary amine salt is cetyl trimethyl ammonium bromide

$$\left[\begin{array}{c} CH_3 \\ | \\ C_{16}H_{33}-N-CH_3 \\ | \\ CH_3 \end{array} \right] Br$$

A more complicated type (useful, for example, in emulsion polymerization) is obtained by quaternizing the tertiary amine obtained by acylating an unsymmetrical dialkyl ethylenediamine with a fatty acid chloride (usually oleyl), i.e., $RCONHC_2H_4N(C_2H_5)_2$.

Many quaternary compounds possess bactericidal activity; thus their presence in a composition may well serve two functions.

The methods used for the production of fatty amines and amine salts are succinctly described by Schwartz and Perry.[15]

Nonionic Surface-Active Emulsfying Agents

This is now the largest and fastest growing group of emulsifying agents. Since these materials are nonionic, they are independent of water hardness and pH. In many cases, the effectiveness of the hydrophobic and hydrophilic portions of the molecule can be modified, so that an emulsifying agent can, in effect, be "tailor-made" for particular applications. This point will be returned to in a later section of this chapter (pp. 190–196).

As indicated in the classification scheme outlined earlier (p. 167), these compounds are best described in terms of the linkage between the hydrophobic group and the solubilizing (hydrophilic) one.

Ether Linkage (Class IIIA). The most prominent members of this class are those compounds formed by the reaction of a hydrophobic hydroxyl-containing compound (e.g., an alcohol or phenol) with ethylene oxide, or, to a lesser extent, propylene oxide. This reaction is a species of polymerization. The ethylene oxide groups, for example, may be added to any desired extent. Obviously, the number of ethylene oxide groups added, for a given hydrophobic group, seriously affects the solubility and surface properties of the resulting agent. As a matter of fact, it can be shown that the ability to stabilize certain types of emulsions depends on the relative amount of ethylene oxide added to a given hydrophobic group (cf. p. 189).

Since these are polymeric products, a compound nominally containing a number n of ethylene oxide units per hydrophobic group, actually contains a spectrum of such groups, with a distribution peaking at the nominal value n.[16] The mode of manufacture dictates what the form of this distribution will be. Thus, in certain critical applications, it may be found that the products of two manufacturers, nominally the same, may have quite different properties. Different lots from the same manufacturer may also vary somewhat in this respect. It must be admitted that this degree of criticality does not often occur, but it must be kept in mind in connection with these useful compounds (as with all compounds based on polyethers).

With fatty alcohols (such as oleyl), useful compounds are obtained with 6 to 8 moles of ethylene oxide. However, compounds have been prepared containing up to thirty such groups. Another important series is based on the alkyl phenols, such as nonyl phenol. Since the alkyl and phenolic groups are in para position relative to each other, the compounds formed are of the form

$$C_9H_{19}\text{---}\langle\ \rangle\text{---}(OCH_2CH_2)_nOH,$$

where n indicates the number of polyoxyethylene groups added. In compounds of this type it has been estimated that the benzene nucleus has the effect of a straight chain of four carbon atoms. Hence the hydrophobic portion of this molecule is roughly equivalent to that derived from lauryl alcohol.

As has been pointed out above, propylene oxide groups also have solubilizing effects when added to appropriate hydrophobes. Propylene oxide itself, however, is polymerized to a fairly high degree of polymerization, the resulting dihydric ether-alcohol is quite hydrophobic, and the subsequent additions of ethylene oxide to one or both of the terminal hydroxyl groups produces a surface-active material which has value as an emulsifying agent. These are actually high-polymeric compounds; commercially available examples range in molecular weight from about 2500 to about 7500. Even the highest molecular weight compounds are appreciably water-soluble. Such polymers are known as *block co-polymers*.[17]

It should be pointed out that compounds of this type form gels at certain aqueous concentrations; this occasionally precludes their use in certain applications.

Ester Linkage (Class IIIB). The esters formed by the reaction of the fatty acids with polyhydric alcohols are an interesting group of nonionic emulsifiers, in that, depending on the nature of the alcohol used, they may be predominantly hydrophobic or hydrophilic, and thus suitable as W/O or O/W emulsifiers, respectively.

A typical example of the hydrophobic type are the monoglycerides, e.g., glyceryl monostearate.

$$CH_2OOCC_{17}H_{35}$$
$$|$$
$$CHOH$$
$$|$$
$$CH_2OH$$

Monoglycerides of other fatty acids are possible, as are monoglycerides of naturally-occurring fatty acids, for example, lard monoglyceride. This group of emulsifiers is especially important since, by a ruling of the Food and Drug Administration (1952), it is the only type which may be used in baked goods. Commercially available monoglycerides contain varying amounts of di- and tri-glycerides; and their performance as emulsifying agents will depend, to some extent, on the amount of such materials present. Various grades are available, and a choice can be made.

Monoesters of ethylene glycol (CH_2OHCH_2OH) and, to a larger extent, propylene glycol ($CH_3CHOHCH_2OH$) are also extensively used.

More hydrophilic esters may be prepared by forming the monoesters of such ether-alcohols as nonaethylene glycol, e.g., the monoöleate

$$C_{17}H_{35}COOCH_2CH_2O(CH_2CH_2O)_8H$$

Similar compounds are produced by the direct addition of varying molar ratios of ethylene oxide to the carboxylic hydroxyl. Again, because of the different distribution curve corresponding to the ethylene oxide ratio per molecule, similar compounds produced by the two different methods may be expected to have slightly different properties.

Another extremely useful group of fatty acid esters are those of the products of the dehydration of sorbitol (the trade-marked series known as *Spans*). The reactions involved in the preparation of these compounds are shown on the opposite page.[18]

It will be noted that the esterification (by long-chain fatty acids) is quite specific, and only certain of the hydroxyl groups are subject to this reac-

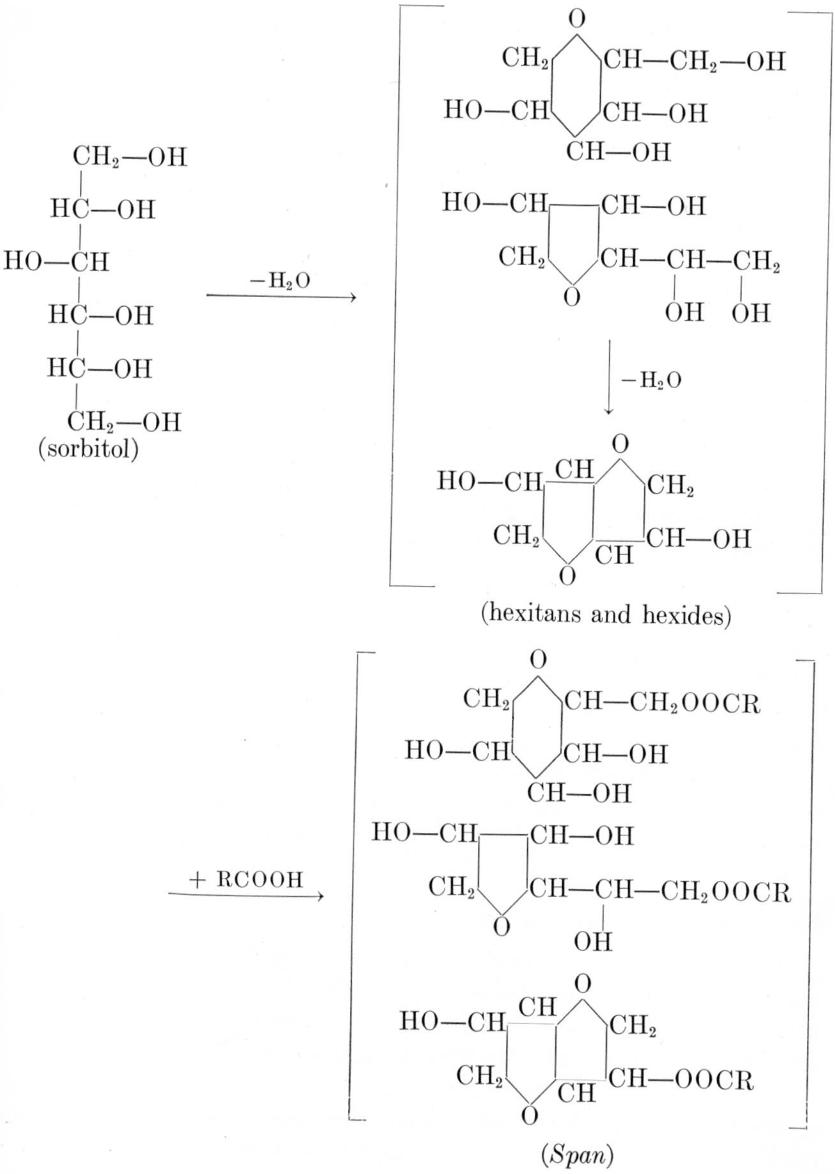

$$CH_2-OH$$
$$HC-OH$$
$$HO-CH$$
$$HC-OH$$
$$HC-OH$$
$$CH_2-OH$$

(sorbitol)

$\xrightarrow{\;-H_2O\;}$

(hexitans and hexides)

$\downarrow -H_2O$

$\xrightarrow{\;+\;RCOOH\;}$

(*Span*)

tion. The remaining hydroxyl groups, however, are reactive with ethylene oxide, which adds in the manner described previously. The resulting compounds (the trade-marked *Tweens*), are thus members of both Classes IIIA and IIIB.

Amide Linkage (Class IIIC). Another interesting series of nonionic compounds are the amides which are derived from the alkylol amines, e.g., diethanolamine, $HN(C_2H_4OH)_2$. A typical member of this group is lauroyl diethanolamide

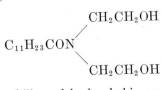

Variations in the hydrophilic and hydrophobic portions of compounds of this type are, of course, possible, and a wide variety of properties can thus be obtained. A more convenient method of varying the hydrophilic/hydrophobic ratio is by allowing unsubstituted amides to react with ethylene oxide to form ether-alcohol amides of the general formula

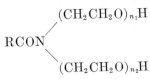

where n_1 and n_2 may be varied as desired.

As indicated earlier, a discussion such as the preceding one can only indicate the broad outlines of the available range of synthetic emulsifying agents. The reader is referred to Appendix B (p. 337) for a list of the commercially available emulsifying agents by their trade-names.

Naturally-Occurring Materials

It must be recognized that many members of the class of emulsifying agents considered under this heading could have profitably been treated under an earlier heading. On the other hand, compounds of this type, derived almost unchanged from natural sources, do represent a somewhat different type of agent. Many of them are relatively inefficient by themselves, but are of considerable value in conjunction with other emulsifiers. Such substances are often called *auxiliary emulsifying agents*. Very often these naturally-occurring agents have the desirable effect of materially increasing the viscosity of the emulsion, being of value in inhibiting creaming. However, they often suffer from the disadvantages of being rather expensive, subject to hydrolysis, and sensitive to variations in pH.

Phospholipids and Sterols. The phospholipids and sterols are found as minor constituents in many fats and oils. An important example of the former group is *lecithin*

$$CH_2OOCR$$
$$CHOOCR \quad OH$$
$$CH_2O\text{------}P\text{---}OCH_2CH_2N(CH_3)_3 \; ,$$
$$\quad\quad\quad\quad \| \quad\quad\quad |$$
$$\quad\quad\quad\quad O \quad\quad\quad OH$$

where the RCOO- groups represent long-chain fatty acid residues. It will be recognized that this compound is similar to ordinary fats. However, instead of being a simple triglyceride, one position on the glycerine is occupied by the substituted phosphoric acid. Interestingly enough, only a few fatty acids are found in lecithin: palmitic, stearic, oleic, linoleic, and arachidonic. Lecithin is present in egg yolk to the extent of about 10 per cent, and is presumably the active material which makes egg yolk effective as the emulsifier for mayonnaise (cf. p. 262). The principal source of commercial lecithin, however, is soybean oil[19]. It is an effective O/W emulsion stabilizer, and has a wide variety of applications.

A typical sterol, also found in egg yolk, although to a minor degree, is *cholesterol*

The principal source of cholesterol is lanolin. Commercial sterols from lanolin may contain dihydrocholesterol (cholestanol), cerebrosterol, lanosterol, dihydrolanosterol, agnosterol, and dihydroagnosterol. The last four of these, however, are not properly sterols (although so named), but are triterpene alcohols.[20] The hydroxyl group attached to the ring imparts its hydrophilic properties to the molecule, but since the hydrophobic portion of the molecule predominates, cholesterol is, as expected, a stabilizer for W/O emulsions.

Lanolin. Lanolin itself has long been known as a stabilizer for water-in-oil emulsions. This natural material is obtained by refining wool wax, obtained from the scouring of the wool of sheep. Although purified lanolin contains small percentages of hydrocarbons, free fatty acids, and free alcohols, the major constituents (about 96 per cent) are esters. These esters are made up of a wide range of saturated, unsaturated, and hydroxy acids, and of sterols, triterpene alcohols, and various aliphatic alcohols It is probable that not all of these constituents are effective as emulsion stabilizers (cf. p. 203).[20]

Water-Soluble Gums. There exist a number of water-soluble gums of vegetable origin which have been used for the stabilization of emulsions since the times of antiquity. These materials are most effective in stabilizing oil-in-water emulsions. They operate principally as auxiliary emulsifiers, through increasing the viscosity, although Serallach and Jones[2] have demonstrated that materials of this sort often concentrate at the interface to form strong interfacial films.

Materials in this class include, e.g., gum arabic, gum tragacanth, gum karaya, guar gum, etc. Many of these gums are derived from the dried exudations of certain varieties of trees. Locust bean gum and guar gum are derived from the seed endosperms from the *Leguminosae* plant family

These gums are principally polysaccharides of a fairly complex sort Mantell[22] describes gum arabic as a mixture of the calcium, magnesium and potassium salts of arabic acid, which is itself pictured by Hirst[23] as consisting of 1-D-glycuronic acid, 3-D-galactose, 2-L-arabinose, and 1-L-rhamnose, arranged as indicated below:

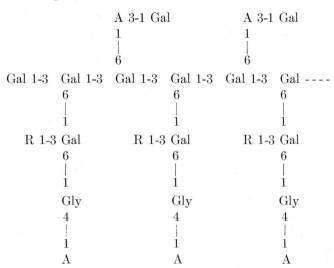

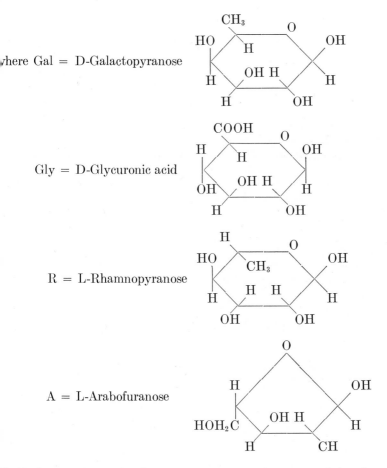

where Gal = D-Galactopyranose

Gly = D-Glycuronic acid

R = L-Rhamnopyranose

A = L-Arabofuranose

Similarly complex molecular structures are to be expected for the other water-soluble gums mentioned.

Mantell[22] has recently described the technology of gum arabic; Goldtein[24] that of karaya; Beach[25] that of tragacanth; Whistler[26] and Deuel and Neukom[26] that of guar and locust bean gums.

Alginates. The alginates are derived from kelp. They are similar to the gums described above.

Although alginates have been known for about seventy years, only in recent years have they been finding wide acceptance. They are used as the sodium, potassium, and ammonium salts, and (more recently) as the propylene glycol esters of alginic acid, with the structure indicated:[27]

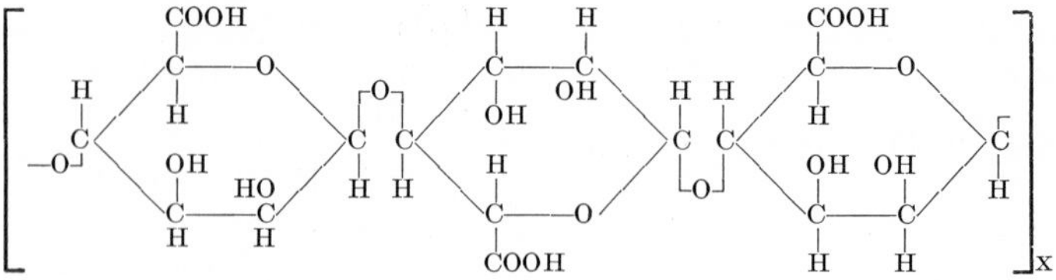

The sodium salt is probably the most widely used form.

The production and properties of alginates have been described b Tseng,[28] Steiner and McNeely,[27] Black and Woodward,[29] and Federici.[29]

Carageenin. A material similar to the alginates, carageenin is derive from another type of seaweed, *Chondrus crispus*, or Irish moss. This ager is apparently a mixed salt sulfate ester of a polysaccharide complex mac up mostly of D-galactopyranose and some L-galactose units, 2-ketoglucon acid units, and nonreducing sugar units, each unit the size of a hexose an combined with one sulfate radical. According to Stoloff,[30] the galactos units are of the following structure

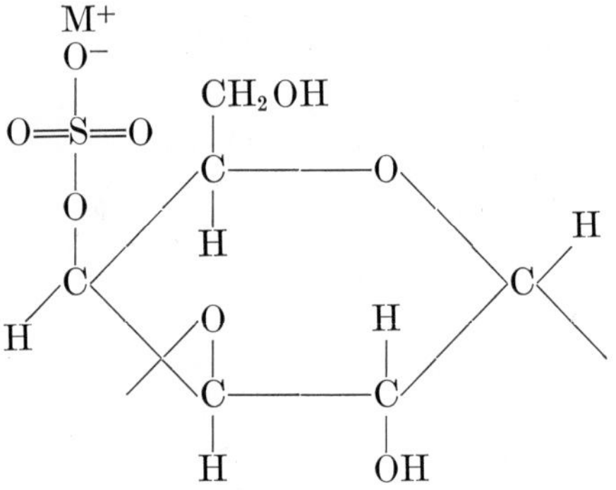

(M = Cation)

They are combined with themselves and the others through the 1- an 3-positions and are sulfated on the 4-position. All the cations in the polyme are ionizable, as monometallic salts are easily prepared through conve tional metathetic reactions.

The work of Serallach and Jones[21] indicates that carageenin forms pa ticularly strong interfacial films.

Cellulose Derivatives. Two useful auxiliary agents, of the type whic has been described in preceding sections, are derived from cellulose, *methy cellulose* and *carboxymethylcellulose*, or CMC (cf. footnote, p. 37). Thes are principally useful in increasing the viscosity of the aqueous phase c oil-in-water emulsions. For example, only 3 per cent of methylcellulos increases the viscosity of water by a factor of ten thousand, thus reducing the velocity of creaming by this same amount (Eq. 5.1).

Methylcellulose is an ether of cellulose formed by interaction of the methyl ester of an inorganic acid with cellulose which has been swollen with a strong base. Materials of this sort have been known for about fifty years.[31]

A more recent type of cellulose derivative has been developed in the last fifteen years: carboxymethylcelluloses, prepared by the action of sodium chloroacetate on alkaline cellulose, and usually employed as the sodium salt. While used in many applications other than emulsion stabilization, some indication of its steadily increasing value is to be found in the fact that production of this compound increased from about 3 million pounds in 1947 to almost 27 million pounds in 1955.[32] The structural formula of CMC has been shown to be[33]

The usual commercial materials contain between 0.5 to 1.2 carboxymethyl groups per anhydroglucose unit, and the average molecular weights vary in the range 50,000–500,000.

Both methylcellulose and carboxymethylcellulose are available in a variety of grades with differing solubilities and viscosities, making possible their use in a wide variety of conditions. Young and Kin[31] give extremely useful data on the variation with concentration in the viscosity of various grades of methylcellulose as compared with the more common water-soluble gums.

Finely-divided Solids. A large number of finely-divided solids have been found to be effective emulsion stabilizers in a variety of applications. These include basic salts of the metals, carbon black, powdered silica, and various clays (principally bentonite). The type of emulsion obtained depends on the extent to which the solid particles are preferentially wetted by either the aqueous or oil phases (cf. p. 127).

Emulsifier Efficiency

The concept of emulsifier efficiency can be approached in two different ways. Unfortunately, the economic, quantity-versus-quality, way is not often considered by the commercial and industrial formulator. As will be seen later (Chapter 8), the quantity of emulsifying agent usually employed varies from one to ten (in some cases, as high as 15) per cent, depending on the particular formulation. For economic reasons, one will wish to formulate with the cheapest emulsifier or with the minimum quantity compatible with the desired stability. Since a given emulsifier may satisfactorily stabilize a given emulsion system at a much lower concentration than another, and hence be more *efficient*, it may be best to employ a relatively expensive material when it is effective at lower concentrations than a cheaper one.

In the second, truly chemical, way of looking at the concept of emulsifier efficiency, one must include the problem of how the proper emulsifying agent for a given system is to be chosen, and how external conditions may tend to modify that choice. The various methods used to determine the proper agent are discussed below.

Emulsion Efficiency and "Balanced" Molecules. Clayton[34] has drawn attention to the concept of *balanced* emulsifying agents as embodied in a series of patents dating back to 1933. What is involved here is the effect on the surface-active properties of the molecule of the relative hydrophile or lipophile "character." For example, palmityl alcohol is compared with palmityl hydrogen sulfate as an anti-spattering agent for margarine, and it is pointed out that (for this application, at least) the hydrophilic character of the —OH group is completely neutralized by the hydrophobic group. On the other hand, the —HSO$_4$ group is a sufficiently strong hydrophil so that the desired surface-active properties are exhibited. It is concluded that in a given homologous series there is a point of range in which the hydrophile and lipophile characteristics are so balanced that an optimum efficiency is reached for the particular application.

While this conclusion is reasonable, and almost certainly correct, it is qualitative. In recent years, a number of schemes designed to put this concept on a quantitative basis have been advanced. The most prominent of these schemes are discussed below.

TABLE 6-1. HLB RANGES AND THEIR APPLICATION[36]

Range	Application
3–6	W/O Emulsifier
7–9	Wetting Agent
8–18	O/W Emulsifier
13–15	Detergent
15–18	Solubilizer

The HLB Method. Probably the most successful approach to this problem, due to Griffin,[35] has been designated the *HLB method*. The letters *HLB* stand for *hydrophile-lipophile balance*.

In this method, an HLB number is assigned to each surface-active agent, and is related by a scale to the suitable applications. Table 6-1 shows the HLB range required for various systems. As can be seen, only those materials with HLB numbers in the range of 4 to 6 are suitable as emulsifiers for W/O emulsions, while only those with HLB numbers in the range 8 to 18 are suitable for the preparation of O/W emulsions. Agents with HLB numbers in different ranges, while possessing important surface-active properties, cannot (according to this classification) be employed as emulsifying agents.

The original method of determining HLB numbers involves a long and laborious experimental procedure.[35] Recently, Griffin[36] has developed equations which permit the calculation of the HLB numbers for certain types of nonionic agents, in particular, polyoxyethylene derivatives of fatty alcohols (cf. pp. 178–180) and polyhydric alcohol fatty acid esters, including those of polyglycols (cf. pp. 180–182).

The formulae for determining HLB numbers may be based either on analytical or composition data. For most polyhydric alcohol fatty acid esters approximate values may be calculated with the aid of the relation

$$\text{HLB} = 20\left(1 - \frac{S}{A}\right), \tag{6.1}$$

where S is the saponification number of the ester and A is the acid number of the acid. Thus, for a glyceryl monostearate with $S = 161$ and $A = 198$, Eq. 6.1 gives an HLB value of 3.8.

Unfortunately, for many fatty acid esters it is difficult to get good saponification number data; for example, esters of tall oil and rosin, beeswax, and lanolin. For these, Griffin gives the relation

$$\text{HLB} = \frac{E + P}{5}, \tag{6.2}$$

where E is the weight percentage of oxyethylene content and P is the weight percentage of the polyhydric alcohol content.

TABLE 6-2. APPROXIMATION OF HLB BY WATER SOLUBILITY[37]

Behavior When Added to Water	HLB Range
No dispersibility in water	1–4
Poor dispersion	3–6
Milky dispersion after vigorous agitation	6–8
Stable milky dispersion (upper end almost translucent)	8–10
From translucent to clear dispersion	10–13
Clear solution	13+

In products where only ethylene oxide is used as the hydrophilic portion and for fatty alcohol ethylene oxide condensation products Eq. 6.2 may be reduced to

$$HLB = E/5, \qquad (6.3$$

where E has the same meaning as above.

These equations cannot be used for nonionic surface-active materials containing propylene oxide, butylene oxide, nitrogen, sulfur, etc., nor can they be used for ionic agents. In these cases, the laborious experimental method[35] must be used.

Attempts have been made to relate HLB to such properties as solubility in water or in other solvents, ratio solubility in two solvents, solubilization behavior both for oils and dyes, surface and interfacial tension data, cloud point behavior, etc. Apparently the cloud-point method offers the most promise.[36]

A rough approximation of the HLB value may be obtained, however, by the water solubility of the agent.[37] Table 6-2, from Griffin's data, shows how this may be done.

Exceptions to these relations are known, but they serve as a quick method for estimating the HLB range.

The assigned HLB numbers for a large number of commercial emulsifiers have been determined or calculated by the more complex methods described above.[36] Table 6-3 gives these values, arranged in order of increasing HLB.

In addition to the materials listed in Table 6-3, Griffin[37] has indicated that the HLB numbers for such well-known emulsifying agents as potassium, sodium, and triethanolamine oleates are 20.0, 18.0, and 12.0, respectively. Oleic acid has an HLB value of approximately unity.

Applications of the HLB Method. The broad ranges in which agents of various HLB values are applicable have been indicated in Table 6-1. More specific data on particular emulsion applications have also been presented, and these are given in Tables 6-4 and 6-5.

The great merit of the HLB numbers resides in the fact that they are algebraically additive; thus, the HLB of a blend of emulsifiers may be predicted

TABLE 6-3. HLB VALUES FOR COMMERCIAL EMULSIFIERS[36]

Name	Mfr.*	Chemical Designation	Type†	HLB††
Span 85	1	Sorbitan trioleate	N	1.8
Arlacel 85	1	Sorbitan trioleate	N	1.8
Atlas G-1706	1	Polyoxyethylene sorbitol beeswax derivative	N	2
Span 65	1	Sorbitan tristearate	N	2.1
Arlacel 65	1	Sorbitan tristearate	N	2.1
Atlas G-1050	1	Polyoxyethylene sorbitol hexastearate	N	2.6
Emcol EO-50	2	Ethylene glycol fatty acid ester	N	2.7
Emcol ES-50	2	Ethylene glycol fatty acid ester	N	2.7
Atlas G-1704	1	Polyoxyethylene sorbitol beeswax derivative	N	3
Emcol PO-50	2	Propylene glycol fatty acid ester	N	3.4
Atlas G-922	1	Propylene glycol monostearate	N	3.4
"Pure"	6	Propylene glycol monostearate	N	3.4
Atlas G-2158	1	Propylene glycol monostearate	N	3.4
Emcol PS-50	2	Propylene glycol fatty acid ester	N	3.4
Emcol EL-50	2	Ethylene glycol fatty acid ester	N	3.6
Emcol PP-50	2	Propylene glycol fatty acid ester	N	3.7
Arlacel C	1	Sorbitan sesquioleate	N	3.7
Arlacel 83	1	Sorbitan sesquioleate	N	3.7
Atlas G-2859	1	Polyoxyethylene sorbitol 4.5 oleate	N	3.7
Atmul 67	1	Glycerol monostearate	N	3.8
Atmul 84	1	Glycerol monostearate	N	3.8
Tegin 515	5	Glycerol monostearate	N	3.8
Aldo 33	4	Glycerol monostearate	N	3.8
"Pure"	6	Glycerol monostearate	N	3.8
Atlas G-1727	1	Polyoxyethylene sorbitol beeswax derivative	N	4
Emcol PM-50	2	Propylene glycol fatty acid ester	N	4.1
Span 80	1	Sorbitan monoöleate	N	4.3
Arlacel 80	1	Sorbitan monoöleate	N	4.3
Atlas G-917	1	Propylene glycol monolaurate	N	4.5
Atlas G-3851	1	Propylene glycol monolaurate	N	4.5
Emcol PL-50	2	Propylene glycol fatty acid ester	N	4.5
Span 60	1	Sorbitan monostearate	N	4.7
Arlacel 60	1	Sorbitan monostearate	N	4.7
Atlas G-2139	1	Diethylene glycol monoöleate	N	4.7
Emcol DO-50	2	Diethylene glycol fatty acid ester	N	4.7
Atlas G-2146	1	Diethylene glycol monostearate	N	4.7
Emcol DS-50	2	Diethylene glycol fatty acid ester	N	4.7
Atlas G-1702	1	Polyoxyethylene sorbitol beeswax derivative	N	5
Emcol DP-50	2	Diethylene glycol fatty acid ester	N	5.1
Aldo 28	4	Glycerol monostearate (self-emulsifying)	A	5.5
Tegin	5	Glycerol monostearate (self-emulsifying)	A	5.5

TABLE 6-3.—(Continued).

Name	Mfr.*	Chemical Designation	Type†	HLB††
Emcol DM-50	2	Diethylene glycol fatty acid ester	N	5.6
Atlas G-1725	1	Polyoxyethylene sorbitol beeswax derivative	N	6
Atlas G-2124	1	Diethylene glycol monolaurate (soap free)	N	6.1
Emcol DL-50	2	Diethylene glycol fatty acid ester	N	6.1
Glaurin	4	Diethylene glycol monolaurate (soap free)	N	6.5
Span 40	1	Sorbitan monopalmitate	N	6.7
Arlacel 40	1	Sorbitan monopalmitate	N	6.7
Atlas G-2242	1	Polyoxyethylene dioleate	N	7.5
Atlas G-2147	1	Tetraethylene glycol monostearate	N	7.7
Atlas G-2140	1	Tetraethylene glycol monoöleate	N	7.7
Atlas G-2800	1	Polyoxypropylene mannitol dioleate	N	8
Atlas G-1493	1	Polyoxyethylene sorbitol lanolin oleate derivative	N	8
Atlas G-1425	1	Polyoxyethylene sorbitol lanolin derivative	N	8
Atlas G-3608	1	Polyoxypropylene stearate	N	8
Span 20	1	Sorbitan monolaurate	N	8.6
Arlacel 20	1	Sorbitan monolaurate	N	8.6
Emulphor VN-430	3	Polyoxyethylene fatty acid	N	9
Atlas G-1734	1	Polyoxyethylene sorbitol beeswax derivative	N	9
Atlas G-2111	1	Polyoxyethylene oxypropylene oleate	N	9
Atlas G-2125	1	Tetraethylene glycol monolaurate	N	9.4
Brij 30	1	Polyoxyethylene lauryl ether	N	9.5
Tween 61	1	Polyoxyethylene sorbitan monostearate	N	9.6
Atlas G-2154	1	Hexaethylene glycol monostearate	N	9.6
Tween 81	1	Polyoxyethylene sorbitan monoöleate	N	10.0
Atlas G-1218	1	Polyoxyethylene esters of mixed fatty and resin acids	N	10.2
Atlas G-3806	1	Polyoxyethylene cetyl ether	N	10.3
Tween 65	1	Polyoxyethylene sorbitan tristearate	N	10.5
Atlas G-3705	1	Polyoxyethylene lauryl ether	N	10.8
Tween 85	1	Polyoxyethylene sorbitan trioleate	N	11
Atlas G-2116	1	Polyoxyethylene oxypropylene oleate	N	11
Atlas G-1790	1	Polyoxyethylene lanolin derivative	N	11
Atlas G-2142	1	Polyoxyethylene monoöleate	N	11.1
Myrj 45	1	Polyoxyethylene monostearate	N	11.1
Atlas G-2141	1	Polyoxyethylene monoöleate	N	11.4
P.E.G. 400 monoöleate	6	Polyoxyethylene monoöleate	N	11.4
P.E.G. 400 monoöleate	7	Polyoxyethylene monoöleate	N	11.4
Atlas G-2076	1	Polyoxyethylene monopalmitate	N	11.6
S-541	4	Polyoxyethylene monostearate	N	11.6

TABLE 6-3.—(*Continued*).

Name	Mfr.*	Chemical Designation	Type†	HLB††
P.E.G. 400 mono-stearate	6	Polyoxyethylene monostearate	N	11.6
P.E.G. 400 mono-stearate	7	Polyoxyethylene monostearate	N	11.6
Atlas G-3300	1	Alkyl aryl sulfonate	A	11.7
		Triethanolamine oleate	A	12
Atlas G-2127	1	Polyoxyethylene monolaurate	N	12.8
Igepal CA-630	3	Polyoxyethylene alkyl phenol	N	12.8
Atlas G-1431	1	Polyoxyethylene sorbitol lanolin derivative	N	13
Atlas G-1690	1	Polyoxyethylene alkyl aryl ether	N	13
S-307	4	Polyoxyethylene monolaurate	N	13.1
P.E.G. 400 monolaurate	6	Polyoxyethylene monolaurate	N	13.1
Atlas G-2133	1	Polyoxyethylene lauryl ether	N	13.1
Atlas G-1794	1	Polyoxyethylene castor oil	N	13.3
Emulphor EL-719	3	Polyoxyethylene vegetable oil	N	13.3
Tween 21	1	Polyoxyethylene sorbitan monolaurate	N	13.3
Renex 20	1	Polyoxyethylene esters of mixed fatty and resin acids	N	13.5
Atlas G-1441	1	Polyoxyethylene sorbitol lanolin derivative	N	14
Atlas G-7596J	1	Polyoxyethylene sorbitan monolaurate	N	14.9
Tween 60	1	Polyoxyethylene sorbitan monostearate	N	14.9
Tween 80	1	Polyoxyethylene sorbitan monoöleate	N	15
Myrj 49	1	Polyoxyethylene monostearate	N	15.0
Atlas G-2144	1	Polyoxyethylene monoöleate	N	15.1
Atlas G-3915	1	Polyoxyethylene oleyl ether	N	15.3
Atlas G-3720	1	Polyoxyethylene stearyl alcohol	N	15.3
Atlas G-3920	1	Polyoxyethylene oleyl alcohol	N	15.4
Emulphor ON-870	3	Polyoxyethylene fatty alcohol	N	15.4
Atlas G-2079	1	Polyoxyethylene glycol monopalmitate	N	15.5
Tween 40	1	Polyoxyethylene sorbitan monopalmitate	N	15.6
Atlas G-3820	1	Polyoxyethylene cetyl alcohol	N	15.7
Atlas G-2162	1	Polyoxyethylene oxypropylene stearate	N	15.7
Atlas G-1471	1	Polyoxyethylene sorbitol lanolin derivative	N	16
Myrj 51	1	Polyoxyethylene monostearate	N	16.0
Atlas G-7596P	1	Polyoxyethylene sorbitan monolaurate	N	16.3
Atlas G-2129	1	Polyoxyethylene monolaurate	N	16.3
Atlas G-3930	1	Polyoxyethylene oleyl ether	N	16.6

TABLE 6-3.—*(continued)*.

Name	Mfr.*	Chemical Designation	Type†	HLB††
Tween 20	1	Polyoxyethylene sorbitan monoleurate	N	16.7
Brij 35	1	Polyoxyethylene lauryl ether	N	16.9
Myrj 52	1	Polyoxyethylene monostearate	N	16.9
Myrj 53	1	Polyoxyethylene monostearate	N	17.9
		Sodium oleate	A	18
Atlas G-2159	1	Polyoxyethylene monostearate	N	18.8
		Potassium oleate	A	20
Atlas G-263	1	N-cetyl N-ethyl morpholinium ethosulfate	C	25–30
		Pure sodium lauryl sulfate	A	App. 40

* 1 = Atlas Powder Company, 2 = Emulsol Corporation, 3 = General Aniline &
Film Corporation, 4 = Glyco Products Company, Inc., 5 = Goldschmidt Chemical
Corporation, 6 = Kessler Chemical Company, Inc., 7 = W. C. Hardesty Company,
Inc.
† A = Anionic, C = Cationic, N = Nonionic.
†† HLB values, either calculated or determined, believed to be correct to ±1.

TABLE 6-4. HLB VALUES FOR VARIOUS EMULSION APPLICATIONS[36]

Application	Emulsion Type	HLB Range
Cream, antiperspirant	O/W	14–17
Cream, cold	O/W	7–15
Cream, emollient	W/O	3–6
Oil, mineral	O/W	9–12
Oils, vitamin	O/W	5–10
Polish, automobile	O/W	8–12

For example, a blend containing 4 parts of *Span 20* and 6 parts of *Tween 60*
(cf. Table 6-3) would have an effective HLB of

$$0.4 \times 8.6 + 0.6 \times 14.9 = 12.3$$

This mixture would then be suitable for preparing an O/W emulsion of
kerosene, as shown in Table 6-5.*

It should be emphasized, however, that this particular mixture of surface
active agents is not the only one which would yield the desired HLB, and
indeed, might not be the one which would yield the most stable emulsion

* An interesting sidelight on these values is suggested by the case of sodium oleate.
It has long been known that the efficacy of this agent is increased by the presence of
free oleic acid. It is usually felt that this serves to inhibit hydrolysis. However, sodium oleate has an HLB number of 18, which puts it at the extreme upper limit for
O/W emulsifiers (Table 6-1). The addition of even a small amount of oleic acid with
its extremely low HLB (approximately 1), will then give total HLB in the middle of
the suitable range.

TABLE 6-5. HLB VALUES REQUIRED TO EMULSIFY VARIOUS OIL PHASES[36]

Oil Phase Material	W/O Emulsion	O/W Emulsion
Acid, stearic	—	17
Alcohol, cetyl	—	13
Kerosene	—	12.5
Lanolin, anhydrous	8	15
Oil		
mineral, heavy	4	10.5
mineral, light	4	10
silicone	—	10.5
Petrolatum	4	10.5
Wax		
beeswax	5	10–16
candelilla	—	14.5
carnauba	—	14.5
microcrystalline	—	9.5
paraffin	4	9

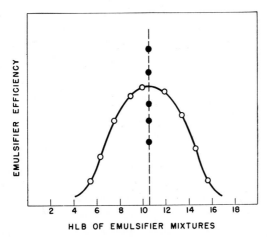

FIGURE 6-1. Determination of appropriate HLB for a given emulsion, according to Griffin.[37] The bell-shaped curve is determined on a particular pair of emulsifying agents; the most efficient pair is then determined by formulating to the HLB corresponding to the peak of the curve.

This point is illustrated by Figure 6-1, adapted from Griffin.[37] In this figure, the stability of a particular emulsion is plotted as a function of HLB. The bell-shaped curve (open circles) represents a series of emulsions made with a given pair of agents at a variety of ratios. As can be seen, this curve peaks at an HLB of about 10.5. The straight line (solid circles) parallel to the stability-axis represents a series of emulsions made up with emulsifier pairs adjusted to give an HLB of 10.5. It is evident that some of these mixtures are less efficient, and some more efficient, than the most efficient ratio of the original pair.

Thus, in formulating a given emulsion, one may take any pair of emulsifying agents and vary their net HLB over the range in which they would be expected to be effective. Having found the most effective HLB, one would then try various pairs of agents until the most effective pair was found.

It should be noted that the HLB concept says nothing about emulsifier concentration. In general, the HLB number required for a stable emulsion does not vary a great deal with the emulsifier concentration. There may, however, be an effect in regions where the emulsion is at best unstable, i.e., very low emulsifier concentrations, high concentration of internal phase.[37]

Other Methods for Hydrophile-Lipophile Balance. Recently Moore and Bell[38] have introduced another method of calculating the hydrophile-lipophile balance of a series of emulsifying agents of the polyoxyethylene type. The hydrophobic groups included in this method are derived from fatty alcohols, saturated and unsaturated fatty acids, alkyl phenols, and castor oil. The number used to classify these agents is calculated from

$$\text{H/L} = \frac{\text{Number EO units} \times 100}{\text{Number C atoms in lipophile}} \qquad (6.4)$$

The range of numbers obtained and their areas of application are summarized in Table 6-6. As can be seen, the H/L numbers are of a different order of magnitude than the HLB numbers described in the previous section. Also, the method of calculation based on Eq. 6.4 does not apparently assign any hydrophilic value to the carboxyl group of a fatty acid group. Thus, a polyglycol ester of stearic acid and a polyoxyethylene derivative of stearyl alcohol containing equal number of ethylene oxide units are assumed to have the same hydrophile-lipophile balance, in disagreement with the results of Griffin[37] (cf. Table 6-3).

Greenwald, Brown, and Fineman[39] have described an experimental method of determining the hydrophile-lipophile character of both surface-

TABLE 6-6. H/L RANGES AND THEIR APPLICATION ACCORDING TO MOORE & BELL[38]

Nature of Agent	H/L	Application
Strongly lipophilic	20	W/O—for paraffins
Moderately lipophilic	40	O/W—for paraffins
Balanced	65	Wetting, dispersing
Moderately hydrophilic	125	O/W—for polar substances
Strongly hydrophilic	150	Solubilizing

TABLE 6-7. WATER NUMBERS OF EMULSIFIERS ACCORDING TO GREENWALD, ET AL.[39]

Compound	Water Number
OPE_0	11.6 ± 0.1
OPE_1	13.8 ± 0
OPE_3	16.7 ± 0.1
OPE_5	18.4 ± 0.1
$OPE_{7.5}$	21.4 ± 0
$OPE_{9.7}$	23.0 ± 0.3
$OPE_{12.3}$	23.6 ± 0.2
$(OPE_{10}-CH_2)_x$	23.6 ± 0.1
$n\text{-}C_{18}H_{37}-C_6H_4-E_8$	13.1 ± 0.1
Branched $C_{12}H_{25}-C_6H_4-E_5$	15.0 ± 0.1
$tert\text{-}C_{18-24}H_{37-49}NE_5*$	9.4 ± 0.2
$tert\text{-}C_{18-24}H_{37-49}NE_{15}*$	12.0 ± 0
$tert\text{-}C_{18-24}H_{37-49}NE_{25}*$	13.7 ± 0.3
n-Hexadecyl + n-octadecyl alcohol	8.4 ± 0.2
Polyethylene glycol (M.W. 300) $(E_{6.4})$	22.8 ± 0
Polyethylene glycol (M.W. 400) $(E_{8.7})$	23.1 ± 0.3
Polyethylene glycol (M.W. 600) $(E_{13.3})$	22.5 ± 0

* Polyethylene oxide adduct of primary amine.

active agents and of oils by means of a water titration. Since this method applied to oils, it suggests directly the type of agent required to emulsify it. In this method, one gram of the material to be tested is dissolved in 30 ml of a solvent consisting of 4 per cent benzene in dioxane, and is then titrated with distilled water until the first persistent turbidity. The number of milliliters of water required for an end-point is defined as the *water number* of the material. The solvent itself has a water number of about 22.6.

Table 6-7 reproduces the values of the water numbers obtained for a number of ethylene oxide derivatives of octyl phenol, primary amines, etc. The symbol *OP* stands for octyl phenol, while *E* represents the ethylene oxide unit (the subscript corresponding to the average number of such units per molecule).

The water numbers of blends are algebraically additive, in analogy to the HLB numbers. Greenwald, Brown, and Fineman have also shown that the water numbers are directly proportional to the HLB numbers; however, the constant of proportionality is apparently different for different families of surface-active agents.

It is unfortunate that these workers should have used the designation "water number" for the property which they are measuring, since pharmaceutical chemists have been using this term as a measure of emulsifier efficiency for some years, with a different definition.

As used in pharmaceutical work, the term is apparently limited to water-in-oil emulsifiers and is defined as the number of milliliters of water, at

20° C, emulsified by 100 grams of the hydrophobic phase containing 5 per cent of the emulsifying agent. As originally defined by Casparis and Meyer,[40] the weight of water imbibed was used as a measure of emulsifying power; but, since the volume is more conveniently determined, it is now generally employed.

Truter[41] describes the determination of this pharmaceutical quantity as follows: Distilled water in 0.3 ml portions is added from a burette to an accurately weighed 5.00 g sample of the material to be studied. The mixture is stirred by hand (apparently mechanical stirrers do not give reproducible results). Addition is continued until it is impossible to emulsify any more water.

The water number, as measured in this manner, has one advantage: if determined as a function of emulsifier concentration, a water number-concentration curve with a peak is often obtained, and this indicates the optimum emulsifier concentration. On the other hand, there does not appear to be a good correlation between water number and emulsion stability; the data of Table 6-8, reproduced from Truter[42] shows this quite clearly.

Nevertheless, the quantity is not without significance, since the work of Halpern and Zopf,[43] Halpern and Wilkins,[44] and Truter[45] show that the water number varies in a systematic way through a homologous series. It is most interesting to note that the water numbers of the n-alcohols from decanol to docosanol (determined at the optimum concentration for each) are directly proportional to their melting points.

Sisley, Reutenauer, and Sicard[46] have adapted the ASTM "Steam Emulsion Test" for the measurement of emulsifier efficiency, while Cockton and Wynn[47] describe a centrifugal method.

Many other methods have been described for evaluating the emulsifying power of various agents. Most of these however, apparently depend on a

TABLE 6-8. EMULSIFYING POWER OF VARIOUS WOOL WAX CONSTITUENTS ACCORDING TO TRUTER[42]

Agent (5% in liquid paraffin)	Water Number	Emulson Stability
Mixed alcohols	650	good
Long-chain fraction (total urea extract)	530	good
Residual fraction (polycyclic alcohols)	360	poor
Monohydric alcohol fraction	110	poor
Dihydric alcohol fraction	565	good
n-Octadecanol	40	good
n-Docosanol	56	good
Hexadecan-1,2-diol (synthetic)	140	good
Cholestan-3,5,6-triol	60	poor
7-Oxycholesterol	320	poor

measurement of emulsion stability (as to some extent do the last three) rather than on some specific property of the emulsifier.

Experimental Data on Emulsifier Efficiency

In addition to the general concepts on emulsifier efficiency described in the preceding section, a great amount of work has been done in the study of the efficiency of particular emulsifiers or groups of emulsifiers. Much of this data deals with specific systems, however, and is thus not of general interest. More recent studies on systems of general interest are described below.

General Observations. Martin and Hermann [48] studied emulsions between $N/30$ sodium oleate solutions and a number of organic liquids: aniline, n-octanol, ethyl oleate, chloroform, benzene, toluene, xylenes, higher paraffins, carbon tetrachloride, carbon disulfide, hexane, and cyclohexane. Emulsions of this sort are divisible into two groups. Those with interfacial tensions less than about 42 dynes/cm gave oil-in-water emulsions with a standard procedure; those above 42 dynes/cm required special procedures (e.g., homogenization). In the standard paraffin and benzene emulsions, the interfacial layer consists of acid soap, whereas in nonyl alcohol emulsions, a composite film of sodium oleate and the alcohol exists.

Loiseleur[49] determined the emulsifying power of the α-amino acids by shaking a 0.5 per cent solution of paraffin in a ketone in a boiling alcoholic solution saturated with the amino acid, then with water, filtering, dialyzing, and determining the optical density, δ, of the emulsion so obtained. The emulsifying power in terms of δ increases with the chain length; i.e., 0.05 for glycine, 0.52 for valine, 1.80 for isoleucine. Amino acids which contain an additional polar group (e.g., aspartic, glutamic, lysine) do not emulsify. Cysteine and tyrosine do not form stable emulsions, but methionine and phenylalanine are very good stabilizers, forming emulsions where $\delta = 1.89$ and 6.0, respectively. The good emulsions are extremely stable and are independent of the pH and salt composition of the medium.

Merill[50] has shown that emulsions of dibutyl phthalate in water, made with a sodium palm-oil soap, were more stable than those made with sodium aleuritate. Similar emulsions prepared with lecithin were found to be very stable, whereas those stabilized with vegetable gums or pectins broke more rapidly than the soap-stabilized emulsions. Vegetable gums, considered as stabilizers for dibutyl phthalate emulsions, can be ranged in the following order of decreasing effectiveness: tragacanth, pectin, karaya, acacia. The paper of Serallach and Jones is of interest in this connection.[21]

In a general discussion of emulsions and emulsifying agents, Lachampt and Dervichian[51] conclude that triethanolamine oleate is a better emulsifier

TABLE 6-9. EMULSIFYING EFFICIENCY OF ALKALI METAL OLEATES
ACCORDING TO KREMNEV AND KAGAN[52]

Soap	Vol. Benzene Emulsified (ml)	Thickness of Interfacial Layer (μ)	Area of Interface per Soap Molecule (Cm²)
Na	178	0.01	50×10^{-16}
K	190	0.005	103
Rb	223	0.0035	147
Cs	247	0.0020	258

than sodium oleate because the triethanolamine tends to inhibit "crystallization" of the interfacial film. The authors also make the broad generalization, with regard to hydrophobic groups, that emulsifying power appears in fatty chain at C_{14} and is strengthened with C_{16} and C_{18}. The benzene ring is not interesting as a hydrophobe; one must reach the naphthalene nucleus for emulsifying power.

Kremnev and Kagan[52] investigated the relative emulsifying efficiency of sodium, potassium, rubidium, and cesium oleates. This was done by noting the volume of benzene emulsified by 1 ml of a $0.32N$ solution of the soap. Microscopic examination of the resulting emulsions allowed calculation of the interfacial area, thickness of interfacial film, and area per soap molecule in the interface. These results are summarized in Table 6-9. The areas per molecule indicate that a complete unimolecular film is not necessary for stabilization. The most frequent drop diameter was 1μ, and the spread of droplet sizes decreased from sodium to cesium soap. It was also found that highly concentrated emulsions did not form in concentrated soap solutions, e.g., in 25 per cent sodium oleate.

Biehn and Ernsberger[53] have reported on the usefulness of polyvinyl alcohol as an emulsifying agent. Polyvinyl alcohols which have been hydrolyzed to 75 to 80 per cent are most effective, and are usable over a wide pH range, but are more efficient at high pH. The presence of salts (e.g., hard water) had an adverse effect.

Jellinek and Anson[54] investigated the mutual increase in emulsifying efficiency by the use of mixed emulsifiers consisting of sodium stearate and α-monostearin. They report interfacial tension studies of systems where α-monostearin in liquid paraffin was measured against water, sodium stearate in water measured against liquid paraffin, and solutions of both agents in the appropriate phase were measured against each. Phase separation studies on emulsions showed that an increase of the α-monostearin concentration from 0.3 to 0.5 per cent caused the emulsion stability to increase rapidly. Emulsions containing sodium stearate alone are less stable than those made with the same molar concentration of the monoglyceride. With a mixture of the two there will be a zone of instability where the emulsion is neither definitely W/O and O/W. The minimal interfacial tension was found between 0.2 per cent α-monostearin in oil and below 0.25 per cent sodium stearate in water at a value of 1.8 dynes/cm, indicating that emulsion formation should be easy. On the other hand, emulsions containing 0.2 per cent monoglyceride and 0.1 per cent soap separate within one hour. Fairly stable W/O emulsions can be formed if the sodium stearate concentration is kept below 0.02 per cent. It is concluded that for each concentration of monoglyceride, there is a particular sodium stearate concentration at which the mechanical properties of the interfacial layer are no longer sufficient to give stable emulsions.

Kaufman, Baltes, and Alberti[55] have studied monoglycerides of saturated and hydroxy fatty acids with respect to their effect on the interfacial tension in the system cottonseed oil/water.

An elaborate study of the emulsifying power of a number of emulsifying agents has been made by Saha.[56] In this investigation a hydrocarbon oil was gradually added, with stirring, to a mixture of water and the emulsifier, to form concentrated emulsions. The interfacial tension and size frequency of the droplets in the emulsion were determined. It was found that the capacity to emulsify oil per unit weight of emulsifier increased, and the specific interface decreased, with dilution of the emulsifier solution.

For a solution of sodium oleate, the specific interface increased from 5.1417×10^3 sq cm/g oil with a 5 per cent sodium oleate solution to 31.7346×10^3 for a 50 per cent solution. The percentage of emulsion droplets below 1μ was increased from 5.8 to 52.0 for the same range of concentrations. At the breaking point of an oil-water emulsion stabilized with sodium oleate, the emulsoid jelly contained 98.32 per cent oil when the percentage by weight of sodium oleate in the jelly was 0.08, and 94.89 per cent oil when the sodium oleate was at 1.70 per cent. Of the other emulsifiers studied, an aqueous mixture of sodium linoleate and rosinate containing excess ammonia had the greatest effect in lowering the interfacial tension and increasing specific interface. For a 5 per cent solution of sodium oleate, it was found that the emulsion had a capacity (expressed as milliliters of oil to form an emulsion jelly per gram of sodium oleate) of 1340; for a 50 per cent solution the capacity was 61.

The emulsifying properties of polyoxyethylated cetostearyl alcohols has been studied by Hadgraft,[57] in connection with emulsions of liquid paraffin and of arachis oil. When used alone, the lower members of the series (two to four ethylene oxide units) are more effective than the higher members in emulsifying these oils. In combination with the free alcohol, however, the higher members (6–10 ethylene oxide units) produce nonionic emulsifying waxes which are equally effective in emulsifying up to 70 per cent oil.

Martynov[58] has studied emulsions of heptane-in-water, stabilized with casein. If the casein solution is cooled to $-15°$ C, its emulsifying efficiency is greatly lowered, and this effect is increased on cooling to $-72°$. Cooling to $-1.5°$ C causes no irreversible changes, but it is believed that lower temperatures must have caused permanent aggregation of the casein protein.

Sisley, Reutenauer, and Sicard,[46] in applying their modification of the ASTM "Steam Emulsion Test," have measured the emulsifying efficiency of a number of agents. For example, it was found that in the series potassium stearate, potassium palmitate, potassium myristate, potassium laurate, sodium laurate, and potassium oleate at concentrations above 2 per cent the emulsifying power (E) was about the same. However, at 0.5 per cent the laurates and myristates formed no emulsion, while the others lost only about half their emulsifying power. One per cent solutions of the following surface-active agents gave E values of: sulforicinate, 41; *Cyclanon* 0, 50; *Melioran*, 51; oxyethylamide, 56; *Igepon T*, 44; *Igepal C*, 47; *Nekal BX*, 0; and *Nacconal NR*, 57. In general, these are to be compared with: potassium stearate, 37; potassium palmitate, 44; potassium myristate, 0; potassium laurate, 16; sodium laurate, 25; and potassium oleate, 46.

Similarly, in connection with their centrifugal method, Cockton and Wynn[37] studied emulsions of peanut oil and liquid paraffin stabilized with hexadecyltrimethylammonium bromide, polyoxyethylene sorbitan monoöleate, stearyl alcohol-ethylene oxide condensate, hexadecyl alcohol-ethylene condensate, and sodium oleate. It was not possible to arrange the surface-active agents in any single order of efficiency that would be applicable at all concentrations. All were less effective for liquid paraffin than for peanut oil.

Agents of Vegetable Origin. Reference has been made earlier to the work of Merill[50] on the efficiency of various of the water-soluble gums (p. 199). Morrison and Campbell[59] have investigated the relative efficiencies of methylcellulose and sodium carboxymethylcellulose (CMC). These workers found that the former agent is more effective at lower concentrations than CMC. A mixture of both is better than either alone. The presence of a high-viscosity CMC reduces the tendency of CMC to foam. Methylcellulose reduces the interfacial tension between water and liquid paraffin to a greater extent than CMC alone. Methylcellulose cannot be used at temperatures in excess of $40°$ C, since its solubility is much reduced. However, an emulsion first made at a lower temperature was stable at $45°$. Emulsions made with methylcellulose are stable over the pH range 2 to 10, whereas in the acid pH range emulsions with CMC break down.

Walsh and Newman[60] have revealed that emulsions of substances of acid, alkaline, or neutral reactions can be prepared by means of an emulsifying agent consisting of the gelatinous material obtained by alkaline extraction of the fleshy growth of agaves and plants of kindred families.

Solid Emulsifying Agents. A large amount of experimental data on solid emulsifying agents has been presented earlier in connection with the stability of such emulsions (p. 125–131). The most general survey of such agents has been made by Bennister, King, and Thomas,[61] and a recent study has been made on the hydrous oxides by Mukerjee and Srivastava.[62]

Fain and Snell[62a] have investigated the effect of pH on the stability of bentonite-stabilized asphalt emulsions.

King[63] has found that emulsions stabilized with inorganic metallic gelatinous hydroxides or oxides can be improved by the addition of a small amount of surface-active material.

Mokrushin and Sheina[64] prepared films of copper and lead sulfides by passing gaseous hydrogen sulfide over solutions containing cupric and lead ions. These films were ground to a powder, and a given weight mixed with 8 ml of benzene, to which was then added 2 ml of water. When the weight of sulfide was varied between 0.02 and 0.1 g, the most stable emulsions (W/O) formed when the weight of solid was 0.08 g. When 2 ml of 0.25 sodium hydroxide solution was used instead of water, it was found that, for the lead sulfide, maximum stability again occurred at the same concentration of solid; but an O/W emulsion formed. With the copper sulfide, an O/W emulsion also formed, but its stability was essentially independent of the amount of solid. Zinc sulfide also gives stable emulsions.

Phospholipids and Sterols. The mode of use of lecithin, especially in paints, has been discussed by Schofield[65] and others.[66] In order to be used successfully as an emulsifying agent, it must be incorporated into the oil phase first, since the hydrate is insoluble in most oils.

The question of the emulsifying power of wool wax, or its refined form, lanolin, is somewhat confused, principally because no one is entirely certain what fraction of the mixture is responsible for the stabilizing action.

Lifshutz,[67] in 1923, reported the isolation of two fractions of lanolin, which he called "metacholesterol" and "oxycholesterol," which he believed represented the emulsifying fractions. Janistyn[68] was able to identify these as mixtures of sterols, only partially precipitated by digitonin. He concluded that the active principle was cholesterol, and that its emulsifying power is somewhat enhanced by the presence of fatty alcohols in the unsaponified portion of wool fat. Since cholesterol blown with oxygen showed increased emulsifying power, various other sterols were tested. However, when lanolin fractions rich in cholesterol were blown with oxygen, no increase in the emulsifying efficiency was noted. The emulsifying action of hexadecanol was found to be superior to hexacosanol, while β-cholestanol was slightly inferior to cholesterol. Pure cholestandiol was a poor emulsifier even when mixed with cholesterol; α-hydroxycholesterol, β-hydroxylcholesterol, and cholestanol were good emulsifiers.

Janistyn concludes that the emulsifying power of cholesterol, and sterols and steroids in general, depends on the spatial arrangement of the hydroxyl groups. Only sterols and steroids which are precipitated by digitonin act as emulsifiers, and always form W/O emulsions. This factor of digitonin-precipitability seems to be limited only by the solubility of the agent in the oil phase. It appears that even bile acids (e.g., hydroxycholanic acid) or triterpene alcohols (e.g., lanosterol) are effective emulsifying agents to the extent that they are precipitable.

Janistyn's observations on the effect of blowing with oxygen may be considered in the light of studies by Muirhead, Oberweger, Seymour, and Simmonite.[69] These investigators found that oxidation of the wool alcohols (the unsaponifiable fraction of lanolin) causes a rise in acid and saponification values and a decrease in acetyl value and cholesterol content. On the other hand, no appreciable change is observed in the interfacial tension reduction afforded by the total wool alcohols, and by some isolated fractions. Nevertheless, it is known that W/O emulsions stabilized by wool alcohols will break on storage, with a coincident rise in the acid value of the oil phase, but that both effects are retarded by the inclusion of antioxidants. The authors conclude that the loss in emulsifying efficiency arises from a loss of solubility of the oxidized material in the hydrocarbon phase of the emulsion.

Janistyn's[68] principal conclusion, however, i.e., that the emulsifying power of lanolin is due to the alcohols, is challenged by Bertram,[70] who has ascribed it to the esters, in particular to the diesters of hydroxy-acids.

An elaborate investigation carried out by Tiedt and Truter[71] would appear to have resolved this problem. Six esters, representing all the types believed to be found in wool wax, were synthesized. These were n-octadecyl acetate, cholesteryl acetate, isocholesteryl acetate, hexadecan-1,2-laurate, cholesteryl-α-palmityl-hexanoate, and cholesteryl-α-hydroxyhexanoate. These were found to be completely devoid of any emulsifying power. Similar conclusions were drawn from lanolin which was fractionated into polar and nonpolar fractions. Thus the emulsifying power of commercial lanolin is attributable to its free alcohol and, to a minor extent, to the free acid content. The stability of emulsions thus stabilized, however, is ascribed to the alcohol content alone. Of the constituents of lanolin, the most powerful emulsifying agents are members of the α,β-glycol series. Measurement of the emulsifying power of cholesterol and isocholesterol leads to the interesting result that the latter compound is an effective agent only in the concentration range of 2 to 4 per cent. At other concentrations it actually decreases the emulsifying power of other alcohols with which it may be mixed.[72]

It should be noted, however, that the possibility does exist that the hydroxy-esters may make some contribution to the emulsifying power of lanolin. Conrad[20] has pointed out that modification of lanolin by causing the hydroxyl groups to react (presumably through acetylation, although this is not explicitly stated) causes its emulsifying power to disappear. If one now adds a surface-active extract of lanolin alcohols to the modified lanolin, the emulsifying power returns, but at a much lower level.

Recently, too, Pickthall[73] has showed that acetylation of beeswax (a substance which has certain similarities to lanolin, considered as an emulsifying agent) reduces the stability of cold creams containing this material.

Hadert[74] and Lower[75] have recently published general discussions of the use of cholesterol and lanolin in emulsions.

Some Miscellaneous Observations. Reference has already been made to the work of Halpern and Wilkins,[44] in which the emulsifying power of a series of n-alkyl alcohols was studied. Halpern and Glasser[76] have found that the fatty mercaptans possess little emulsifying ability, while Halpern and Squeglia[77] found that the fatty acid esters of ethylene glycol, propylene glycol, and glycerol did not form W/O emulsions of petrolatum at low concentrations, with the exception of glyceryl monoöleate. The monostearates of ethylene and propylene glycol gave some emulsification at concentrations higher than 10 per cent. The fatty acid esters of sorbitan and mannitan were the best emulsifying compounds in the group studied.

Janistyn[78] has reviewed the subject of nonionic emulsifiers, and Sisley[79] has considered their application in the textile industry. General reviews which include considerations on emulsifying agents are due to Nolla and Diaz,[80] Monson,[81] and Ruemele.[82, 83] Trusler has discussed emulsifiers for waxes.[84]

An important contribution to formulation has recently been made by De Navarre and Bailey.[85] These workers have found that certain nonionic emulsifiers completely or partially inactivate some common microbial preservatives. This is of considerable importance in compositions which are prone to bacterial or mold infection.

Bibliography

1. Schwartz, A. M. and Perry, J. W., "Surface Active Agents," pp. 15–17, New York, Interscience Publishers, Inc., 1949.
2. Sisley, J. P., "Encyclopedia of Surface-Active Agents," (trans. from French and revised by P. J. Wood), pp. 31–38, New York, Chemical Publishing Co., 1952.
3. Schwartz, A. M. and Perry, J. W., op. cit., pp. 18–20.
4. Hilditch, T. P., "Chemical Constitution of the Natural Fats," 2nd Ed., New York, John Wiley, 1947; Markley, K. S., "Fatty Acids," New York, Interscience Publishers, Inc., 1947.

5. Goodyear, G. H. and Hathorne, B. L., *J. Soc. Cosmetic Chemists* **5,** 96 (1941).
6. Van Zile, B. S. and Borglin, J. N., *Oil & Soap* **22,** 331 (1945).
7. Levinson, H. and Minich, A., *Soap* **14,** No. 12, 24, 69 (1938).
8. Burton, D. and Robertshaw, G. F., "Sulfated Oils & Allied Products," Ch. 2 and 3, New York, Chemical Publishing Co., 1940.
9. Rueggeberg, W. H. C. and Sauls, T. W. (to Tennessee Corp.), U. S. Pat. 2,743,288, Apr. 24, 1956.
10. Glicher, S., *Paint Manuf.* **14,** 248 (1944).
11. Trask, R. H., *J. Am. Oil Chemists' Soc.* **33,** 568 (1956).
12. Reed, R. M. and Tartar, H. V., *J. Am. Chem. Soc.* **57,** 570 (1935).
13. Schwartz, A. M. and Perry, J. W., *op. cit.*, p. 88.
14. Van Antwerpen, F. J., *Ind. Eng. Chem.* **31,** 64 (1939).
15. Schwartz, A. M. and Perry, J. W., *op. cit.*, pp. 151–200.
16. Mayhew, R. L. and Hyatt, R. C., *J. Am. Oil Chemists' Soc.* **29,** 357 (1952).
17. Vaughn, T. H., Jackson, D. R., and Lundsted, L. G., *J. Am. Oil Chemists' Soc.* **29,** 240 (1952).
18. Schwartz, A. M. and Perry, J. W., *op. cit.*, p. 209.
19. Stanley, J. in Alexander, J. (ed.), "Colloid Chemistry," **VI,** pp. 263–267, New York, Reinhold Publishing Corp., 1946.
20. Conrad, L. I., *J. Soc. Cosmetic Chemists* **5,** 11 (1954).
21. Serallach, J. A. and Jones, G., *Ind. Eng. Chem.* **23,** 1016 (1931).
22. Mantell, C. L. in "Natural Plant Hydrocolloids," Advances in Chemistry Series, No. 11, pp. 20–32, Washington, American Chemical Society, 1954.
23. Hirst, E. L., *J. Chem. Soc. (London)* **1942,** 70.
24. Goldstein, A. M., in "Natural Plant Hydrocolloids," Advances in Chemistry Series, No. 11, pp. 33–37, Washington, American Chemical Society, 1954.
25. Beach, D. C., in "Natural Plant Hydrocolloids, "Advances in Chemistry Series, No. 11, pp. 38–44, Washington, American Chemical Society, 1954.
26. Whistler, R. L., in "Natural Plant Hydrocolloids," Advances in Chemistry Series, No. 11, pp. 45–50, Washington, American Chemical Society, 1954; Deuel, H. and Neukom, H., *ibid.*, pp. 51–61.
27. Steiner, A. B. and McNeely, W. H., in "Hatural Plant Hydrocolloids," Advances in Chemistry Series, No. 11, pp. 68–82, Washington, American Chemical Society, 1954.
28. Tseng, C. K., in Alexander, J. (ed.), "Colloid Chemistry," **VI,** pp. 629–734, New York, Reinhold Publishing Corp., 1946.
29. Black, W. A. P. and Woodward, F. N., in "Natural Plant Hydrocolloids," Advances in Chemistry Series, No. 11, pp. 83–91, Washington, American Chemical Society, 1954.
29a. Federici, N. J., *Soap Sanit. Chemicals* **23,** No. 5, 35, 90 (1947).
30. Stoloff, L., in "Natural Plant Hydrocolloids," Advances in Chemistry Series, No. 11, pp. 92–100, Washington, American Chemical Society, 1954.
31. Young, A. E. and Kin, M., in Alexander, J. (ed.), "Colloid Chemistry," **VI,** pp. 926–933, New York, Reinhold Publishing Corp., 1946.
32. Anon., *Chem. Eng. News* **35,** 78 (1957).
33. Ott, E. and Spurlin, H. M., "Cellulose and Cellulose Derivatives," 2nd Ed., pp. 927ff., New York, Interscience Publishers, 1954.
34. Clayton, W., "Theory of Emulsions," 4th Ed., p. 127, Philadelphia, The Blakiston Co., 1943.

35. Griffin, W. C., *J. Soc. Cosmetic Chemists* **1,** '311 (1949).
36. Griffin, W. C., *J. Soc. Cosmetic Chemists* **5,** 249 (1954).
37. Griffin, W. C., *Offic. Dig. Federation Paint & Varnish Production Clubs* **28,** 466 (1956).
38. Moore, C. D. and Bell, M., *Soap, Perfumery, & Cosmetics* **29,** 893 (1956).
39. Greenwald, H. L., Brown, G. L., and Fineman, M. N., *Anal. Chem.* **28,** 1693 (1956).
40. Casparis, P. and Meyer, E. W., *Pharm. Acta Helv.* **10,** 163 (1935).
41. Truter, E. V., "Wool Wax," p. 91, New York, Interscience Publishers, Inc., 1956.
42. Truter, E. V., *op. cit.*, p. 98.
43. Halpern, A. and Zopf, L. C., *J. Am. Pharm. Assoc.* **36,** 101 (1947).
44. Halpern, A. and Wilkins, W. J., *J. Am. Pharm. Assoc.* **38,** 283 (1949).
45. Truter, E. V., *J. Appl. Chem. (London)* **1,** 254 (1949).
46. Sisley, J. P., Reutenauer, G., and Sicard, P., *Teintex* **14,** 361 (1949).
47. Cockton, J. R. and Wynn, J. B., *J. Pharm. Pharmacol.* **4,** 959 (1952).
48. Martin, A. R. and Hermann, R. N., *Trans. Faraday Soc.* **37,** 25 (1941).
49. Loiseleur, J., *Compt. rend.* **213,** 351 (1941).
50. Merill, R. C., Jr., *Ind. Eng. Chem., Anal Ed.* **15,** 743 (1943).
51. Lachampt, F. and Dervichian, D., *Bull. soc. chim.* **1946,** 491.
52. Kremnev, Ya. and Kagan, R. N., *Kolloid. Zhur.* **10,** 436 (1948); *C.A.* **43,** 7775i.
53. Biehn, G. F. and Ernsberger, M. L., *Ind. Eng. Chem.* **40,** 1449 (1948).
54. Jellinek, H. H. G. and Anson, H. A., *J. Soc. Chem. Ind.* (London) **68,** 108 (1949).
55. Kaufmann, H. P., Baltes, J., and Alberti, G., *Fette u. Seifen* **55,** 670 (1953).
56. Saha, A. N., *J. Indian Chem. Soc.* **28,** 23 (1951).
57. Hadgraft, J. W., *J. Pharm. Pharmacol.* **6,** 816 (1954).
58. Martynov, V. M., *Kolloid. Zhur.* **18,** 443 (1956); *C.A.* **51,** 803b.
59. Morrison, R. I. and Campbell, B. *J. Soc. Chem. Ind. (London)* **68,** 333 (1949).
60. Walsh, C. L. and Newman, A. A., Brit. Pat. 551,298, Feb. 18, 1943.
61. Bennister, H. L., King, A., Thomas, R. K., *J. Soc. Chem. Ind. (London)* **59,** 185 (1940).
62. Mukerjee, L. N. and Srivastava, S. N., *Kolloid-Z.* **147,** 146 (1956).
62a. Fain, J. M. and Snell, F. D., *Ind. Eng. Chem.* **31,** 48 (1939).
63. King, A., Brit. Pat. 519,769, April 5, 1940.
64. Mokrushin, S. G. and Sheina, Z. G., *Kolloid. Zhur.* **9,** 285 (1947); *C.A.* **47,** 943i.
65. Schofield, M., *Paint Manuf.* **15,** 45 (1945).
66. Anon., *Paint, Oil & Chem. Rev.* **116,** No. 13, 27, 30, 36 (1953).
67. Lifshutz, I., *Pharm. Zentrh.* **64,** 305 (1923).
68. Janistyn, H., *Fette u. Seifen* **47,** 351 (1940).
69. Muirhead, G. S., Oberweger, K. H., Seymour, D. E., and Simmonite, D., *J. Pharm. Pharmacol.* **1,** 762 (1949).
70. Bertram, S. H., *J. Am. Oil Chemists' Soc.* **26,** 454 (1949).
71. Tiedt, J. and Truter, E. V., *J. Appl. Chem. (London)* **2,** 633 (1952).
72. Truter, E. V., *op. cit.*, p. 94.
73. Pickthall, J., *J. Soc. Cosmetic Chemists* **6,** 263 (1955).
74. Hadert, H., *Seifen-Öle-Fette-Wachse* **80,** 502 (1954).
75. Lower, E. S., *Drug & Cosmetic Ind.* **74, 52,** 127 (1954).
76. Halpern, A. and Glasser, A. C., *J. Am. Pharm. Assoc.* **38,** 287 (1949).
77. Halpern, A. and Squeglia, N., *J. Am. Pharm. Assoc.* **38,** 290 (1949).
78. Janistyn, H., *Seifen-Öle-Fette-Wachse* **75,** 26 (1949).
79. Sisley, J. P., *Am. Dyestuff Reporter* **38,** 513 (1949).

80. Nolla, J. M. and Diaz, E., *Afinidad* **21,** 540 (1944); **22,** 31, 76, 168, 205 (1945); **23,** 541 (1946).
81. Monson, L. T., *Am. Perfumer* **46,** No. 5, 35; No. 6, 51, 53; No. 9, 37; No. 10, 39 (1944).
82. Ruemele, T., *Seifen-Öle-Fette-Wachse* **75,** 416 (1949).
83. Ruemele, T., *Mfg. Chemist* **25,** 76 (1954).
84. Trusler, R. B., *Proc. Chem. Specialties Mfrs. Assoc.*, Dec., **1951,** 127.
85. De Navarre, M. G. and Bailey, H. E., *J. Soc. Cosmetic Chemists* **7,** 427 (1956).

CHAPTER 7

Technique of Emulsification

The present chapter considers the actual details of putting an emulsion together. The first section is devoted to such problems as the mode of incorporating the emulsifying agent, the order of addition of materials, the effect of mixing technique, etc. The subsequent sections of the chapter consider the various types of equipment available for the actual mechanical production of emulsions.

Techniques for Producing Emulsions

Cobb[1] has pointed out that there are, in general, two ways in which emulsification can be made to take place. These are:

1. Brute force.
2. Persuasion.

This classification is, no doubt, made with tongue in cheek; nevertheless there is considerable truth in it. One is tempted to think that the best emulsions could be made by the "persuasive" method, and this is, to a large extent, quite true. It is most practical, however, to make most emulsions by the "brute force" method, i.e., vigorous mixing.

Mode of Addition. In connection with the above classification, the author[2] has previously noted that only one of the four standard techniques for incorporating emulsifying agent, the agent-in-water method, is probably to be classed as an instance of the "brute force" technique. In most cases, other standard methods for incorporating emulsifying agent will yield emulsions with only moderate treatment.

The four modes of addition are as follows:

1. Agent-in-water method.

In this method, the emulsifying agent is dissolved directly in the water, and the oil is then added, with considerable agitation. This procedure makes O/W emulsions directly; should a W/O emulsion be desired, the oil addition is continued until inversion takes place.

2. Agent-in-oil method.

The emulsifying agent is dissolved in the oil phase. The emulsion may then be formed in two ways:

209

a. By adding the *mixture* directly to the water. In this case, an O/W emulsion forms spontaneously.

b. By adding *water* directly to the mixture. In this case, a W/O emulsion is formed. In order to produce an O/W emulsion by this method, it is necessary to invert the emulsion by the addition of further water.

3. Nascent soap (in situ) *method.*

This method is suitable for those emulsions which are stabilized by soaps, and may be used to prepare either O/W or W/O types. The fatty acid part of the emulsifying agent is dissolved in the oil, and the alkaline part in the water. The formation of the soap (at the interface), as the two phases are brought together, results in stable emulsions.

4. Alternate addition method.

In this method, the water and oil are added alternately, in small portions, to the emulsifying agent. This method is particularly suitable for the preparation of food emulsions, e.g., mayonnaise, and other emulsions containing vegetable oils.

Cosmetic chemists recognize two general procedures in preparing emulsions: the so-called English and Continental Methods.[3] The English method is essentially the alternate addition method described above, and is probably quite satisfactory. The Continental Method is the agent-in-oil method. According to de Navarre,[3] the Continental Method gives generally somewhat better emulsions; however, a higher concentration of stabilizer is required for a completely satisfactory product.

The agent-in-water technique usually results in quite coarse emulsions, with a wide range of particle size. As has been indicated earlier (p. 46), such emulsions tend to be somewhat unstable. This can be corrected by passing the emulsion through a homogenizer or colloid mill, following the initial mixing.

On the other hand, the agent-in-oil method usually results in uniform emulsions, with an average droplet diameter of about 0.5 μ, which probably represents the most stable type of emulsion.

In all cases where soaps are employed as emulsifying agents, the nascent soap method, resulting in *in situ* formation of the emulsifying agent, is the preferred method. The data of Dorey[4] on olive oil emulsions stabilized with sodium oleate illustrates this point, as well as the need for homogenization. Table 7-1, adapted from the data of Dorey, shows the particle-size distributions of four different emulsions. All of the emulsions had the same composition, olive oil 10 per cent, soap 0.5 per cent, water 89.5 per cent, but different methods of preparation were used. Emulsion I was prepared by the agent-in-water method, using simple mixing with a high-speed mixer; Emul-

TABLE 7-1. PARTICLE SIZE DISTRIBUTION FOR OLIVE OIL EMULSIONS[4]

Size Range of Particles (microns)	Per Cent of Particles in Range			
	Emulsion I	Emulsion II	Emulsion III	Emulsion IV
0–1	47.5	71.8	68.5	80.7
1–2	41.1	26.4	28.4	17.1
2–3	7.4	1.4	2.0	2.0
3–4	2.1	0.3	0.5	0.2
4–5	0.1	—	0.1	—
5–6	0.7	0.1	0.3	—
6–7	0.1	—	—	—
7–8	0.6	—	0.1	—
8–9	0.1	—	—	—
9–10	0.2	—	—	—
10–11	—	—	—	—
11–12	0.1	—	—	—

sion II was also prepared by the agent-in-water method, but with homogenization following mixing. Emulsion III was prepared by the *in situ* formation of soap, with simple mixing; Emulsion IV was also prepared by the nascent soap method, but with mixing and homogenization. As can be seen, the nascent soap technique yields, by mixing, an emulsion (III) very nearly as uniform as that obtained by homogenizing an agent-in-water type (II), while the homogenized nascent soap emulsion (IV) is as uniform, and consequently as stable, an emulsion as could be required.

If mixing is omitted, and the two liquids fed directly to the homogenizer quite uniform emulsions are obtained by either technique; even great homogeneity is obtained by multiple passes. Similar data are reported for emulsions of arachis oil.

De Navarre[5] has indicated that, when the Continental Method is employed, the optimum ratios of water to oil to stabilizing gum are 4:2:1. This has been verified in a study by Husa and Becker,[6] who determined the quality of emulsions obtained by microscopic measurement of the dispersed droplets, as well as by observing the rate of creaming. Castor oil and cod-liver oil give excellent emulsions in systems formulated on the quoted ratios; it was found necessary to increase the amount of gum (acacia) employed to make satisfactory emulsions of linseed oil and mineral oil.

In producing these emulsions by trituration, it was also found that the use of an initially dry mortar and pestle was extremely important, as was the time of trituration of the preliminary emulsion. On the other hand, the following factors were without influence: use of dried acacia, rate of dilution of the primary emulsion, excessive trituration of the oil and acacia when making the primary emulsion, and direction of trituration, i.e., clockwise, counterclockwise, or in both directions.

Stanko, Fiedler, and Sperandio[7] have also reported on a study of the effect of physical factors on the formation of emulsions. These workers used a basic emulsion of the following composition:

Heavy mineral oil	35.0%
Lanolin anhydrous	1.0
Cetyl alcohol	1.0
"Span 80"	2.1
"Tween 80"	4.9
Distilled water	56.0

As shown below, seven types of addition and mixing techniques were employed:

1. Water to oil—slow.
2. Water to oil—fast.
3. Water to oil—intermittent.
4. Oil to water—slow.
5. Oil and water combined.
6. Water to oil cold.
7. Oil to water cold.

For the exact experimental details the reader is referred to the original paper. Numerous physical properties were measured, the most important of which, in the present connection, is the particle size distribution. This data is given in Table 7-2, for the different addition methods. While there are differences, they are surprisingly small. The authors rated the emulsions prepared on a scale of one-to-seven, for the various properties measured; these ratings are given in Table 7-3. The general conclusion is that the

TABLE 7-2. PARTICLE SIZE DISTRIBUTION OF EMULSIONS[7]

Emulsion	Particle Size Ranges (μ)						
	Below 1.9	2.0–3.9	4.0–5.9	6.0–7.9	8.0–9.9	10.0–11.9	12.0 and above
1	58.4	20.3	5.5	4.1	2.6	1.3	7.8
2	53.8	23.9	7.5	4.2	2.7	1.4	6.5
3	42.2	24.0	10.1	5.9	3.8	2.7	11.3
4	46.3	29.3	10.0	5.0	3.3	1.3	4.8
5	57.0	22.6	5.2	3.0	2.6	1.8	7.8
6	37.3	24.5	12.3	8.8	4.9	2.5	9.7
7	23.6	43.6	18.4	6.5	3.3	0.7	3.9

TABLE 7-3. RATING OF EMULSIONS FROM PHYSICAL FACTORS[7]

Factor	Emulsion						
	1	2	3	4	5	6	7
Original viscosity	1	2	3	4	4	5	6
Viscosity change after 60 days	7	1	6	3	5	4	2
Surface tension	2	1	7	2	3	5	4
Total creaming, %	3	1	5	2	6	4	3
Creaming rate	5	1	6	2	7	4	3
Particle size distribution,							
0–3.9 microns	2	3	6	4	1	7	5
0–5.9 microns	4	2	5	1	3	6	7
General appearance	4	1	4	3	2	6	5
Ease of redispersion	2	1	6	3	4	6	5

best method of preparing O/W emulsions is method 2, i.e., addition of the aqueous phase to the oil phase with rapid and continuous stirring.

Effect of Mixing Time and Technique. A certain amount of data has already been introduced on the question of mixing technique; this matter will now be discussed in more detail. Generally, emulsion mixing can be achieved by three techniques: simple stirring (propellor or turbine mixers), homogenization, or colloid milling. These will be discussed from the point of view of the mechanics of the process in a later section of the present chapter.

It has already been shown, in connection with the data of Dorey,[4] that the mixing technique is at least as important as the mode of addition of materials in defining the final particle size distribution. This is shown more clearly in additional experiments with the same emulsion, i.e., olive oil 10 per cent, soap 0.5 per cent, water 89.5 per cent. In these experiments Emulsion V consists of an emulsion prepared by the agent-in-water method, in which the oil and water phases were added individually to the homogenizer; Emulsion VI is Emulsion V after another pass. Emulsion VII is an emulsion prepared by *in situ* formation of the soap, with the individual phases being fed to the homogenizer. Emulsion VIII represents an additional pass through the equipment. The particle-size distributions thus obtained are shown in Table 7-4.

It is evident that the *in situ* addition of emulsifying agent does represent a slight advantage. In any case, it is evident that straight homogenization is more efficient than homogenization following a prior mixing step.

Griffin[8] has presented data for a number of oil-in-water emulsions showing the broad particle-size ranges obtained with different mixing tech-

TABLE 7-4. PARTICLE SIZE DISTRIBUTION FOR OLIVE OIL EMULSIONS[4]

Size Range of Particles (microns)	Per Cent of Particles in Range			
	Emulsion V	Emulsion VI	Emulsion VII	Emulsion VIII
0–1	80.8	87.7	88.6	97.3
0–2	18.1	11.6	10.7	2.5
2–3	0.8	0.7	0.5	0.2
3–4	0.2	—	0.1	—
4–5	0.1	—	—	—

TABLE 7-5. PARTICLE SIZE DISTRIBUTION AS A FUNCTION OF AGITATION[8]

Type of Agitation	Particle Size Range (microns)		
	1% Emulsifier	5% Emulsifier	10% Emulsifier
Propellor	No emulsion	3–8	2–5 (0.1–0.5)*
Turbine	2–9	2–4	2–4
Colloid Mill	6–9	4–7	3–5
Homogenizer	1–3†	1–3	1–3

* Proper choice of emulsifying agent and careful technique.
† 50% oil, nonionic emulsifier.

niques, as a function of emulsifier concentration (Table 7-5). As can be seen, these data are consistent with those of Dorey.[4]

Husa and Becker[9] extended their investigation of the Continental and English methods of preparing emulsions to consider various types of mechanical stirrers and hand homogenizers. With efficient mixing or homogenization, it was found that the ratios (cf. page 211) required for the Continental method were far less critical.

Dvoretskaya[10, 11] has published papers in which the emulsion type obtained is related to the properties of the vessel in which the mixing occurs. When air was bubbled through a mixture of one volume of water and one volume of oil, an O/W emulsion was formed in a glass vessel which was well wetted with water. However, a W/O emulsion formed when a plastic vessel, not wetted by water, was used.[10]

In further studies, mixtures of transformer oil or kerosene with aqueous solutions of sodium oleate, sodium naphthenate, sodium butyrate, or petroleum resins gave oil-in-water emulsions if they were agitated with a glass piston in a glass cylinder. If hydrophobic plastic pistons and cylinders were used, W/O emulsions were obtained. Sodium napthenesulfonates caused formation of W/O emulsions in plastic vessels, if the concentration was less than 0.5 to 1.0 per cent, whereas 2 per cent solutions gave rise to O/W emulsions. Dvoretskaya concludes that the selective wetting of the mixing vessel determines the type of emulsion, while the emulsifier determines its stability.[11]

On the other hand, in an investigation of the formation of petroleum emulsions, Ben'kovskii and Zavorokhin[12] found evidence contrary to Dvoretskaya's. Three samples of crude oil (resinous, paraffinic, and mixed) were agitated with water or aqueous sodium chloride by means of an iron piston in an iron vessel or a glass piston in a glass vessel. In all instances, save one, W/O emulsions resulted. The time required to achieve stable emulsification, however, was greater in the glass container, i.e., seventeen minutes as against six. The time required increased with the water-oil ratio, and, in the iron vessel, rose linearly with the concentration of sodium chloride in the aqueous phase.

Jürgen-Lohmann[13] has reported experiments on the emulsification of xylene in aqueous turkey-red oil and of petroleum in aqueous sodium oleate, in an all-glass emulsifying apparatus, fitted with a glass stirrer. Operating speeds of 4000 rpm are attainable. The results indicate that the radii of the droplets in the emulsion decrease with increasing stirrer speed and with increased emulsifying time, and depends on the composition of the emulsifying agent. This is in agreement with the results of other workers described above. A limiting radius is obtained after a definite time (10 to 20 minutes in these experiments).

Kremnev and Soskin[14] have studied the homogenization of highly concentrated emulsions by forcing them through capillaries under pressure. For example, an emulsion of 40 ml of benzene in 1 ml of 5 per cent sodium oleate solution was repeatedly forced, by a pressure of one meter of water, through a glass capillary 10 cm long and 0.5 mm in diameter. Under such treatment droplets in excess of 10 μ disappeared, and the number of particles of the order of 1 μ increased. After twelve passages, the viscosity of the emulsion increased in the ratio 2.5:1.

Kremnev and Kuprik[15] extended this study to other emulsions which, although still highly concentrated, covered a wider range of concentrations. Between 80 and 97.6 volume per cent of benzene was dispersed in 5 per cent sodium oleate solution. The most frequent particle diameter remained about 1 μ after passage through the capillary, but emulsions became less polydisperse. The specific surface (i.e., interfacial area per milliliter of disperse phase) increased under such treatment from, e.g., 500 sq m/ml to 2500 sq m/ml when an emulsion was forced through a capillary at a linear flow rate of 12 cm/sec. The increase in specific surface was found to depend only on the linear velocity and was independent of length (25 to

100 cm) and diameter (0.36 to 1.48 mm) of the capillary employed. It was also independent of the pressure, over a range of 0.3 to 4.5 atmospheres.

It was also found that the increase in the specific interface was greater when 90 per cent emulsions were treated than when 80 per cent emulsions were used. When, instead of benzene, mixtures of benzene and liquid paraffin were used, the effect of passage through the capillary was greater with higher percentages of benzene. This corresponds to lower viscosity. Emulsions stabilized with saponin, gum arabic, and gelatin behaved similarly.

Fundamental Studies. Much of the older literature on the physics of emulsification deals with very special cases of little general interest; however, a large amount of useful and interesting information is embedded in these papers. Summaries of this older work will be found in Berkman and Egloff[16] and Clayton.[17]

Taylor[18] has deduced some useful relations between the radii of droplets formed and such parameters as rate of shear, interfacial tension, and the viscosities of the two phases. At low speeds, the following relation is found to hold approximately

$$(L - B)/(L + B) = F$$

where L is the radius of the largest droplet which can exist under the conditions of shear, B the radius of the corresponding smallest droplet, and F is a non-dimensional quantity proportional to the speed of flow, which also involves the interfacial tension, viscosity, and droplet radius.

Under shear, a droplet undergoes distortion, elongates into threadlike filaments, and subsequently break up into smaller drops. This is evidenced by photographic studies of oil drops suspended in syrup under stress. Mathematical studies based on this sort of model were made by Tomotika[19] and Oka.[20]

Recently, Kremnev and Ravdel[21] have considered this problem anew. Assuming that the large droplet takes on the shape of a cylinder (whose ratio of height to radius is K) under shear, it will spontaneously break into two spherical drops (whose radii are in the ratio $n:1$) when the total surface area of the drops is smaller than that of the cylinder; i.e., when the parameter

$$\beta = [2(n^2 + 1)/(1 + K)] [3K/4(n^2 + 1)]^{2/3}$$

is greater than unity.

The formation of the two drops is least probable when $K = 2$ and when $n = 1$; its probability is almost independent of n when $0.2 > n > 5$. An ellipsoid (a more realistic picture) which has an axial ratio K spontane-

TABLE 7-6. VALUES OF CRITICAL VELOCITY FROM OHNESORGE'S EQUATION[22]

Liquids	V_0 (cm/sec)		
	$D = 0.1$ cm	$D = 0.15$ cm	$D = 0.275$ cm
Aniline/water	70	50	19
Aniline/brine	89	70	26
Benzene/water	100	70	25
Paraffin/water	150	110	90

ously gives two drops when

$$(n^2 + 1)K^{1/3}/(n^3 + 1) > 1$$

for the oblate shape and when

$$(n^2 + 1)/(n^3 + 1)K^{1/3} > 1$$

for the prolate form.

Also considered are the cases where the ellipsoid breaks down into many drops and where there is successive separation of many small droplets from an ellipsoid.

Richardson[22] has considered the situation when an emulsion is prepared by injection of one liquid phase in another. Under these circumstances the velocity of flow is extremely important, and breakup of the liquid jet is controlled by inertial and viscous forces. In order to get good breakup a certain critical velocity V_0 is required. It is calculable from Ohnesorge's[23] relation

$$\eta_1/(\rho_1 \gamma_i D)^{1/2} = 2000(\eta_1/V_0 \rho_1 D)^{4/3} \tag{7.1}$$

where D is the nozzle diameter, ρ_1 and η_1 are the density and viscosity, respectively, of the internal phase (i.e., of the liquid being forced through the nozzle), and γ_i is the interfacial tension.

Table 7-6, from Richardson,[22] gives values of the critical velocity as calculated from Ohnesorge's equation (Eq. 7.1) for a number of systems. Figures 7-1 and 7-2, after Richardson, show quite clearly the difference in breakup at velocities below and above the critical.

If one constructs summation curves from drop distributions, such as may be obtained from examination of Figures 7-1 and 7-2, they will be as shown in Figure 7-3. In this figure the distribution corresponding to supercritical values is to the left; subcritical to the right. On such curves the mean droplet diameter can be read on the 50 per cent intercept. The Reynolds number Vd/ν, where V is the linear efflux velocity, d the mean drop-

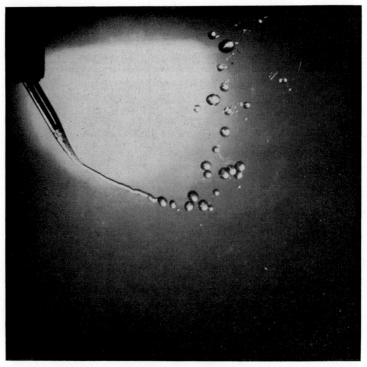

FIGURE 7-1. Breakup in a liquid jet at velocities below the critical.[22] *Courtesy: Prof. E. G. Richardson and Journal of Colloid Science.*

let diameter, and ν the *kinematic* viscosity of the liquid in the drop, can be plotted logarithmically against the ratio ν/ν_0, ν_0 being the viscosity of the continuous phase. Under these circumstances, a straight-line relation is obtained for a number of systems. This is fitted by the empirical formula[24]

$$(Vd/100 \, \nu)^{5/4} = 100(\nu/\nu_0) \qquad (7.2)$$

Richardson[24] points out that if it is desired to produce an emulsion of a given mean drop size by such a process, Eq. 7.2 can be used to calculate the velocity of injection which should be used, if the viscosities of the two components are known. In most commercial injection processes the velocity is well above the critical, so that this formula should apply.

The experiments of Clay[25] may also be cited in this connection. In this work, emulsions were "refined" (i.e., the mean droplet size decreased) by pumping in turbulent motion through a glass pipe of four-inch diameter, and in the Reynolds number region 5×10^4 to 5×10^5. On changing the pumping rate, and hence the Reynolds number, a change in droplet-size

FIGURE 7-2. Breakup in a liquid jet at velocities above the critical.[22] *Courtesy: Prof. E. G. Richardson and Journal of Colloid Science.*

distribution was found to occur. This change in distribution is proportional to the thickness of the boundary layer along the glass and to the number of drops in it. Increasing the Reynolds number promoted further breakup of drops; while decreasing it favored coalescence.

Shevyakhova and Smirnov[26] have studied the movement of drops of variable mass in a liquid medium. While this work applies strictly to systems in which the droplet is *dissolving*, the mathematical techniques will undoubtedly be found to have application in emulsion theory.

The subject of mixing and emulsification has been reviewed by Belani,[27] Brancker[27a], and Gerö.[27b] A recent review of mixing (confined principally to propellor mixers) by Rushton[28] is also of considerable value.

Emulsification Equipment

As has been pointed out earlier (p. 86), the formation of an emulsion requires the expenditure of a certain amount of energy, this being partly the work required to form the interface between the two phases. For ex-

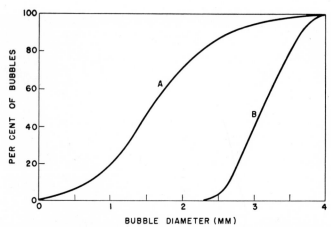

FIGURE 7-3. Summation curves of drop distributions of Figures 7-1 and 7-2. The lefthand curve is obtained with supercritical values, the righthand curve with subcritical values of the jet velocity.[22]

ample, it was calculated that, in the presence of a suitable emulsifying agent, the emulsification of one hundred pounds of olive oil requires the introduction of 0.75 kg-cal into the system. If the operation is to take a reasonable time, say five minutes, 0.14 horsepower will be required. This is certainly a reasonable power requirement.

Unfortunately, this is not the only factor involved. The creation of the interface will require that mechanical work also be done, for example, against the internal friction (i.e., viscosity) of the system. Thus, practical emulsion processing equipment with a power rating of several horsepower may be needed.

Types of Emulsification Equipment. For practical emulsion manufacture there are three types of emulsification equipment which can be considered:

1. Simple mixers
2. Homogenizers
3. Colloid mills

At this time it is possible to add a fourth:*

4. Ultrasonic devices

The simple stirrers include both propellor and turbine stirrers, as well as paddle devices and whisks.

Unfortunately, there is a certain semantic confusion with regard to these emulsification devices, particularly as between homogenizers and colloid

---

* This is truer in England and on the Continent than in this country, but considerable progress is being made in this regard.

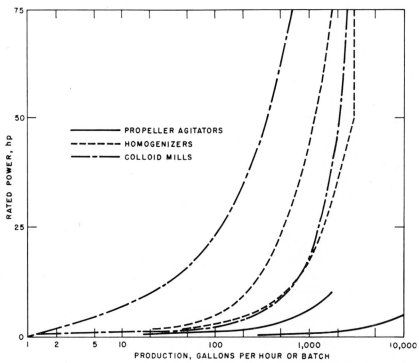

FIGURE 7-4. The relation between rated horsepower and production rate for different types of emulsators.[29]

mills, the latter often being incorrectly referred to as homogenizers. In addition, the term "emulsifier" has often been applied to these machines, resulting, on occasion, in confusion as to whether the mechanical device or the surface-active agent is intended. Since emulsifier is most often used in this way, the writer has suggested (p. 3) that "emulsifier" be reserved for the emulsifying agent, while the word "emulsator" be used as a generic term for all emulsification machinery.

Energy Requirements for Emulsators. The data of Griffin[8] has already been presented, showing how the particle-size of an emulsion is affected by the type of emulsator employed. Figure 7-4, adapted from Griffin,[29] shows the relation between rated horsepower and production rate, for the different types of emulsators. According to Griffin, power requirements for the different types of equipment increases in the order:

1. Propellor stirrers
2. Turbine stirrers
3. Homogenizers
4. Colloid mills

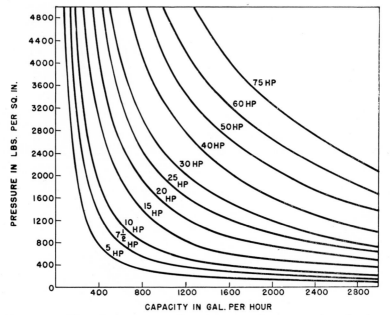

FIGURE 7-5. The relation between horsepower requirements and production rate as a function of homogenizer pressure. *Courtesy: Manton-Gaulin Manufacturing Co.*

Griffin[30] has also conveniently tabulated the power requirements, as based on manufacturer's data, for various types of emulsators as a function of production rate and other operating conditions. These range from 1/100 hp requirement for a thin liquid, propellor-stirred, with rate of 0.5 to 2 gph, to the 75 hp required of a colloid mill producing at the rate 2500 to 3000 gph.

In a homogenizer, the pressure under which the liquid is forced through the orifice has an effect on the eventual form of the emulsion (cf., p. 215), hence this may be a controlling factor. Figure 7-5, based on data from one manufacturer,[31] shows how the horsepower requirements vary with production rate and homogenizer pressure.

Similar considerations apply in the case of colloid mills, except that the controlling factor is the setting of the gap between the stator and rotor. Decreasing the clearance at a constant velocity decreases the horsepower requirement and production capacity of the mill. Increasing the viscosity while maintaining the same gap has the same effect. Thus, the overall efficiency of a homogenizer may well be greater than that of a colloid mill of the same horsepower rating.

Simple Mixers. Simple mixers may be of various types, ranging from high-powered propeller shaft stirrers immersed in a small tank or drum to

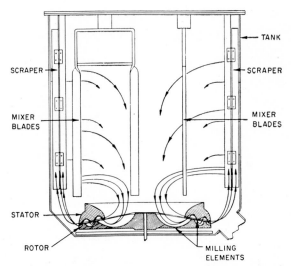

FIGURE 7-6. Simple mixer. The walls of the vessel may be jacketed, and steam or refrigerants circulated to maintain constant temperature. *Courtesy: Abbé Engineering Co.*

large self-contained units with propellor or paddle systems and jacketed tanks through which heating or cooling materials may be circulated (Figure 7-6). For many emulsions, e.g., floor waxes, such simple mixers will produce entirely satisfactory processing. For many systems, such mixers will only produce relatively coarse emulsions, and in such cases the simple mixer may be employed only as a pre-mixer before using a homogenizer or colloid mill.

A somewhat unusual type of mixing unit, depending on centrifugal action and continuous in operation, is illustrated in Figure 7-7.

Reference has been made to the work of Rushton[28] on propellor stirrers. Roerich[32] has offered recommendations on the use of high-speed stirrers, and Magnuson[33] has suggested general principles for evaluating their emulsifying performance.

Münzel[34] has studied the Eppenbach Homo-mixer (a turbine stirrer) rather extensively. Using raw milk as a test material, he was able to decrease the mean diameter of the fat particles from 3.05 to 1.18 μ, and to increase the percentage of particles in the range 0.5 to 1.2 μ diameter from 38.2 to 71.7 per cent.

Hailstone[35] has described the uses of the Watten high-speed mixer in the preparation of chloroform-water emulsions, and in the solubilizing of sodium alginate, "Promulsin," and gums. Kempson-Jones[36] has discussed the techniques of emulsification and emulsators with particular attention to the manufacture of cosmetics.

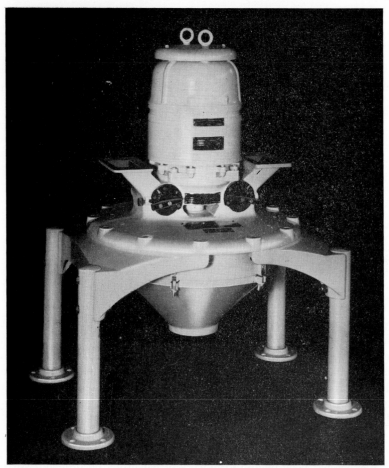

FIGURE 7-7. A less usual type of mixing unit, depending on centrifugal action, and capable of being operated continuously. *Courtesy: Troy Engine & Machine Co.*

Colloid Mills. In the standard colloid mill emulsification is carried out by means of a shearing action imparted to the liquid by a rotor, revolving at speeds of from 1000 to 20,000 rpm, and a stator. The emulsion passes between these two opposing faces through a clearance which may be as small as 0.001 inch. The faces of the rotor and stator may be completely smooth, or may be roughened by a series of concentric or radial corrugations.

Figure 7-8 illustrates a large-capacity horizontal operation colloid mill, as produced by Chemicolloid Laboratories, Inc. Figure 7-9 illustrates a slight modification of this type (Manton-Gaulin), with provision for recycling the liquid; Figure 7-10 shows the rotor-stator unit demounted.

FIGURE 7-8. A large-capacity horizontal operation colloid mill, shown disassembled. Note the corrugations in the rotor. *Courtesy: Chemicolloid Laboratories, Inc.*

A vertical unit, as produced by the Troy Engine and Machine Co., is illustrated in Figure 7-11; a cross-sectional view of the operation of the unit is given in Figure 7-12.

A number of different rotors, employed in the Tri-Homo colloid mill, are illustrated in Figure 7-13A, B, C, and D. The rotor in Figure 7-13D is made of porcelain.

A radically different type of rotor-stator combination is employed in Kinetic-Dispersion mills. Notches in the rotor and stator impart an angular impulse to the liquid under milling, and this, reportedly, gives finer dispersion. The principle underlying this particular mill, a production model of which is shown in Figure 7-14, is primarily intended for pigment dispersions, but it should prove effective in the production of emulsions.

Another type of colloid mill (the "Versator") dispenses with the stator. Instead, this machine consists of a vacuum, or pressure chamber, mounted on a base. Inside the chamber there is a 26-inch diameter disc, shaped like an open shallow bowl. This disc may be driven at speed between 900 and 2000 rpm. The material being emulsified enters the unit through a feed line at the center of the disc and is impounded by either a fixed or spring-

FIGURE 7-9. A slightly modified colloid mill, with provision for recycling the emulsion. *Courtesy: Manton-Gaulin Manufacturing Co.*

loaded film-forming ring, which can be preset to determine the initial thickness of the entering film.

As the material to be processed is fed onto the spinning disc, centrifugal force drives the mass away from the center of the disc, the film decreases in thickness (thus increasing the surface), and the entire mass takes on the characteristics of a two-dimensional surface. Turbulent flow results from the forces acting on the film of liquid, and this, combined with the shape, area, and speed of the disc, develops great shearing force, resulting in a reduction in size of the dispersed droplets. The liquid is collected at a discharge line at the periphery of the rotor. A cross-sectional view of this unit is shown in Figure 7-15.

Colloid mills are available in units varying in power requirements from 3 to 75 horsepower, and varying in capacity from 20 to 4500 gallons per

FIGURE 7-10. The colloid mill of Figure 7-9, with the rotor-stator unit demounted. *Courtesy: Manton-Gaulin Manufacturing Co.*

hour. Higher horsepower mills have been constructed for special applications.

Homogenizers. A homogenizer is a device in which dispersion is effected by forcing the mixture to be emulsified through a small orifice under very

FIGURE 7-11. A vertical-operation colloid mill. *Courtesy: Troy Engine & Machine Co.*

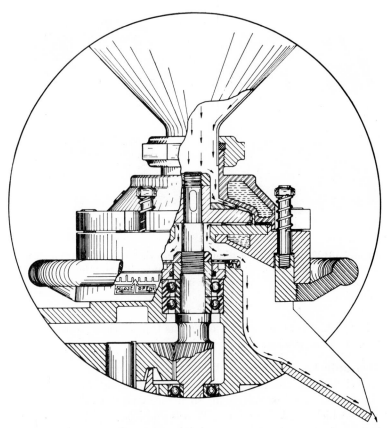

FIGURE 7-12. The vertical colloid mill of Figure 7-11 shown in cross-section. *Courtesy: Troy Engine & Machine Co.*

high pressure. A convenient hand-powered homogenizer, suitable for many laboratory preparations, is shown in Figure 7-16.

A production homogenizer will consist of a pump which provides the required pressure of between 1000 and 5000 psi, and a special spring-loaded valve which constitutes the orifice. The various commercially available homogenizers differ principally in the design of the valve.

Homogenizers are produced as either single- or double-stage units, the double-stage homogenizer being so constructed that the liquid passes through *two* orifice valves arranged in tandem. The pressure-drop through the second valve is, of course, somewhat lower than that through the first. Single-stage and double-stage valves for the well-known "Gaulin" homogenizers are illustrated in Figures 7-17 and 7-18. A production homogenizer is shown in Figure 7-19.

FIGURE 7-13. Various forms of rotors. (A) Smooth: for general purpose emulsions. (B) Serrated, medium: for ointments and lubricants. (C) Serrated, coarse: for very viscous products. (D) Vitrified stone: for ink, paint, and pigment dispersions. *Courtesy: Tri-Homo Corp.*

As is well known, a primary use for this type of emulsator is in the production of "homogenized" milk.

Ultrasonic Emulsification. While the use of mixing and shearing devices obviously represents a simple way of introducing the energy required for the formation of an emulsion, such mechanical methods are not the only ones which can be employed. Woods and Loomis,[37] in 1927, were able to produce emulsions by the use of ultrasonic vibrations, i.e., acoustic waves of such high frequency as to be inaudible. In this work, vibrations of a frequency of about 200,000 cps were employed (the audible limit, for persons of average hearing, is about 15,000 cycles).

Alexander[38] has pointed out that there are four general methods whereby acoustic waves of the requisite energy can be generated:

1. *Piezoelectric effects.* This method depends on the fact that certain crystals contract in an electric field. If an alternating current of the same frequency as the natural mode of vibration of the crystal is applied across the crystal's faces, extremely powerful oscillations can be produced.

2. *Electromagnetic effects.* In principle, this method is, that involved in the production of sound waves by means of the familiar moving-coil loudspeaker.

3. *Magnetostriction effects.* Certain ferromagnetic metals, particularly nickel, are found to change in length when put in a magnetic field. If an alternating field of the natural frequency of the metal rod is imposed, large

FIGURE 7-14. A production model employing the notched rotor-stator principle. The notches in the rotor impart an angular impulse to the liquid. *Courtesy: Kinetic Dispersion Corp.*

amplitude oscillations can be obtained. This has the advantage over the previous two methods in that the oscillating element itself carries no current, and can therefore be placed directly in water.

4. *Mechanical effects.* The principle involved here is similar to that of an organ pipe. Obviously, however, the problem of setting up acoustic waves in this way in a liquid is by no means simple. One solution has been by means of the so-called Pohlman whistle (cf. below, p. 236).

The energies obtained by such devices are sufficiently high so that Bull and Sollner[39] were able to obtain stable mercury-in-water emulsions without the use of any additional emulsifying agent. This effect was further inves-

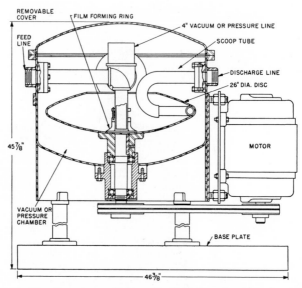

FIGURE 7-15. The Versator colloid mill. Centrifugal action is used to obtain breakup of the emulsion particles. *Courtesy: Cornell Machine Co.*

FIGURE 7-16. A convenient, hand-powered homogenizer, suitable for laboratory applications. *Courtesy: Central Scientific Co.*

tigated by Sollner and Bondy,[40] who concluded that the mechanism involved in the formation of such metal-in-water emulsions is different from that required for the more usual liquid-in-liquid ones.

Sollner[41] has pointed out the significant fact that oil-water emulsions are

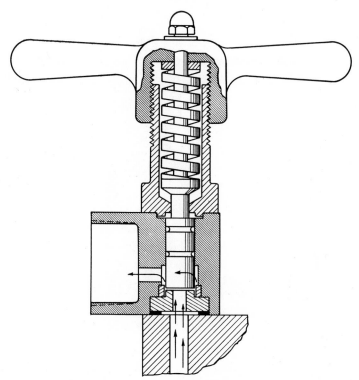

FIGURE 7-17. Valve assembly for a single-stage homogenizer. *Courtesy: Manton-Gaulin Manufacturing Co.*

not formed by acoustic waves when the operation is carried out *in vacuo* or under a sufficiently high external pressure. This is true even when the system contains an efficient emulsifying agent. Also, emulsions which are formed at fairly low frequencies (about 30 kc per sec) are broken by subsequent exposure to higher frequencies (about 400 kc per second) at a lower power level.

Sollner and Bondy[40] explain the phenomena involved in the formation of liquid-liquid emulsions as being due to cavitation. A sound wave travelling through a liquid compresses and stretches it. When the stretch is moderate and the liquid contains no gas nothing happens, but if the liquid is saturated with gas, bubbles appear. What actually seems to happen is that the liquid disrupts under the action of the vibrations, and this disruption forms actual cavities in the liquid.

Rayleigh calculated the pressure which can occur when a bubble collapses in a liquid; these are of the order of several thousand atmospheres. The forces thus engendered are, of course, capable of all sorts of mechanical effects, including the desired shearing.

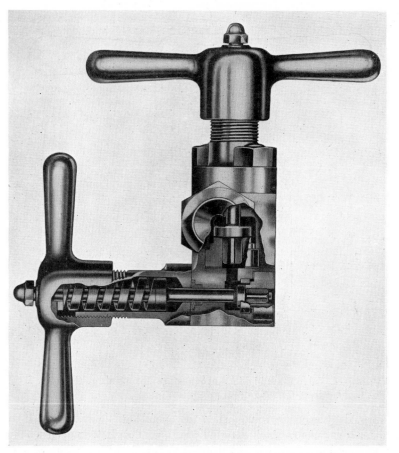

FIGURE 7-18. Valve assembly for a double-stage homogenizer. This type is commonly used in homogenizing milk. *Courtesy: Manton-Gaulin Manufacturing Co.*

On the other hand, the intense agitation which is brought on by these effects has the property of increasing the number of collisions between the dispersed droplets, and hence increasing the possibility of coalescence. In fact, Roth[42] has pointed out that the actual process represents a competition between these opposing forces, and it is necessary to choose operating conditions and frequencies in which the disruptive effect predominates. Obviously, however, other conditions can be found favoring coalescence, and ultrasonic devices can be employed this way.

Ultrasonic Emulsators. A diagramatic representation of the emulsification vessel of a piezoelectric and of a magneto-strictive emulsator are

FIGURE 7-19. A production homogenizer of the single-stage type. *Courtesy: Manton-Gaulin Manufacturing Co.*

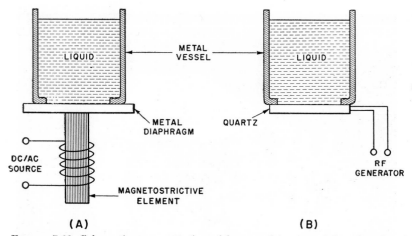

FIGURE 7-20. Schematic representation of homogenizing vessel for piezoelectric and magnetostrictive ultrasonic emulsators.[42a]

given in Figure 7-20.[42a] A commercially available piezoelectric laboratory device of this type is illustrated in Figure 7-21.

There are difficulties involved in commercial applications of these devices, although these are evidently not insurmountable. One problem is the difficulty of securing large enough piezocrystals, for example, to drive sufficiently large units. This may be largely surmounted by the use of barium titanite transducers (these are technically magnetostrictive). However, both magnetostrictive and piezoelectric transducers suffer the disadvantage that it is difficult to transmit the ultrasonic energy from them to liquid loads.

An ingenious solution to this problem has been the invention of the so-called Pohlman "liquid whistle," in which the principle of the organ-pipe has been applied to liquids.[42b] The basic construction of the unit is shown diagramatically in Figure 7-22. A jet of liquid is forced through the orifice and impinges on a blade, which is thus forced into vibration at its resonant frequency. As it vibrates, the stream of liquid is alternately forced up and down, and, if the frequency is sufficiently high (this depends solely on the dimension and physical characteristics of the blade), powerful oscillations are set up in the liquid. These are strongest near the blade, and it is in this region where emulsification takes place. An idea of the physical characteristics of the blade may be gleaned from the fact that in early experiments by Alexander[38] an ordinary razor blade was employed.

A commercial unit of this type is presently available in the United States, under the trade-name "Rapisonic." It is reported that the unit is capable

FIGURE 7-21. A laboratory ultrasonic emulsator. This device is driven by a piezo-crystal. *Courtesy: Brush Electric Co.*

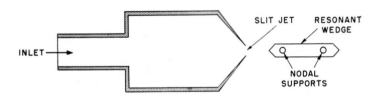

FIGURE 7-22. Schematic diagram of a Pohlman whistle. The nodal supports of the blade are separated by a distance equal to half the wave-length of the characteristic vibration of the system.

of production rates of 5 to 7 gpm at 2 horsepower. What is apparently a similar unit, developed in France, is available in this country, under the name "Multiwhistle RD Air-Jet Ultrasonic Generator."

A discussion of ultrasonic whistles has been published by Crawford,[43] while an extremely useful discussion of ultrasonic transducers in general is found in the paper of Mattiat.[44] For complete discussion of the theory underlying the production of ultrasonic energy, the reader is referred to the monographs of, among others, Richardson,[45] Crawford,[45a] and Hueter and Bolt.[46]

It should, perhaps, be pointed out that cavitation, as an effect, may be produced by other than ultrasonic generators, and that these techniques are frequently applicable to emulsions. Thus, Auerbach[47] has considered cavitation which occurs when a liquid moving through a pipe at high speeds undergoes a sudden change in cross-section, or in the neighborhood of a rapidly-rotating propellor. He found that quite stable O/W and W/O emulsions could be obtained by pumping coarse dispersions through a Venturi tube at high speeds.

The experiments of Richardson[22, 24] and Clay,[25] cited earlier (pp. 217–220), should be recognized as similar to this in many ways.

Polotskii[48] has reported on cavitation caused by superheated steam. Quite stable emulsions, even in the absence of emulsifying agent, can be prepared with this technique.

Applications of Ultrasonics. In addition to the early exploratory work reported by Sollner[41] some recent studies may be noted. Audouin and Levavasseur[49] have investigated the emulsification of vegetable oils by the use of ultrasonic waves. They found that cavitation, and therefore emulsification, takes place when the input energy is sufficiently high and the pressure sufficiently low; otherwise demulsification may occur. The frequency used determines the type of the emulsion, regardless of emulsifying agent. As indicated by earlier work, certain frequencies tend to break the emulsion. Progressive waves are more effective than stationary waves, and emulsification is favored by low temperatures and the presence of emulsifiers.

Polotskii's[48] steam-cavitation techniques have been employed in the emulsification of such systems at 10 per cent camphor oil, castor oil, peach-seed oil, or paraffin in water; 60 and 74.2 per cent benzene-in-water; 2 per cent lecithin, 5 per cent cresol in water, etc.

A recent, and most elaborate study, of the use of ultrasonic methods in preparing emulsions is due to Beal and Skauen.[50] In this study, the effect of various parameters were measured principally as they related to the droplet-size distribution obtained. An important factor considered was the effect of the ultrasound on the emulsifying agent itself. It was discovered that solutions of some agents suffered a marked decrease in viscosity after several minutes exposure. This presumably has nothing to do with any chemical breakdown, but rather with a breakdown of some sort of gel-structure in the solution. This decrease in viscosity may often be undesirable, however, and should be kept in mind.

The effect of the type of chamber in which the emulsification takes place, and of the orientation of the sample with respect to the transducer (in this case a piezocrystal) was also considered. As might be expected, it was found that most intense emulsification occurred at distances which were semi-multiples of the crystal's characteristic frequency.

In addition, the effect of exposure time and the type of emulsifier used were also considered. Interestingly enough, hard soap, USP, was found to be the most efficient emulsifying agent in the systems studied, with complete dispersion of the liquid occurring within three minutes. The data obtained with this agent is shown in Figure 7-23, adapted from Beal and Skauen. In this figure the root mean cube diameter (a measure of the average droplet diameter) is plotted as a function of the exposure time.

Grutzner[51] has discussed the use of ultrasonics in the preparation of photographic emulsions, while Tröger[52] has reviewed applications in the lacquer industry.

General discussions of applications of ultrasound include the recent ones of Arnold,[53] Speyer,[54] Sollner,[55] Alexander,[38] and Roth.[42] An extensive discussion is to be found in Crawford.[45]

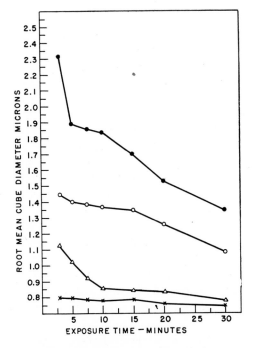

FIGURE 7-23. The effect of exposure time to ultrasound on root mean cube droplet diameter of emulsions stabilized with hard soap, USP.[50]

Emulsification Equipment in the Patent Literature

The older patent literature pertaining to emulsators has been extensively surveyed by Philip[56] (with a bias to German equipment), Clayton,[57] and Berkman and Egloff.[58] Slightly more recent data is to be found in a review by Appell.[59] Some more recent patents of interest are discussed below.

A number of devices for mixing and homogenization have been patented by Facks,[60] Plauson,[61] and others.[62-64] Schmidt[65] has described an apparatus employing an annular disc impeller. Hawes[66] has patented portable agitating and emulsifying apparatus suitable for attachment to tanks, and particularly applicable to the production of food products. Schöfer[67] describes an apparatus for the preparation of oil-in-water emulsions.

A mixing and emulsifying mill suitable for the manufacture of mayonnaise, salad dressings, or insecticidal emulsions has been described by McLean.[68] A mixing apparatus, on the colloid-mill principle, is revealed by Hoffman,[69] and a complete batch emulsifying plant, with a low-pressure valve is described in a patent issued to Alfa-Laval.[70] A later patent describes the adaptation of the plant to the production of mixtures of liquids and solids.[71]

Gerhold[72] has revealed a method for the emulsification of two liquids by carrying out the mixing operation under varying pressure, i.e., between one at which the material is completely liquid and one at which it is partially vaporized while in a turbulent state.

More conventional homogenizers are the subjects of patents by Sussex Bitumen, Ltd.[73] and Colony.[74]

Zarecki[75] has described an apparatus for the production of solid or semisolid emulsions of the oil-in-water type, especially edible ones, e.g., margarine. Another emulsator specifically for use in margarine production is revealed by Grasso.[76] In this latter unit there is a rapidly rotating stirring apparatus, and a slower-rotating one behind it. This serves to deaerate the mix.

Indrikson[77] describes a unit for the emulsification of bitumen, pitch, tar, etc. This is a mixer in which lengths of chain are attached to a rotating shaft.

Nooteboom[78] has devised an emulsator for emulsifying and thickening the cream from reconstituted milk, and for mixtures of liquids of widely differing densities. In this device, the two liquids are mixed and then compressed by means of a reciprocating pump.

Denier[79] produces emulsions by leading the oil phase into the aqueous one by injection through hollow needles which are subjected to a potential of 30,000 to 200,000 volts.

A patent issued to Hochholzer[80] describes an apparatus which produces emulsification by the simultaneous action of impact and tearing forces. Two patents issued to Müller[81] describe an apparatus the stirring unit of which is a horizontal plate provided with conical holes. Vertical vibration of the plate causes thorough mixing of the liquids.

An arrangement of propellor blades on a mixing unit, which is particularly applicable to the production of emulsions, have been patented by Gallenkamp[82] and Lamoreaux.[83]

In situ formation of emulsifying agent is used in a patent issued to McMillan and Sullivan.[84] In this method, the solution containing the acid is heated and then injected countercurrently into a flowing stream of the other solution, held at a lower temperature.

A few patents have been issued in recent years covering the use of ultrasonics. Japanese patents have been issued to Wada[85] and Inada.[86] The use of a magnetostriction vibrator of tubular construction is revealed by Schöfer.[87] Magnetostriction transducers for this purpose have also been treated by Hertz and Wiesner.[88] Kolthoff and Carr[89] have patented the use of ultrasonic vibrations to cause the emulsification required for the emulsion polymerization of e.g., styrene-butadiene copolymer. A jet disperser, involving cavitation as the emulsifying principle, is described by Hjort and Jansa.[90]

Bibliography

1. Cobb, R. M. K., in "Emulsion Technology," pp. 7–32, Brooklyn, Chemical Publishing Co., 1946.
2. Becher, P., "Principles of Emulsion Technology," p. 110, New York, Reinhold Publishing Corp., 1955.
3. de Navarre, M. G., "Chemistry and Manufacture of Cosmetics," p. 191, New York, D. Van Nostrand Co., 1941.
4. Dorey, R., in "Emulsion Technology," pp. 119–126, Brooklyn, Chemical Publishing Co., 1946.
5. de Navarre, M. G., *op. cit.*, p. 192.
6. Husa, W J., and Becker, C. H., *J. Am. Pharm. Assoc.* **30**, 83 (1941).
7. Stanko, G. L., Fiedler, W. C., and Sperandio, G. J., *J. Soc. Cosmetic Chemists* **5**, 39 (1954).
8. Griffin. W. C., in Kirk-Othmer "Encyclopedia of Chemical Technology," **5**, p. 709, New York, Interscience Encyclopedia, Inc., 1950.
9. Husa, W. J. and Becker, C. H., *J. Am. Pharm. Assoc.* **30**, 141 (1941).
10. Dvoretskaya, R. M., *Kolloid. Zhur.* **11**, 311 (1949); *C.A.* **44**, 903b.
11. Dvoretskaya, R. M., *Kolloid. Zhur.* **13**, 432 (1951); *C.A.* **46**, 2374a.
12. Ben'kovskii, V. G. and Zavorokhin, N. D., *Kolloid. Zhur.* **14**, 15 (1952); *C.A.* **46**, 4773i.
13. Jürgen-Lohmann, L., *Kolloid-Z.* **124**, 41 (1951).
14. Kremnev, L. Ya. and Soskin, S. A., *Kolloid. Zhur.* **10**, 209 (1948); *C.A.* **43**, 7775h.
15. Kremnev, L. Ya. and Kuprik, V. S., *Kolloid. Zhur.* **14**, 98 (1952); *C.A.* **46**, 5931b.

16. Berkman, S. and Egloff, G., "Emulsions and Foams," pp. 161–167, New York, Reinhold Publishing Corp., 1941.
17. Clayton, W., "Theory of Emulsions," 4th Ed., pp. 351–360, Philadelphia, Blakiston, 1943.
18. Taylor, G. I., *Proc. Roy. Soc. (London)* **A138,** 47 (1932).
19. Timotika, S., *Proc. Roy. Soc. (London)* **A150,** 322 (1935); **A153,** 302 (1936).
20. Oka, *Proc. Physico-Math. Soc. Japan* **18,** 524 (1936).
21. Kremnev, L. Ya. and Ravdel, A. A., *Kolloid. Zhur.* **16,** 17 (1954); *C.A.* **48,** 6780c.
22. Richardson, E. G., *J. Colloid Sci.* **5,** 404 (1950).
23. Ohnesorge, W. V., *Z. angew. Math. u. Mech.* **16,** 355 (1936).
24. Richardson, E. G., in Hermans, J. J. (ed.), "Flow Properties of Disperse Systems," pp. 42–46, New York, Interscience Publishers, 1953.
25. Clay, P. H., *Proc. Acad. Sci. Amsterdam* **43,** 852, 979 (1940).
26. Shevyakhova, I. P. and Smirnov, N. I., *J. Appl. Chem. U.S.S.R.* **29,** 207 (1956).
27. Belani, E., *Fette u. Seifen* **48,** 230 (1941).
27a. Brancker, A., *Petroleum (London)* **8,** 88 (1945).
27b. Gerö, F., *Mitt. chem. Forsch. Inst. Ind. österr.* **5,** 50 (1951).
28. Rushton, J. H., *J. Am. Oil Chemists' Soc.* **33,** 598 (1956).
29. Griffin, W. C., *loc. cit.*, p. 706.
30. Griffin, W. C., *loc. ci.*, p. 708.
31. Manton-Gaulin Mfg. Co., Everett, Mass.
32. Roerich, H., *Pharm. Ind.* **15,** 27 (1953).
33. Magnusson, K., *Chem. Process Eng.* **35,** 276 (1954).
34. Münzel, K., *Schweiz. Apoth.-Ztg.* **90,** 317 (1952).
35. Hailstone, W. N., *Pharm. J.* **165,** 268 (1950).
36. Kempson-Jones, G., *Mfg. Chemist* **27,** 310 (1956).
37. Wood, R. W. and Loomis, A. L., *Phil. Mag.* (7) **4,** 417 (1927).
38. Alexander, P., *Paint Manuf.* **21,** 157, 175 (1951); *Mfg. Chemist* **22,** 5, 12 (1951).
39. Sollner, K. and Bull, H. B., *Kolloid-Z.* **60,** 263 (1932).
40. Sollner, K. and Bondy, C., *Trans. Faraday Soc.* **32,** 1119 (1936).
41. Sollner, K., in Alexander, J. (ed.), "Colloid Chemistry," **VII,** pp. 337–373, New York, Reinhold Publishing Corp., 1944.
42. Roth, W., *J. Soc. Cosmetic Chemists* **7,** 565 (1956).
42a. Yeager, E. and Hovorka, F., in Kirk-Othmer "Encyclopedia of Chemical Technology," **14,** p. 409, New York, Interscience Encyclopedia, Inc. 1955.
42b. Janovsky, W. and Pohlman, R. *Z. angew. Phys.* **1,** 222 (1948).
43. Crawford, A. E., *Research (London)* **6,** 106 (1953).
44. Mattiat, O., *J. Acoust. Soc. Am.* **25,** 291 (1953).
45. Richardson, E. G., "Ultrasonic Physics," *passim*, New York, Elsevier Publishing Co., 1952.
45a. Crawford, A. E., "Ultrasonic Engineering," *passim*, London, Butterworth Scientific Publications, 1955.
46. Hueter, T. F. and Bolt, R. H., "Sonics," *passim*, New York, John Wiley & Sons, Inc., 1955.
47. Auerbach, R., *Chem. Tech.* **15,** 107 (1942).
48. Polotskii, I. G., *Farmatsiya* **9,** No. 1, 18 (1946).
49. Audouin, A. and Levavasseur, G., *Oleagineaux* **4,** 95 (1949).
50. Beal, H. M. and Skauen, D. M., *J. Am. Pharm. Assoc.* **44,** 487, 490 (1955).
51. Grutzner, E., *Gelatine, Leim, Klebstoffe* **8,** 3 (1940).

52. Tröger, G., *Fette u. Seifen* **52**, 115 (1950).
53. Arnold, M. H. M., *Chem. Process Eng.* **35**, 15 (1954).
54. Speyer, K., *Seifen-Öle-Fette-Wachse* **78**, 153 (1952).
55. Sollner, K., *Chem. Eng. Progress Symposium Series*, "Ultrasonics," **47**, No. 1 30 (1951).
56. Philip, C., "Technisch Verwendbare Emulsionen," **I**, pp. 7–78; **II**, pp. 7–82, Berlin, Allgemeiner Industrie-Verlag, 1938 (Ann Arbor, Edwards Bros., 1944).
57. Clayton, W., *op. cit.*, pp. 371–402.
58. Berkman, S. and Egloff, G., *op. cit.*, pp. 199–211.
59. Appell, F., *Chimie & Industrie* **57**, 241, 341 (1946).
60. Facks, W., Brit. Pat. 513, 197, Oct. 5, 1939.
61. Plauson, H., Ger. Pat. 709,704, July 17, 1941 (Cl. 50c. 18.01).
62. Coöperative condensfabriek "Friesland," Fr. Pat. 850,901, Dec. 29, 1939.
63. Hannotte, J. (to Eberhard Hoesch & Söhne), Ger. Pat. 681,943, Sept. 14, 1939 (Cl. 12e. 4.01).
64. British Emulsifiers Ltd., Brit. Pat. 524,728, Aug. 13, 1940; British Emulsifiers, Ltd. and Curzon, T., Brit. Pat. 530,191, Dec. 6, 1940.
65. Schmidt, E. (to Lawrence Pump & Engine Co.), U. S. Pat. 2,267,341, Dec. 23, 1941.
66. Hawes, D. M. A. G. (to Joe Lowe Food Products Co., Ltd.), U. S. Pat. 2,247,439, July 1, 1941.
67. Schöfer, R. (to Siemens-Schuckertwerke A.-G.), Ger. Pat. 727,155, Sept. 24, 1942 (Cl. 12e. 4.50).
68. McLean, W. A. (to Geneva Processes, Inc.) U. S. Pat. 2,313,760, March 16, 1943.
69. Hoffman, M. P. (to The C. O. Bartlett & Snow Co.), U. S. Pat. 2,321,599, June 15, 1943.
70. Alfa-Laval Co., Ltd. and Hollister, A. H., Brit. Pat. 553,644, May 31, 1943.
71. Alfa-Laval Co., Ltd. and Hollister, A. H., Brit. Pat. 554,573, July 9, 1943.
72. Gerhold, C. G. (to Universal Oil Products Co.), U. S. Pat. 2,382,871, Aug. 14, 1945.
73. Sussex Bitumen & Taroleum Ltd., Hatt, E. R., and Norton, C. W., Brit. Pat. 568,742, April 18, 1945.
74. Colony, M. W. (to Donald G. Colony and John K. Colony), U. S. Pat. 2,389,486, Nov. 20, 1945.
75. Zarecki, Y. and Palestine Oil Industry "Shemen" Ltd., Brit. Pat. 569,731, June 6, 1945.
76. Grasso, H. A. M., Dutch Pat. 58,497, Nov. 15, 1946.
77. Indrikson, G. P., U.S.S.R. Pat. 66,882, Aug. 31, 1946.
78. Nooteboom, F., Belg. Pat. 476,167, Inv. Sept. 19, 1947.
79. Denier, A., Fr. Pat. 941,365, Jan. 10, 1949.
80. Hochholzer, H., Austr. Pat. 168,282, May 10, 1951 (Cl. 12c4).
81. Müller, H., Swiss. Pat. 278,280, Jan. 3, 1952 (Cl. 36e); 286,342, Oct. 15, 1952 (Cl. 36e); cf. also U. S. Pat. 2,681,798, June 22, 1954.
82. Gallenkamp, N. B., U. S. Pat. 2,585,925, Feb. 19, 1952.
83. Lamoreaux, F., Can. Pat. 495,498, Aug. 25, 1953.
84. McMillan, F. M. and Sullivan, R. D. (to Shell Development Co.), U. S. Pat. 2,684,949, July 27, 1954.
85. Wada, S. (to Tokyo Shibaura Electric Co.), Jap. Pat. 2364 ('50), Aug. 16, 1950.
86. Inada, T., Jap. Pat. 3358 ('50), Oct. 10, 1950.

87. Schöfer, R. (to Siemens-Schuckertwerke A.-G.), Ger. Pat. 718,744, Feb. 26, 1942 (Cl. 12e 4.50).
88. Hertz, G. and Wiesner, R. (to Siemens & Halske A.-G.), Ger. Pat. 712,216, Sept. 18, 1941; 716,231, Dec. 18, 1941 (Cl. 12e 4.50).
89. Kolthoff, I. M. and Carr, C. W. (to Phillips Petroleum Co.), U. S. Pat. 2,606,174, Aug. 5, 1952.
90. Hjort, C. I. F. and Jansa, O. V. E., U. S. Pat. 2,460,884, Feb. 8, 1949.

CHAPTER 8

Emulsion Applications

A complete discussion of the formulation of emulsions would be a most formidable effort. It would, of course, be possible to give a long list of tried-and-true formulations in various fields of application, but such a listing would be inordinately long without in any way exhausting the possibilities. Rather than taking such a jejune course, the present chapter concerns itself with a detailed discussion of various topics in the field of emulsion manufacture. The methods and general principles thus illustrated should serve as guides in specific problems.

Cosmetic Emulsions

Historically, cosmetic emulsions represent the oldest members of the class. Emollient and cosmetic creams have been known for literally thousands of years; indeed, the invention of cold cream is ascribed to Galen in the second century.

In the not too distant past, the formulation of cosmetic emulsions was an art, the exercise of which was limited to an initiated few. Today, of course, broad generalizations can be made which will guide the uninitiated; an attempt is made to do this in the present section.

Basic Cosmetic Emulsions. The advantages of the emulsion form in the field of cosmetics have been stated by Leslie.[1] These include:

1. Economy and ease of application.
2. Increase in rate and extent of penetration into the skin.
3. Possibility of applying both water- and oil-soluble ingredients simultaneously (e.g., deodorants).
4. Efficient cleansing action.

Leslie supports the second item by the statement that the hydrophilic nature of the human skin permits greater penetration by an O/W emulsion than by a straight oil or fat. The validity of this is denied by some authorities,[2] but the other three advantages are real, their importance varying, to be sure, from case to case.

Hollenberg[3] classifies cosmetic emulsions as being either "nonvehicle" or "vehicle" emulsions. A *vehicle* emulsion is one whose principle purpose is to supply a medium whereby some agent (e.g., astringents, deodorants,

TABLE 8-1. NON-VEHICLE EMULSIONS[3]

	Water	Alkali (e.g., Borax)	Oil (Mineral or Vegetable)	Beeswax	Stearic Acid	Cetyl Alcohol	Lanolin	Absorption Base	Gum	Glyceryl or Glycol Stearate	Humectant	Petrolatum
	\multicolumn{12}{} Per cent of ingredients											

	Water	Alkali (e.g., Borax)	Oil (Mineral or Vegetable)	Beeswax	Stearic Acid	Cetyl Alcohol	Lanolin	Absorption Base	Gum	Glyceryl or Glycol Stearate	Humectant	Petrolatum
Cold Cream												
A	29	1	55	15	—	—	—	—	—	—	—	—
B	35	1	50	10	—	1	—	1	—	—	—	2
C	55	—	20	—	—	—	—	2	—	15	3	5
Vanishing Cream												
A	65	2	—	—	25	—	—	—	—	—	8	—
B	67	2	—	—	20	1	1	—	—	3	5	1
C	55	1	5	—	—	2	—	10	—	25	2	—
Night Cream												
A	40	—	25	5	—	3	10	10	—	—	—	7
B	40	—	27	5	—	3	—	25	—	—	—	—
C	55	1	5	—	—	2	—	10	—	25	2	—
Cream Lotion												
A	88	1	1	—	3	1	—	—	1	—	5	—
B	90	1	—	—	2	—	1	—	1	—	5	—
C	94	—	—	—	1	—	—	1	—	1.5	2.5	—

etc.) may be brought into intimate contact with the skin. Table 8-1, after Hollenberg, lists some schematic formulae for a number of non-vehicle emulsions, i.e., cold creams, vanishing creams, night creams, and cream lotions. Table 8-2 lists a number of formulae for vehicle emulsions of the astringent and deodorant cream types. It should be pointed out that, in some respects, vehicle emulsions have much in common with pharmaceutical emulsions (cf. pp. 248–249).

Cold Creams. According to Leslie,[1] cold creams are cosmetic emulsions of high oil content. These originally were W/O emulsions, but are now being replaced by O/W emulsions. Harry[4] states that the majority of cold creams being produced at present are of the latter type. The cooling sensation characteristic of cold creams occurs through the fact that the emulsion breaks on being rubbed into the skin, liberating the water, whose evaporation has the desired effect.

TABLE 8-2. VEHICLE EMULSIONS[3]

Ingredient	Percentage		
Emulsifier	18*	5†	11††
Stearic Acid	—	15	—
Cetyl Alcohol	1.5	3	6
Petrolatum	1	3	—
Glycol Stearate	1	—	8
Mineral Oil	2.5	—	—
Carbamide	5	—	—
Humectant	3	3	—
Water	53	52	59
Aluminum Salt	15	18	15
Titania	1	1	1

* Fatty alcohol sulfate
† Fatty amide
†† Nonionic

Leslie[1] presents two basic formulae for cold creams, one of each type:

W/O Cold Cream

Ingredient	Per Cent
Oils	50–70
Borax/waxes	0–20
Lanolin	0–15
Water	30–35
Preservative, perfume	q.s.

O/W Cold Cream

Ingredient	Per Cent
Oils	40–50
Borax/waxes	0–15
Soaps/esters	0–5
Water	33–45
Preservative, perfume	q.s.

Hollenberg's Cream "C" is probably a W/O cream, but is interesting because no beeswax and no borax (or other alkali) are used. Instead, a nonionic emulsifying agent and a small amount of absorption base§ are present.

§ An "absorption base" is a mixture of an oil and a surface-active agent which will readily absorb several times its own weight of water. De Navarre[7] describes the usual bases as consisting of mixtures of lanolin or lanolin isolates with hydrocarbon oil. Others are based on the use of nonionic emulsifiers. Absorption bases are characterized by a "water number" equal to $100B/(A - B)$, where B is the maximum weight of water absorbed by a weight A of base. For example, petrolatum containing 4 per cent cetyl alcohol has a water number in the range 39 to 51. Petrolatum plus lanolin has a water number of 79, independent of the lanolin concentration in the range 5 to 15 per cent.

Classical cold cream, however, is based on the use of beeswax. The addition of an alkali, notably borax, has also become standard, because it increases the stability. This presumably arises through the neutralization of the free fatty acid to form soaps. Indeed, it is recommended that the amount of borax addition be based on the amount of free fatty acid present.[5] Aside from the emulsifier supplied by the borax neutralization, the beeswax itself plays an important part in stabilizing the emulsion. Pickthall[6] has demonstrated that the hydroxy-compounds in beeswax are significant in this regard. Cold creams prepared with acetylated beeswax proved unstable; stability was improved somewhat by the addition of cetyl alcohol, and improved more by the addition of a cetyl alcohol-ethylene oxide condensate.

The work of Salisbury, Leuallen, and Chavkin[8] on the effect of phase-volume on emulsion type in beeswax-borax cream systems has already been referred to (cf. p. 147). Mullins and Becker[9] have studied the effect of the adjustment of the specific gravity of the internal phase to equal that of the continuous phase in similar systems. It was found that such adjustment decreased the specific interfacial area, but that, in spite of this, the stability was increased in some instances.

Vanishing Creams. A vanishing cream is an O/W emulsion of much lower oil content than cold creams. In addition, vanishing creams contain a small quantity of a strongly hydrophilic compound (e.g., glycerine) to increase penetration. The vanishing creams cited by Hollenberg[3] contain no waxes; however, Leslie[1] has indicated that a vanishing cream may contain up to 10 per cent of oil and wax combined, as shown in the typical vanishing cream formula.

Vanishing Cream

Ingredient	Per Cent
Oil/waxes	0–10
Stearic acid/esters	10–25
Alkali or Amine	0–2
Glycerine	0–5
Water	60–80
Preservatives, etc.	q.s.

Even more dilute O/W emulsions are employed as hand lotions, facial milk, etc. Because of their low viscosity, emulsions of this type may require efficient homogenization for maximum stability.

Vehicle and Pharmaceutical Emulsions. Deodorant creams are the major types which may be classified as vehicle emulsions. Table 8-2 gave some representative formulae for this type. Pharmaceutical emulsions may also be considered to be in this class, the emulsion serving as a vehicle for transport of the medical agent to the skin. General formulae for O/W bases have been reported; the medicament can be readily incorporated into these bases. For example, a simple base emulsion of this type is prepared as follows:[10]

Basic Pharmaceutical Emulsion

Ingredient	*Per Cent*
Polyethylene Glycol (200) Stearate	15.0
Magnesium Aluminum Silicate	5.0
Carbowax (1000) Monoöleate	1.0
Methyl *p*-Hydroxy Benzoate	0.1
Water	78.9

To an ointment base such as this may be added small percentages of bacteriostats (e.g., hexachlorophene), fungicides (e.g., undecylenic acid salts), etc.

Harry[11] has discussed the formulation of pharmaceutical emulsions in detail. Ointment bases are discussed by Martin,[12] with special consideration of fat and paraffin type bases. Münzel and Amman[13] have made an extensive study of washable oil-in-water ointments, with respect to the effect of various emulsifying agents.

Recent studies have been made on the effect of formulation on the release of medication to the skin from ointment bases. Coran and Huyck[14] have studied the effect of water/oil ratio on the diffusion of sulfathioazole from cold-cream type bases, and have found that this material diffuses more rapidly from O/W emulsion bases than from the W/O type. Barker, deKay, and Christian[15] investigated the effect of different nonionic emulsifying agents on the release of mercuric oxide and iodine. Optimum release was obtained when the emulsifying agent was employed in a 1 per cent concentration, almost irrespective of type.

Wood and Rising[16] found that O/W emulsions were more effective for antiseptics (in agreement with Coran and Huyck), but found that no one emulsifier could be recommended for all antiseptics.

Halpern and Hartwell[17] have found that in antiseptic ointments of the W/O type, the presence of water is not necessary for the development of antiseptic action. However, the degree of action is influenced by the water concentration.

Literature on Cosmetic Emulsions. For detailed data on the formulation of cosmetic emulsions, much of the older literature is covered by de Navarre.[18] A recent, current survey of the field has been undertaken by Harry.[19]

General discussions of the broad applications of emulsions in cosmetic fields include the previously cited papers of Leslie[1] and Hollenberg.[3] Lewinson[20] and Pickthall[21] have also contributed general discussions relating theories of emulsification to practical applications. Hollenberg[22] has treated of emulsifier selection and methods of preparation, and a later paper by the same author discusses the evaluation and development of emulsifiers.[23]

The application of surface chemistry to cosmetic problems is treated by Davies;[24] criteria for stability in terms of the times of coalescence of model systems are included. Rothemann[25] also surveys the field of emulsifiers in cosmetics.

A number of formulations to illustrate the application of nonionic emulsifying agents in series of mineral-oil-based O/W emulsions have been presented by King.[26] Formulation is discussed from the practical point of view by Hollenberg[27] and Ruemele.[28]

The use of specific materials are treated in a number of instructive papers. Schweisheimer[29] has dealt with the use of beeswax, spermaceti, 1-hexadecanol, 1-octadecanol, carnauba wax, paraffins, ozocerite, and lanolin. The use of *cera emulsificans*, an oil-in-water emulsifying wax consisting of partially sulfated cetyl and stearyl alcohol, is the subject of two papers by Wells.[30] Hilfer[31] has published on the use of newer emulsifiers.

Currie and Francisco[32] have listed the properties of silicones which suggest their usefulness as protective ingredients in cosmetic creams, lotions, ointments, etc. Macias-Sarria[33] has discussed the application of sodium lauryl sulfate as the sole emulsifier in cold creams, while Federici[34] has treated of the applications of the alginates. Ruemele[35] has made a contribution to the formulation of self-emulsifiable oils, with a view to promoting highly stable O/W emulsions in alkaline media. Collier[36] discusses the composition, properties, and uses of self-emulsifying monostearate.

Specific formulation problems are also covered in a number of papers. Vanishing creams are discussed by Winter.[37] Janowitz discusses the formulation of water-free skin creams,[38] water-in-oil emulsion creams,[39] and creams based on gels and suspensions.[40] Peel[41] treats of the constitution, formulation, and production of water-in-oil face creams; later papers deal with cleansing creams.[42] Hilfer[42a] discusses the formulation of emollient and lubricating creams, and Keithler[42b] has presented illustrative examples of lotion formulations.

Peel[43] has also discussed the preparation of brilliantine hair creams of the W/O type. An interesting type of hair-cream emulsion has been developed which is unusual in that it is intended to yield an unstable emulsion. The mixture of vegetable oil and water emulsifies when the bottle is shaken, but is otherwise in separate layers.[44]

Sollazzo[45] has considered the problems connected with the preparation of emulsions of oil in petrolatum, and suggests a number of alginate and nonionic emulsifying agents which are useful in this connection. Avis,[46] in a discussion of the preparation of a lotion having an acid pH, has brought out in a very precise way the problems involved in the production of stable emulsions. A large number of formulae for cosmetic and pharmaceutical emulsions are given by Swallow.[47]

The effect of humectants on cosmetic emulsions has been the subject of a number of interesting papers. Cessna, Ohlmann, and Roehm[48] studied the effect of propylene glycol, glycerol, and sorbitol on rates of evaporation. While in aqueous solution the evaporation rate decreases until all but the equilibrium water has evaporated, in emulsions of the W/O type propylene glycol has little effect on evaporation rate, while glycerol and sorbitol increase the rate.

A much more elaborate study was made by Griffin, Behrens, and Cross.[49] These authors tabulate hygroscopicity and other physical data for a large number of humectants. The differences between equilibrium and dynamic hygroscopicity are discussed, and a chart for the rapid selection of humectants, based on equilibrium data, is included.

Rate of weight loss from a number of O/W cosmetic creams was studied at various humidities and different humectant concentrations. In a stearic acid-soap cream sorbitol was found to retard moisture loss to an increasing extent with increasing concentration, in disagreement with the observations of Cessna, Ohlmann, and Roehm.[48] However, in agreement with this earlier work, the addition of small amounts of glycerol and propylene glycol increased the evaporation rate. This effect was reversed for these materials at higher concentrations. In creams stabilized with nonionic agents, the differences between these agents is less pronounced. A method for measuring the extent of "crust" formation in dried-out creams is also described.

Formulation of creams containing humectants is discussed by Vasic.[50]

Preservatives for cosmetic emulsions have been recently discussed.[51] De Navarre and Bailey[52] have pointed out that certain preservatives may have an adverse effect on the stability of cosmetic emulsions (cf. p. 205). Wynne[53] has treated of the effect of perfume oils on emulsion stability. He found that essential oils did not have an effect on the stability of lotions. However, a potassium stearate-quinceseed emulsion was separated by geraniol and terpineol. Pickthall[54] has also treated of this subject recently.

Hilfer[55] has discussed the entire problem of shelf-life stability of cosmetic emulsions, and the use of accelerated tests to evaluate this property.

Bruening[56] has reported on methods for the analysis of cold creams and vanishing creams.

Polishes

Polishes are oil or wax emulsions in water. They have to be formulated with the end-use in mind, i.e., after application the emulsion must be broken so that the polishing ingredient can spread to a smooth, even film. Very often this imposes a set of contradictory requirements on the formulator. For example, once applied, the wax in a floor polish should not be readily emulsifiable, or water-spotting will occur. On the other hand, if some degree of emulsifiability is not retained, removing the wax coating at a later date will be attended with considerable difficulty.

Oil Base Polishes. Oil base polishes may be readily formulated with nonionic emulsifiers. For example, one manufacturer proposes the following as an oil furniture polish:[57]

Oil Base Furniture Polish

Ingredient	Per Cent
Mineral Oil	40.0
"Span 85"	3.5
"Tween 81"	3.5
Water	53.0

This particular formula is interesting in that the preparation involves a procedure which is, perhaps, more important in food emulsions (p. 261): initial preparation of the wrong type of emulsion and inversion to the desired type. The emulsifying agents are added to the oil phase, and the water then added. This leads to the formation of a W/O emulsion; continued addition of water finally produces inversion to the desired O/W type.

Similar results may be obtained with other nonionic materials, as shown in the formula[58] below:

Oil-Base Furniture Polish

Ingredient	Per Cent
Mineral Oil	40.0
"Ethofat 60/15"	2.5
"Ethofat 60/20"	2.5
Water	55.0

Both of these oil-base furniture polish formulae involve the use of mixed emulsifying agents. This involves the application of the HLB concept (or equivalent techniques) in formulation.

Silicone-based polishes have recently become important; a typical formula is[59]

Silicone Polish

Ingredient	Per Cent
Silicone (DC-200, 300–500 cps or SF-96)	2.00
Kerosene or Mineral Spirits	8.00
"Ethomeen S/12"	0.25
"Arquad 2C"	0.25
Water	89.50

Polishes which include wax in addition to the silicone oil have also been proposed.

Floor Polishes. In the case of floor polishes, the preferred emulsifying agent has been the amine soaps. Kroner[60] has given a useful and complete description of the manufacturing technique involved in such preparations. The formulation described in this work involves a product containing a wax emulsion together with an aqueous shellac solution, the shellac being solubilized by ammonia. The use of shellac as an extender for wax polishes is quite usual. A low shellac content gives a polish that is water-resistant, elastic, and easily repolishable, on the other hand, high shellac content gives a harder film, and is less slippery, but is also less water-resistant and not repolishable. Obviously, a good formulation must strike a happy medium.

Kroner formulates the wax emulsion portion of the floor polish as follows:

Floor Polish (Wax Emulsion Portion)

Ingredients	Parts
Carnauba Wax	100
Oleic Acid	20
Morpholine	13
Borax	4
Water	363

This is compounded with a shellac solution of the following composition:

Floor Polish (Shellac Solution Portion)

Ingredients	Parts
Bleached Shellac	25
Ammonia (28%)	3.6
Water	171.4

The solutions are made up separately; the final product containing 100 parts of the wax emulsion and 10 to 20 parts of the shellac solution.

It is instructive to consider, in some detail, the method of producing this emulsion. A steam-jacketed mixing vessel, fitted with a mixing apparatus whose arms reach to the walls of the vessel (to prevent deposition of the wax) is employed. A mixer such as that illustrated in Figure 7-6 is (p. 223) suitable for this purpose.

The wax is heated to 95° C, and the mixer started as soon as the wax is soft enough to mix. When the wax is nearly all melted, the oleic acid is added. After the melting is complete, the morpholine is added gradually, and no further addition made for about three minutes. It is important that the temperature of 95° C not be exceeded, since morpholine boils at about 128°. Following this, the 4 parts of borax, dissolved in 40 parts of *boiling* water is added slowly, in small portions. Agitation is continued until the mixture is completely homogeneous. During the addition of the borax solution the liquid wax is changed into a thick, syrupy, transparent mass. On further addition of increasingly large amounts of boiling water, the solution becomes thinner. After the total amount of water is added, the solution is slowly cooled to room temperature, the water lost by evaporation is replaced, and the mixture filtered. From the time of the melting of the wax until the end of the cooling, the mixing must be continuous.

The duration of this standard mixing technique is approximately one-half hour. The process may be speeded up somewhat, following the borax addition, by pouring the mix rapidly into the remaining quantity of boiling water with vigorous mixing. Another method of speeding up this process is to add, to the mixture of wax, oleic acid, and morpholine, approximately 0.5 per cent of an additional emulsifying agent (Kroner suggests *Emulphor O*) just prior to the borax addition. This, according to Kroner, has the effect of lowering the interfacial tension further, and the water addition can be accelerated.

The shellac solution may be prepared by heating two-thirds of the required water and the ammonia to about 55° C. To this warm solution, the shellac is slowly added, with mixing. At the same time the temperature is slowly raised to 90° C in order to bring about complete solution of the shellac, and the remainder of the water added. The solution is then cooled with further stirring, the evaporated water replaced, and the mixture filtered. The emulsion and the shellac solution are then mixed in the desired proportions.

In this type of polish, morpholine is the emulsifier of choice. However, it has the disadvantage of boiling at a low temperature, making careful control of processing temperature important, and it is relatively expensive. Other amines, such as the various ethanolamines, have the disadvantage of an excessively high boiling point.

Recently the compound 2-amino-2-methyl-1-propanol has been introduced as the emulsifier.[61] This boils somewhat higher than morpholine (165° C), which is not entirely disadvantageous, and has a molecular weight of about the same order of magnitude, which makes it as efficient on a weight basis. Indeed, according to Frump,[61] this alcohol-amine is even more efficient than morpholine on a weight basis.

However, it is also possible to formulate polishes of this type with nonionic emulsifiers. A simple formula is as follows:[62]

Dry-Bright Floor Polish (Nonionic Agent)

Ingredients	Per Cent
Carnauba Wax	10
"Tween 80"	3
Water	87

Leslie[1] has indicated that in addition to carnauba wax (which, in the past, has been most widely used), beeswax, candelilla, ceresin, paraffin, and montan waxes may be used. In addition to amine and alkali soaps, sulfated oils, glyceryl or glycol fatty esters, glycol ethers and sulfonates, and petroleum sulfonates may be conveniently employed as emulsifying agents.

Literature on Polishes. In addition to the paper by Frump,[61] Eaton and Hughes[63] have considered the use of amine emulsifiers in carnauba wax emulsions. These workers measured the effect of the type and amount of amine emulsifying agent on the particle size as measured by the turbidity of the emulsion. Among the amines studied were ethylamine, ethylmonoethanolamine, ethyl-diethanolamine, isopropyl-monoethanolamine, isopropyl-diethanolamine, diethyl-monoethanolamine, triethanolamine, and morpholine. The data indicate that, with respect to the ratio of amine to oleic acid, there is a minimum which must be exceeded and a maximum which cannot be exceeded. Between these extremes is an optimum ratio for finest particle size. The optimum ratios are different for different amines, varying between 0.7 to 1.5.

Trusler[64] discusses the entire question of emulsifiers for waxes, including statistical and economic aspects. Behrens and Griffin[65] treat of the application of the HLB principle (cf. pp. 189–196) to the formulation of wax emulsions.

Schoenholz and Kimball[66] have shown, by means of the electron microscope, that the size of the wax particles in a bright-drying polish emulsion is of the order of 0.05 to 0.1 μ, and find that the particle size decreases as the emulsifier/wax ratio is increased.

Welch[67] discusses the use of silicones in wax polishes in some detail, giving illustrative examples. Another possible substitute for waxes in such emulsions, i.e., polyethylene, is the subject of papers by Kselik[68] and Clark.[69] The latter author emphasizes the advantages, in terms of durability, high gloss, toughness, etc., to be gained by the use of this material.

Treffler[70] has carried out an extensive investigation of self-polishing wax emulsions, by studying the effect of variation in the wax and the emulsifying agent. In the case of the latter, he has varied the amine and the fatty acid employed.

The general subject of the formulation of wax emulsions has been considered by Figliolino,[71] Lesser,[72] and, more recently, by Wasserman.[73] Italian practice is treated by Camperio.[74]

Emulsion Paints

An emulsion paint may be considered as a relatively simple system consisting of a *vehicle* (the liquid phase) and *pigments*, which are the materials which impart the color and opacity to the painted surface. To this may be added a volatile *thinner*, which serves principally to make the application easier. Basically, it is only the vehicle portion (and, to some extent, the thinner) which is of interest in the present discussion.

For many years, the only types of paints available were the so-called oil and water paints, in which the names are descriptive of the vehicle employed.* The important distinction between oil and water paints is in the mode of drying which the vehicle imparts. Oil paints can be considered to dry in two stages: in the first, the volatile thinner evaporates; in the second, the oil (usually a bodied linseed oil) undergoes a complex oxidation-polymerization to form a hard glossy film. On the other hand, a water paint dries by simple evaporation of the aqueous vehicle.

Both types have certain disadvantages. Water paints are inexpensive and easily applied, but generally do not impart a glossy finish, and are easily marred. On the other hand, oil paints give hard, glossy coatings which are not easily marred; but they are expensive, and the strong odor which accompanies the drying process is frequently objectionable, and often irritating.

Formulation of Emulsion Paints. In recent years, it has been found possible to combine the best elements of both oil and water paints, while diminishing their disadvantages. This has been done by the development of emulsion paints. In their simplest form, these are water paints to which a certain amount of drying oil is added. A more sophisticated approach

---

* Water paints are of the greatest antiquity. The ancient Egyptians used a paint composed of sour milk and freshly burned lime.

involves the use of true emulsions of either O/W and W/O type, although the O/W type is generally preferred.

Werthan[75] has described the formulation of a vehicle which amounts to the production of a more-or-less standard oil paint vehicle in emulsion form. The formula is as follows:

Emulsion Paint (Vehicle)

Ingredients	Parts
Ester Gum	93
Bodied Linseed Oil	75
Casein	42
Oleic Acid	36
Ammonia (28%)	16
Mineral Spirits	93
Pine Oil	32
Liquid Drier	8
Preservative	1
Water	604

The principal emulsifier is ammonium oleate (formed *in situ*), with the casein imparting additional high stability. The high total emulsifier content of the above formulation indicates the importance attached to stability in this type of emulsion. Preservative (e.g., chlorinated phenyl phenol) is needed to inhibit bacterial decomposition which may be a problem in systems containing so much proteinaceous and oily matter.

The use of the ammonium soap as the main emulsifier is predicated on the gradual evaporation of the ammonia, during the drying process, leading to a breaking of the emulsion.When the emulsion breaks, the now tightly-knit oil film can dry by the usual oxidation process.

Another type of emulsion paint which has found some application is essentially an oil paint in which a quantity of an oil-soluble emulsifier has been dissolved. Thus, these paints may be "thinned" by stirring in water to form an O/W emulsion; or, if desired, the paint can be thinned with mineral spirits in the conventional way.

Werthan[75] has also reported on the use of materials other than casein in the basic vehicle formulation. Materials such as soya protein, sodium alginate, gelatin, bentonite, and methyl cellulose were studied. While substitution was possible, Werthan feels that casein produces the most satisfactory type of emulsion.

Other types of emulsifying agents may also be employed. Allen[76] describes a vehicle in which the principal emulsifying agent is a sodium lauryl sulfate, with sodium alginate present for additional stabilization. It may be of interest to see how such an emulsion vehicle is combined with pigments to produce a finished paint. A white paint can be formulated as listed below.[76]

White Emulsion Paint

Ingredients	Parts
Pure Titanium Dioxide	100
Lithopone	400
China Clay	180
Mica	100
Emulsion Vehicle	620

A more recent approach to the formulation of emulsion paints has involved the use of synthetic resin emulsions in the vehicle, the synthetic high polymer constituting, in whole or part, the "oil" phase. These resin emulsions are produced principally by emulsion polymerization techniques, and will be discussed in more detail in the treatment of emulsion polymers (cf. pp. 272–274).

The first such resin emulsions used in formulating emulsion paints were the so-called alkyd resins. According to Levesque,[77] an alkyd resin is a polymeric ester prepared by the condensation of a polyhydric alcohol and a polybasic acid. When the fatty acid used is one from a drying oil, a drying resin is produced. The chemistry and production of such resins is described by Levesque[77] and by Ferguson and Sellers.[78] The manufacture of alkyd resin-based emulsion paints is discussed in considerable detail by Cheetham and Pearce.[79]

More recently, emulsion polymer latices of various types have been used to formulate the vehicle. Such emulsions are used in formulation in much the same way as the earlier drying-oil-based types. However, it is possible to produce paints with even better properties by use of these materials; indeed, polymer latices can be "custom-made" to serve as vehicles for specific applications.

Literature on Emulsion Paints. In addition to some of the basic references already cited, the following literature is of interest. General papers on the properties and formulation of emulsion paints include the work of Elm and Werthan,[80] who consider in some detail the problems connected with the stability of such emulsions. The general problem of emulsion paints is treated by Iddings[81] and Payne.[82]

Musgrave[83] describes the applications of emulsion technology to the production of emulsion paints, while formulation problems are covered in papers by Paxon,[84] Burr and Matvey,[85] Steig,[86] and Chatfield.[87] McLean[88] covers formulation problems connected with the use of polymer latices, including poly(vinyl acetate), polystyrene, styrene copolymers, etc. A thorough discussion of problems connected with the formulation of interior wall paints of this type is given by Davis.[89]

The literature on materials for emulsion paints is extensive. A recent paper by Cogan and Clarke[90] is useful. Palmer and Cass[91] have discussed vinyl acetate copolymer vehicles; Agabeg[92] has treated styrene emulsions; Hand[93] polyacrylate emulsions; and Hurd[94] has given the most recent data on alkyds.

A useful paper by Ford[95] concerns itself with the use of colloid mills in the production of emulsion paints.

Nitrocellulose Lacquer Emulsions. Although not strictly to be characterized as emulsion paints, nitrocellulose lacquer emulsions may be briefly discussed here. Normally, nitrocellulose lacquers are formulated in volatile solvents, but in certain applications (notably the treatment of porous surfaces) the use of the lacquer in the emulsion form has advantages.

Emulsifying agents to be used in these preparations must be free of electrolytes and essentially neutral. In fact, since lacquers are themselves somewhat acidic, the emulsifying agent used should be stable under acid conditions.

Such emulsifying agents as a mixture of sodium lauryl sulfate and sulfonated castor oil are satisfactory with nitrocellulose lacquers, while potassium oleate is useful with ethyl cellulose, and sodium oleate with "Parlon." Small amounts of protective colloid, e.g., CMC, casein, etc., added to the aqueous phase may be useful in increasing stability.

Simpson[96] has published an interesting study of the viscosity of such emulsions. In the case of nitrocellulose emulsions stabilized with sodium poly(vinyl acetate-phthalate), Newtonian flow is found only at low phase volumes of the disperse phase. At higher values, pseudo-plastic behavior is observed. The "residual" viscosity (as obtained by extrapolation to zero residual shear) increases with stabilizer content, the phase volume of the disperse phase, and continuous phase viscosity.

A general discussion of the formulation and production of nitrocellulose emulsions has been presented by Campbell.[97]

Agricultural Sprays

The use of chemicals as insecticides for plants and animals is quite old; the application of these materials in the form of emulsion sprays has itself been in practise for nearly a century. Aside from the commercial importance of this application of emulsion technology, the operation presents certain formulation problems of broad interest.

Formulation of Agricultural Sprays. Agricultural sprays are usually sold in the form of so-called "emulsifiable concentrates," or "soluble oils."

Such concentrates consist of a solution of the toxicant in an organic solvent, to which is added an oil-soluble emulsifier. In use, water is added to the concentrate, and the desired emulsion (usually, but not always of the O/W type) is prepared by hand-mixing.

Obviously, in order to form an emulsion under these conditions, considerable care has to be made in the choice of emulsifier. In particular, since waters of all degrees of hardness may be used, a calcium-insensitive agent is required. This rules out many otherwise attractive anionic agents. Cationics, largely unaffected by hard water, may be used, but the majority of formulations have involved nonionic agents. Recently, mixtures of nonionic and anionic agents have found application.

In the older types of insecticidal sprays, the active material consisted of mineral oils or tar distillates, which operate mostly by suffocation of the undesirable insects. A typical spray oil of this type is formulated as follows:[98]

Self-Emulsifying Agricultural Spray Oil

Ingredient	Per Cent
Agricultural Spray Oil	97.5
"Ethofat 142/15"	2.0
Mahogany Soap	0.5

In recent years, toxicants of the type of DDT, Chlordane, Toxaphene, etc., have been used in these applications. An emulsifiable concentrate containing DDT and stabilized by cationic agents may be formulated as follows:[99]

DDT Emulsifiable Concentrate

Ingredient	Per Cent
DDT	25.0
"Arquad 2C"	1.0
"Ethofat 142/20"	0.5
Xylol	73.5

A highly concentrated Chlordane preparation may be produced, using nonionic agents exclusively.[99]

Chlordane Emulsifiable Concentrate

Ingredient	Per Cent
Chlordane (Technical)	85
"Ethofat 142/20"	15

Concentrates of these types disperse readily to form reasonably stable emulsions. In fact, the stability of such emulsions must be nicely balanced.

The emulsion should remain intact in the spray tank but should break almost instantly on contact with the foliage, etc., upon being sprayed. When this happens the toxic agent forms a film on the surface, and is not easily removed. If the emulsion does *not* break, much of the material drips off and serves no useful purpose. By the same token, the particle size of the spray emulsion is important, as emphasized by Smith and Goodhue.[100] If the emulsion droplets are too fine, breaking on contact with the foliage tends to be incomplete; while, if the droplets are large, creaming will take place, causing uneven distribution of the material on the leaves.

A complete discussion of the older literature with respect to the oil-type emulsions is given by Woodman.[101] Felber[102] has reported on studies of the films formed from sprayed agricultural emulsions. There is a fairly extensive literature on DDT emulsions in particular; the papers of Jones and Fluno,[103] Plante,[104] and Hackman[105] may be cited. The paper of Plante, in particular, has valuable data on formulations involving a number of different organic solvents.

An extensive study on emulsifying agents for use in military insecticide formulations is due to Sparr and Bowen.[106] In studying Lindane, DDT, and NBIN, these workers found that when using polyoxyethylated nonylphenols of different ethylene oxide mole ratios there appeared to be an optimum mole ratio for each of the three toxicants. Brown and Riley[107] have discussed the evaluation of emulsifying agents for these formulations, and a recent publication gives data on types, consumption, and markets of emulsifiers for pesticide formulations.[108]

The spreading properties of the oil phase are significant with regard to the efficacy of the applied agent. Clayton[109] discusses some of the literature in this regard.

Food Emulsions

A most important field of emulsion technology is that involving food preparations. Some of the oldest emulsions known to man, i.e., butter, salad dressings, etc., are included in this group, while the applications of emulsion theory to the baking process and to therapeutic feeding represent modern developments.

Mayonnaise and Salad Dressings. Probably the most instructive of all the food emulsions which might be considered is mayonnaise. Mayonnaise is essentially an oil-in-water emulsion in which the disperse (oil) phase represents from 60 to 80 per cent of the total. From the point of view of the phase-volume theory of emulsion stability (cf. pp. 90–93), this represents a considerable challenge. This is particularly true, since, in edible emulsions, one is sharply limited in the choice of emulsifying agent. In fact, the emulsifying agents which are used in mayonnaise are those which have been known for hundreds of years: egg yolk and mustard.

Corran[110] has given a remarkably complete discussion of the problems connected with the production of mayonnaise. According to Corran, a typical formula for a commercial mayonnaise is

Mayonnaise

Ingredient	Per Cent
Oil	75.0
Salt	1.5
Egg tolk	8.0
Mustard	1.0
Water	3.5
Vinegar (6% acetic acid)	11.0

In addition, other flavoring or coloring materials may be added. It will be noted that most of the aqueous phase is introduced with the vinegar.

Corran discusses in detail the various factors which influence the stability of the mayonnaise emulsion, i.e.,

1. Egg yolk.
2. Phase volumes.
3. Emulsifying effect of the mustard.
4. Method of mixing.
5. Hardness of water.
6. Viscosity.

Not all of these factors are, of course, equally significant. From many points of view, indeed, the first of these is most critical, since egg yolk is far from a satisfactory emulsifier in the system under consideration. This can be made clear if one examines the composition of egg yolk. Egg yolk has the approximate composition:

Egg Yolk

Components	Per Cent
Fats	22.5
Proteins	16
Lecithin	10
Cholesterol	1.5
Salts	2
Water	48

The surface-active components of the egg yolk are seen to be lecithin and cholesterol. Lecithin, the major surface-active component, is known to be a good O/W emulsifier; however, cholesterol is an efficient W/O emulsifying agent, and, as might be expected, this is a source of difficulty. Corran's investigation of the effect of the lecithin-cholesterol ratio on the type of emulsion produced has already been cited (cf. p. 146 and Table 5-1). From these data, it may be concluded that in approximately 50–50 oil-water emulsions, inversion to W/O emulsions will occur when the lecithin/cholesterol ratio falls lower than 8:1. (In naturally-occurring egg yolk, it is seen to be only 6.7:1).*

Indeed, with this consideration in mind, it is probable that the only reason that a stable mayonnaise emulsion is possible is due to the presence of the mustard. The mustard presumably operates as a finely-divided solid in stabilization of the mixture (cf. pp. 125–131).

Corran has demonstrated the effectiveness of this agent by a series of experiments in which emulsions of olive oil in lime-water were studied. Equal volumes of the two liquids were shaken together, with the result that the calcium ions of the lime-water reacted with the free fatty acid of the oil to form calcium soaps, which, as is well known, stabilize water-in-oil emulsions. The effect on the emulsion type of additions of mustard flour was observed, both fine and coarse mustard being investigated. Some of Corran's observations are summarized in Tables 8-3 and 8-4.

As can be seen, the presence of as little as 2 per cent of the fine mustard flour causes inversion to the O/W type; the coarse mustard is less efficient, as might well be expected.

As has been pointed out, the phase volume situation in mayonnaise is also not favorable to the formation of the desired emulsion type; this is further complicated by the usual processing technique. It is customary to dry-mix the solid components (i.e., egg yolk, mustard, seasoning) and then add only a small amount of water. Finally, the oil is added gradually, with vigorous mixing, to form a quite viscous "nucleus." The additional aqueous components are then added to dilute the nucleus to the desired composition. It will be recognized that the conditions under which the nucleus is prepared favor the formation of W/O emulsion, rather than the desired O/W, although the emulsifying agent will correct this to some extent.

The method used, however, has a perfectly rational explanation. Unless the processing is carried out in such a way as to give a product of high consistency rather early in the process, the mixing efficiency is quite low, and satisfactory homogenization is not obtained. The use of colloid mills obviates this problem to some extent, although in this case care must be exer-

---

* This observation may serve to explain the known fact that fresh egg yolk is more satisfactory than materials which have been stored for some time; the lecithin suffers some breakdown, and thus the lecithin/cholesterol ratio becomes even less favorable.

TABLE 8-3. EFFECT OF FINE MUSTARD ON LIME WATER-OLIVE OIL EMULSION[110]

Per Cent Mustard	Emulsion Type
0	W/O
0.17	W/O
0.3	Mainly W/O (unstable)
0.5	Mainly W/O (unstable)
0.7	Mainly W/O (unstable)
1.0	Multiple (unstable)
1.25	Multiple (unstable)
1.7	Mainly O/W (unstable)
2.1	O/W
3.3	O/W
4.2	O/W

TABLE 8-4. EFFECT OF COARSE MUSTARD ON LIME WATER-OLIVE OIL EMULSION[110]

Per Cent Mustard	Emulsion Type
0	W/O
0.4	W/O
0.8	W/O
1.25	W/O
1.7	W/O
2.5	Multiple (unstable)
3.3	Multiple (unstable)
5.9	Multiple (unstable)

cised to keep the particle size from becoming too *small*, for in that case the interfacial area may become larger than can be stabilized by the quantity of emulsifier present. Figure 8-1 shows a modern mayonnaise plant, which uses a colloid mill in the processing.

Corran has also investigated the effect of the time and temperature of mixing on the viscosity of the final product. When the acid component is not added in the initial stage of preparation, the viscosity of the product (as measured with a mobilometer) increases steadily with mixing time. On the other hand, when even a portion of the acetic acid (as vinegar) is added in the initial stage, the consistency of the final product varies in an indeterminate manner with mixing. Stability of the mayonnaise may be increased by the addition of gums, etc., while hard water may have an undesirable effect on stability.

The related emulsion, i.e., the so-called salad dressing, is relatively simple to prepare, in contradistinction to mayonnaise. This, too, is an O/W emulsion largely stabilized by egg yolk. The oil concentration is very much lower, however, usually around 45 per cent, and additional stabilizers in the form of gums are quite common. Thus, completely stable emulsions can be formed by almost any technique.

FIGURE 8-1. A modern mayonnaise plant. A colloid mill is used to homogenize the product. *Courtesy: Chemicolloid Laboratories, Inc.*

Baked Goods. The use of emulsifying agents in baked goods is widespread. Generally speaking, the emulsifying agent appears to serve two purposes. In the actual preparation of the dough, the presence of the agent serves to improve the intimate mixing of the components, and hence has a desirable effect on the quality of the final product. Secondly, in the final product, the presence of the emulsifying agent has the effect of retarding staling.

The exact mode of action in this latter regard is somewhat obscure, but it appears to take place as a result of the inhibition of the formation of hydrogen bonds between adjacent starch chains in the finished cake or bread.

Thus, for example, Jooste and Mackey[111] found that when glyceryl monostearate was added to a cake formula, the fat, together with the air bubbles, became more highly dispersed. Improved fat and air dispersal was brought about by the emulsifier concentrating at the fat-liquid and air-liquid interfaces. The stabilizing effect of the emulsifier on fat and air distribution was most effective when the proteins were coagulated early in the baking process. No difference was noted when the emulsifying agent was used at a 3 per cent or 6 per cent level, nor did it make any difference if the emulsifier were dispersed in the fat or in the water.

Similarly, Skovholt and Dowdle[111a] have studied bread containing 3 and 2 per cent of shortening plus 0.5 per cent each of polyoxyethylene stearate and vegetable oil monoglyceride. Both were found to retard staling effectively, as measured by compressibility. It was noted that breads firmed more quickly in temperatures such as prevail indoors during winter than during the warmer months of summer.

The use and application of emulsifiers in the baking industry is discussed in detail by Pyler.[112] Coppock[113] has discussed a large number of potential new compounds for use in the baking industry, including emulsifiers. This paper, written in 1950, indicates the effectiveness of polyoxyethylene stearates as baking emulsifiers, but emphasizes the need for information on their physiological action.

In view of some doubts as to their physiological effects, the Food & Drug Authority banned the use of polyoxyethylene esters in food products in 1952, and restricted the use of emulsifiers to monoglyceride esters, such as, for example, hydrogenated lard monoglyceride. Longnecker has shown that the ethylene oxide derivatives are not safe under all conditions,[114] while in the case of glycerides there is no evidence to question their safety.[115]

Coppock and Cookson[116] have presented evidence to support the latter conclusion. They point out that bread prepared without emulsifier still contains monoglycerides. As much as 5 per cent of monglycerides are found in flour oils extracted from 81 per cent extraction flours, and in oils extracted from breads baked from severally treated flours. Similar results are due to Sager,[116a] who points out that these materials in the diet serve to promote the bodies' absorption of fat.

Margarine. In American practise, margarine consists of highly refined vegetable oils emulsified in fresh pasteurized milk. To a small extent, edible meat fats are used, usually in admixture with vegetable oils. On the Continent, marine oils, in particular oils derived from the whale, are employed.

Clayton[117] gives as a typical fatty composition for margarine:

Margarine: Fatty Composition

Ingredient	Per Cent
Premier Jus	15
Coconut Oil	15
Palm-kernel Oil	50
Peanut Oil	10
Cottonseed Oil	10

This is a characteristic meat fat-vegetable oil mixture.

According to Clayton, the usual procedure in America (and, to some extent, in England) is to melt the mixture of oils and fats, and then intimately mix them with milk, which has previously been "ripened" by inoculation with certain strains of lactic acid microorganisms.

The basic problem in margarine manufacture is the preparation of a stable W/O emulsion containing from 77 to 84 per cent of fat and 23 to 16 per cent of water (in the form of milk serum). The milk is used at a temperature of approximately 7° C, while the molten fatty phase is added at a temperature of about 40° C. The emulsion is cooled down, and worked at the lower temperature. The cooling, by suddenly increasing the viscosity, stabilizes the emulsion, and the high viscosity of the finished product is, of course, desirable. The working at the lower temperature serves to squeeze out excess water, and to occlude water in the plastic mass.

By proper choice of fats, a margarine which will retain its consistency at high temperatures can be prepared, and such formulations were found to be valuable in tropical areas during World War II. A recent trend has been the incorporation of varying percentages of butter, which serves to improve the spreading properties.

Miscellaneous Food Emulsions. Pattison[118] has discussed the use of various nonionic emulsifying agents in food. Cox[119] treats of the physiological and pharmacological properties of the various surface-active agents used in food products, and discusses their desirability. Ferri[120] gives an extensive discussion of the use of vegetable gums in stabilizing food emulsions.

Lecithin has been used for some time as an ingredient in chocolate to retard the unsightly fat "bloom." Easton, Kelly, Bartron, Cross, and Griffin[121] have discussed the use of other surface-active agents in this regard.

Snyder[122] has pointed out that ice cream mixes do not ordinarily contain the emulsifying constituents (principally lecithin) of the original dairy products, owing to losses in preparing the mix. Added emulsifiers are accordingly useful, providing an ice cream of superior texture. Snyder discusses the theory of emulsifier action in ice cream, and the use of lecithin, mono- and diglycerides, and sorbitan derivatives.

Potter and Williams[123] discuss these emulsifiers with respect to ice cream, as well as such materials as gelatin, sodium alginate, tragacanth, karaya, ground psyllium-seed husks, locust bean gum, quince-seed extract, Irish moss extract, pectin, oat gum, and cellulose gum.

The use of emulsifiers in ice cream is also the subject of a review by Redfern and Arbuckle.[124]

The manufacture of flavoring emulsions is discussed by Saul,[125] while the application of emulsion technology to the manufacture of cheese is treated by Faivre.[126]

Romanoff and Yushok[127] have found that treatment of shell eggs with an emulsion of stearic and lactic acids in mineral oil decreased the permeability of the egg shell to water vapor to about 1 per cent of its original value.

A recent application of emulsions in therapeutic feeding has been the oral or intravenous administration of emulsions consisting primarily of about 40 per cent of peanut oil in cases of malnutrition.[128] The preparation of such emulsions (requiring aseptic conditions) is described by Geyer, Olsen, Andrus, Waddell, and Stare.[129]

Asphalt Emulsions

Emulsions of asphaltic bitumen have wide application in fields where the principal requirement is the production of water-repellent surfaces. Thus, emulsions of asphaltic bitumen are employed in road construction (probably the most important application), in roofs and floors, for paper and fabric impregnation, electrical and heat insulation, and as binders for other insulating materials, such as cork or asbestos. They are also used as binders for coal briquets and carbon electrodes.

Road Surfacing. Probably the most thoroughly investigated application of asphalt emulsions is that of road surfacing. The principal advantage of applying asphalt to road surfaces in the form of an emulsion lies in the ease of application. The fluidity of the emulsion, much greater than that of molten asphalt, permits quicker application, and results in a smoother coating than would direct application. Furthermore, it is possible to use it on damp surfaces, which is impossible with the straight asphaltic material.

The formulation of asphalt emulsions involves a complex stability problem. In the first place, a reasonably stable mixture is desired so that it can be transported from the manufacturing point to the place of application, and, in some cases, stored for a period of time, without breaking. On the other hand, it is required that the emulsion break soon after application, so that the film of asphalt forms and sets quickly.

The number of formulations for this type of emulsion is enormous; hundreds of patents for various types have been granted. Berkman and Egloff[130] and Clayton[131] give extensive listings of the older literature in this regard.

In the simplest formulations, the aqueous phase contains a quantity of alkali; upon homogenization, this reacts with the free naphthenic acids in the asphalt to form naphthenate soaps *in situ*, which act as emulsifying agents. Obviously, the stability of such emulsions depends to a large extent on the nature of the particular asphalt employed, and additional soap (or other emulsifier) may be required. A typical formulation is

Asphalt Emulsion

Ingredient	Per Cent
Asphaltic Bitumen	55
Water	44
Sodium Linoleate/Alkali	1

Such emulsions are usually prepared in homogenizers or colloid mills; in some cases strong agitation is sufficient.

Literature on Asphalt Emulsions. The earliest colloid-chemical studies are probably those of Nellensteyn.[132] In particular, Nellensteyn has pointed out that the mechanism by which the emulsion breaks on being sprayed onto the stone sub-surface is two-fold. The breaking occurs as a result of

1. Adsorption of the emulsifier.
2. Changes in interfacial tension and interfacial electrical conditions.

The emulsions may be classified by their stability as unstable, semi-stable, and cold-stable (the last being suitable for road repairs in cold weather). The variation in stability which is possible is exemplified by the studies of Keppler, Blankenstein, and Borchers,[133] who studied the "break-ability" of asphalt emulsions as a function of asphalt concentration, the concentration of emulsifying agent in the aqueous phase remaining constant. Breakability was measured by determining the weight of the particular emulsion which will break when poured on 100 g of a standard stone. Some of the data obtained by these workers is given in Figure 8-2.

Reference has been made earlier (p. 73) to the studies by Lyttleton and Traxler[134] on the viscosity of asphalt emulsions; they conclude that commercial asphalt emulsions possessing Newtonian flow characteristics are a rarity. Most such emulsions are non-Newtonian liquids which show a decrease of consistency with increasing rate of shear.[135]

The preparation of such emulsions has been more recently discussed by Nüssel.[136] He discusses the important effect of the hardness of water used in the preparation on the stability of the emulsion, particularly with respect to transport. Becker[137] gives extensive directions for the manufacture of bituminous emulsions, and Rick[138] gives data on their production, properties, and uses. The older, but useful, paper by Gabriel[139] discusses the measurement of the physical properties of such emulsions.

Avetikyan and Gol'denberg[140] have discussed the effect of the nature of the bitumen employed on the stability of the emulsion. It was found that asphalts rich in naphthenes were often most stable when emulsified with sodium naphthenate, while asphalts rich in aromatic compounds were most often stable when emulsified with sodium benzoate.

These same workers have studied the effect of the nature and concentration of the emulsifier.[141] They report that, for example, 0.0035 M sodium oleate was as efficient as 0.35 M sodium stearate. The efficiency of sodium naphthenates increases with molecular weight, and exceeded that of sodium oleate when the molecular weight was equal to 394.

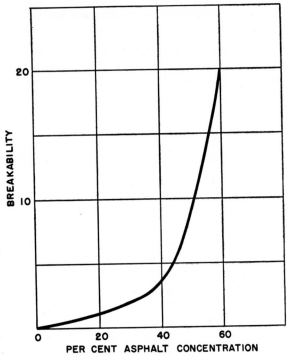

FIGURE 8-2. "Breakability" of asphalt emulsions as a function of asphalt concentration.[133]

Arriano[142] has summarized some of the methods for defining an index of coagulation for bituminous emulsions, and the methods for determining stability are critically reviewed by Klinkman.[143]

Gabriel and Peard[144] divide the tests for road emulsions into two classes, i.e., "imitative" and "empirical." "Imitative" tests seek to reproduce, in miniature, all factors operating on the road; their value is to act as a substitute for actual road trial of the emulsions.

Harsch and Spotswood[145] have described a new apparatus for the determination of the rate of break of asphalt emulsions, and have used it in an investigation of mechanism of breaking. They conclude that breaking results from loss of water by evaporation, the chemical effect of soluble salts from the rock aggregate, and physical effects of the aggregate surfaces.

Evans and Keiller[146] describe an apparatus for measuring the permeability of water of bituminous emulsion films. Blakely[147] has discussed the application of asphalt emulsions in preventing leakage under and through dams.

Patents. A number of recent patents of interest will be summarized here. The use of chlorinate tall oil as an emulsifying agent is revealed by Grader.[148, 149] Buckley[150] describes the preparation of an emulsion having increased stability and improved drying properties as a result of the addition of *Vinsol* resin. Chadder, Spiers, and Arnold[151] describe a process for making asphalt emulsions which involves the initial preparation of an emulsion of the W/O type, and subsequent inversion to the desired O/W form.

Barth[152] reveals that the wax acid soap of aluminum, copper, iron, or lead may be treated with sodium hydroxide, and that the resulting product is a satisfactory emulsifying agent. Carr[153] has described a process for producing an asphalt emulsion of low viscosity, while Johnson and Brown[154] reveal the use of a mixture of a zinc-tall oil reaction product with an oleic acid amide as an emulsifying agent, for the formation of bituminous emulsions of the water-in-oil type.

Another type of emulsifying agent has been patented by Mayfield.[155] This consists of the sulfonation product of a substantially petroleum-hydrocarbon insoluble pine wood resin. Gabriel and Rawlinson[156] reveal the use of bentonite to produce low viscosity emulsions containing not more than 25 per cent of bituminous material.

A patent issued to Borglin,[157] covering the use of sodium caseinate as a stabilizer, emphasizes the importance of the use of soft water. Mayfield,[158] by the use of saponified B-wood-rosin oil, has produced asphalt emulsions which combine high demulsifiability with good physical stability. In another patent, Mayfield[159] has revealed the use of 0.01 to 0.1 per cent methylcellulose to control the viscosity of such emulsions.

Arnold[160] treats of the use of alkali-starch mixtures to form, initially, W/O emulsions. The emulsion is inverted to the desired type by the further addition of more alkali-starch solution, alkali-soluble casein, and ammonium hydroxide. The rate of break is governed by varying the emulsifier content.

Fratis and Oakley[161] classify the acids present in the asphalt as "naphthenic" if they have a molecular weight below 400, and as "asphaltic" if they are above. They reveal that acids in the molecular-weight range 500 to 900 form salts which emulsify asphalt properly. Asphalts which contain acids either outside of the optimum molecular weight range or outside the optimum content range may be made emulsifiable by the addition of the correct amount of acids of the right molecular weight.

Allen[162] has patented the use of a saturated or unsaturated straight-chain aliphatic amine containing 5 to 20 carbon atoms as an antistripping agent, while Johnson[163] has revealed the use of small amounts of amine soaps (1 per cent) to improve the adhesion and ageing properties.

Worson[163a] describes the use of neutral salts of alkaline earth metals, such as calcium, barium, magnesium, or strontium, together with reemulsifiable bentonite, to form asphalt emulsions whose dried film is nonreemulsifiable.

Treatment of bitumen at 100 to 120° C with 0.5 to 3.0 per cent of sulfuric acid or oleum improves its emulsifiability, according to an Italian patent.[164]

McCoy[165] has patented a non-alkaline asphalt emulsion. This is stabilized by the addition of up to 10 per cent of sulfonated tall oil in the presence of 0.2 to 0.3 per cent sodium dichromate.

Craig[166] has described the use of clay as the major emulsifying agent in asphalt emulsions, while Cross[167] has revealed the use of a nonionic emulsifier prepared by the addition of ethylene oxide to a fatty secondary amine.

For information on the standards for bituminous emulsions, and for methods of testing, the publications of the ASTM should be consulted.[168]

Emulsion Polymerization

As is well known, the production of high polymeric materials (plastics, synthetic fibers, and synthetic rubbers) is an increasingly important technological problem; the production of such materials is measured in the millions of tons annually.

The earliest polymerization techniques involved the use of simple batch processes. The monomer, or mixture of monomers, was polymerized in bulk to the desired degree of polymerization through the action of appropriate catalysts. However, unless the degree of polymerization is sharply limited, a not particularly desirable arrangement, considerable processing difficulty is experienced. This arises from the fact that the polymerization reaction is highly exothermic, and, since high reaction temperatures lead to undesirable side effects, efficient cooling is necessary. This, however, is hindered by the high viscosity which results even at moderate conversions, rendering efficient stirring, and hence efficient heat transfer, extremely difficult.

Some of these difficulties were removed by the use of suspension polymerization, but the introduction of emulsion polymerization, although not without its own problems, represents a considerable improvement. The heat of reaction is readily dissipated into the aqueous phase and, since the viscosity ordinarily does not increase much during reaction, efficient stirring is possible. Furthermore, since the catalysts employed are water-soluble, their effectiveness is enhanced, and polymerizations can be carried out at lower temperatures, which is often desirable. Finally, for certain purposes (such as the formulation of coatings) it is desirable to use the polymer as is, i.e., as a latex, which is, of course, directly produced by this technique.

Most of the polymerization reactions are carried out under alkaline conditions, and soaps are especially suitable as emulsifying agents. Cationic agents (notably dodecylamine hydrochloride) are also used, however, and there is no reason why, under appropriate conditions, nonionic surface-active agents may not be used. A typical emulsion polymerization formula (for GR-S rubber) would be:

Emulsion Polymerization Formula (GR-S)

Ingredients	*Per Cent*
Butadiene	25
Styrene	8.5
Emulsifying Agent	1
Catalyst	0.5
Water	65

The reaction is ionically catalyzed. The usual catalyst is potassium persulfate, although hydrogen peroxide and ferrous ions or a water-soluble organic hydroperoxide may be used.

Theory of Emulsion Polymerization. Although the literature on the theory of emulsion polymerization is extensive, the basic theory is due to Harkins,[169] while the experimental verification of this theory was performed by Smith and Ewart.[170, 171]

According to Harkins' theory it may be considered that the monomer is distributed in three different phases, i.e., dissolved to a small extent in the aqueous phase, suspended in the form of emulsion droplets, and solubilized in the detergent micelles. By the same token, the surface-active molecules are distributed in the same manner. Clearly, the concentration of emulsifying agent must be sufficient to satisfy all these needs.

The process of initiation occurs in the aqueous phase, and the monomer radical thus formed diffuses into a micelle, where polymerization begins to take place. Little or no polymerization occurs in the emulsion droplets themselves, which merely serve as a reservoir to supply the micelles as the monomer concentration therein is depleted by reaction. The shrinking of the emulsion droplets and the corresponding growth of the solubilizing micelles has actually been observed microscopically.[169]

As the micelle grows, it also adsorbs the free emulsifier out of solution, and eventually from the surface of the emulsion droplets. The emulsifier thus serves to stabilize the polymer droplets. Although, on first consideration, this mechanism appears almost improbable, the experimental results of Smith and Ewart bear it out. It is found that the rate of polymerization depends principally on the number of particles present, and this, in turn, is largely dependent on the emulsifier concentration.

In spite of the success of this approach, the theoretical study of emulsion polymerization is far from complete, and much remains to be done. For additional details, the works of Flory[172] and Williams[173] should be consulted. The latter is of considerable value in giving the practical details of manufacture.

Miscellaneous Applications of Emulsions

In addition to the applications of emulsion technology outlined in the previous sections of this chapter, there are many more equally worthy of attention. However, since these applications very often do not offer much in the way of new principles, limitations of space dictate a more condensed discussion. In many cases, extended discussions do exist, and, wherever possible, reference is made to them.

Leather Treatment. Following the tanning of leather or fur pelts, it is customary to incorporate fat or oil in order to lubricate the fibers, and improve the general feel of the leather with respect to such properties as softness, stretchiness, and elasticity. Such oils or fats may be applied directly, in which case the process is called *stuffing* or *oiling*. An alternative procedure is to apply the fatty material in the form of an emulsion. This is known as *fat-liquoring*, and the emulsion employed is known as fat-liquor.

A typical fat-liquor formula would be:[174]

Fat-Liquor

Ingredients	Parts by Weight
Paraffin Oil	60
Sulfonated Cod Oil	40
Borax	2
Water	160

Sulfonated oils, particularly marine oils, are widely used as emulsifying agents for fat-liquor. Other agents, including egg yolk, lecithin, petroleum sulfonates, fatty alcohol sulfates, and quaternary ammonium salts, are also the subject of patents.

The particle-size distribution in fat-liquor emulsions is moderately critical, since coarse emulsions will deposit oil on the surface of the leather, while fine emulsions will penetrate to deeply, with consequent excessive pliability. The effect of these parameters, of the concentration and nature of the emulsifier, etc., have been investigated intensively. Clayton[175] and McLaughlin and Theis[176] discuss the older work in considerable detail. Recent publications in this regard include the papers of Roux[177] and Smith,[178] on the subject of new emulsifiers for fat-liquor. Smith states that fat-liquors made with mixtures of cationic and nonionic emulsifying agents combine the advantages of soap and sulfonated oil fat-liquors. Poujade and Poujade,[179] in a French patent, reveal the use of bentonite as a stabilizer for fat-liquor emulsion, while Kroch[180] suggests the use of activated gelatinous aluminum oxide for the same purpose.

Textile Industry. Emulsions in many forms are widely used in the textile industry, as a method of incorporating fatty agents of one sort or another into the fibers. A particularly important application of this sort is in the use of lubricant emulsions which, applied to the fibers, control the friction developed between the fibers and solid surfaces, e.g., the metal parts of the spinning and weaving machinery. This has the desirable effect of reducing abrasion, as well as improving the handling properties of the fabric.

Similarly, anti-static and waterproofing agents may be applied to fabrics in the form of emulsions.

Sisley[181] and Smith[182] have discussed the use of surface-active materials in the textile industry, including emulsifying agents. The former paper is restricted to nonionic agents.

Emulsions also occur in the textile industry in the disposition of so-called scouring liquors. This is a problem of demulsification, however, and is discussed in the next chapter.

Metal Dispersions. The well-known disadvantages of handling metallic sodium may largely be obviated by using the fact that, at temperatures above 97.5° C, a mixture of sodium and a liquid hydrocarbon will exist as two immiscible liquids, which may be emulsified in much the same way as oil and water. If the emulsification is properly carried out, a sodium-in-hydrocarbon emulsion results; on subsequent cooling, the sodium solidifies as tiny spheroids suspended in the hydrocarbon.

In addition to the increased ease of handling, the reactivity of the sodium is increased by virtue of the large surface area. Sittig[183] presents data showing that for the reaction of sodium with octanol, a given weight of sodium in one piece results in less than 25 per cent reaction in 15 minutes, whereas if the sodium is dispersed in spheroids of 5 μ in diameter complete reaction occurs in under 2 minutes.

Although several methods are available for the preparation of such emulsions, methods based on mechanical mixing techniques, including the use

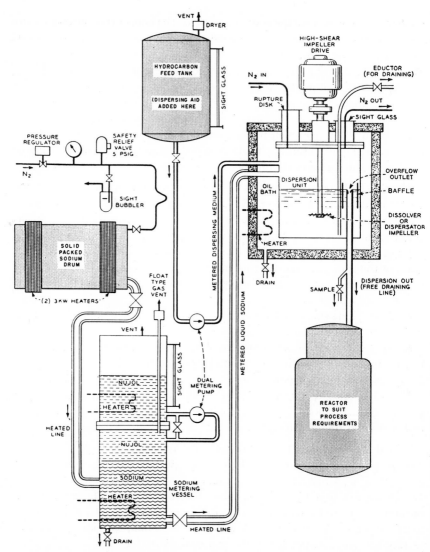

FIGURE 8-3. Flow diagram of a recently-developed continuous process for the production of sodium dispersions. *Courtesy: U. S. Industrial Chemicals Co., Div. of National Distillers & Chemical Corp.*

of colloid mills, are preferred.[184] A recently-developed continuous process is illustrated in Figure 8-3.

A series of patents by Hansley[185, 186, 186a] and Hansley and Hilts[187] over the use of alkali metal soaps and binary mixtures of activated carbon and alkali metal soaps, or alkali metal salts of hydroperoxides, as stabilizing agents. Hansley and Hilts discuss similar dispersions of the other alkali metals or alloys of these metals.

Emulsions as Transport Media. A not entirely unrelated application of emulsions is in the transport of dangerous materials other than alkali metals. Aqueous emulsions of nitroglycerine and similar liquid explosives are employed to transport the explosive from the manufacturing point to storage depots. Bryce and Williams[188] have patented a technique which prevents a possible detonation from being transported through the system.

Similarly, Schutte[189] has revealed a technique for rendering a normally solid wax-containing hydrocarbon pumpable; this involves the formation of an emulsion of the molten hydrocarbon in water.

Drilling Fluids. Drilling fluids are used in the petroleum industry as aids in the boring of wells. Their principal function is to transport bit cuttings to the surface. At the same time, the liquid exerts a hydrostatic pressure on exposed fluid-bearing formations, so that entry of these fluids into the hole is prevented. Larsen[190] has discussed the colloid chemistry of these materials in considerable detail. Numerous materials are used for the purpose, including oil-water emulsions. These are usually oil-in-water, although the inverse type have been proposed.

Parkins,[191] in a discussion of oil emulsion drilling fluids, points out that such a fluid consists of three components, e.g., an emulsifying agent (soaps, lignosulfonates, starch, CMC, finely-divided solids), oil (crude oil, gas oil, Diesel oil), and an aqueous mud slurry. The fluid is made by emulsifying 5 to 30 per cent of oil in a slurry containing the emulsifying agent. Van Dyke[192] also treats of the preparation of these fluids; Lummus[193] has patented a drilling fluid of the W/O type.

Detergency. Although the exact mechanism of detergency is not well understood, there is no doubt that emulsification of the fatty part of the soil represents an important part of the process. Rosano and Weill[194] state that, for aqueous detergents, the mechanical work appears to be a work of emulsification. Adam and Stevenson[195] have reviewed the mechanism whereby a detergent removes oil films. Niven[196] should be consulted for these and other aspects of the detersive process.

Isolated Applications. A few isolated applications may be cited. Milne[197] has discussed the preparation of cutting-oils. The use of W/O emulsions as lubricants (as a substitute for oil) is treated by Beuerlein,[198] Marke,[199] and Baum.[200]

Powell[201] has pointed out how emulsions may be used in oil-fire fighting. If the burning surface is sprayed with dispersed water in the form of fine droplets, and if the droplets are of the correct size and driven with the correct velocity, an emulsion is formed with the oil, and the fire is extinguished.

Nitrobenzene, in the form of an emulsion in water, may be readily reduced by electrolytic means by the use of an iron cathode. Yields are reported to be 90 per cent.[202]

Pollack[203] has described the formulation of solvent emulsions used in metal cleaning. Kadmer[204] gives considerable detail on the preparation of emulsions for use in metal-working.

Aqueous emulsions of vitamins are described in a patent issued to Zentner,[205] while Wallenmeyer, McDonald, and Henry[206] describe a method of spray-drying emulsions of vitamins A and D.

Bibliography

1. Leslie, R., *Mfg. Chemist* **18**, 494 (1947).
2. Rothman, S., "Physiology and Biochemistry of the Skin," pp. 46–49, Chicago, University of Chicago Press, 1954.
3. Hollenberg, I. R., *J. Soc. Cosmetic Chemists* **1**, 368 (1949).
4. Harry, R. G., "Cosmetics: Their Principles and Practices," p. 118, New York, Chemical Publishing Co., 1956.
5. Harry, R. G., *op. cit.*, pp. 117–118.
6. Pickthall, J., *J. Soc. Cosmetic Chemists* **6**, 4 (1955).
7. De Navarre, M. G., "Chemistry and Manufacture of Cosmetics," pp. 65–69, New York, D. Van Nostrand Co., 1941; cf. also *Schimmel Briefs* **230**, (May, 1954).
8. Salisbury, R., Leuallen, E. E., Chavkin, L. T., *J. Am. Pharm. Assoc.* **43**, 117 (1954).
9. Mullins, J. D. and Becker, C. H., *J. Am. Pharm. Assoc.* (Sci. Ed.) **44**, 110 (1956).
10. Glyco Products Co., "Cosmetic and Drug Manual."
11. Harry, R. G., *Mfg. Chemist* **10**, 366 (1939); **11**, 13 (1940).
12. Martin, B. K., *Mfg. Chemist* **25**, 242 (1954).
13. Münzel, K. and Ammann, R., *Pharm. Acta Helv.* **30**, 1, 49 (1955); **31**, 140 (1956).
14. Coran, A., and Huyck, C. L., *J. Soc. Cosmetic Chemists* **7**, 20 (1956).
15. Barker, D. V., DeKay, H. G., and Christian, J. E., *J. Am. Pharm. Assoc.* (Sci. Ed.) **45**, 527 (1956).
16. Wood, J. A. and Rising, L. W., *J. Am. Pharm. Assoc.* **42**, 481 (1952).
17. Halpern, A. and Hartwell, L. M., *Am. J. Pharm.* **120**, 386 (1948).
18. De Navarre, M. G., *op. cit.*, *passim.*
19. Harry, R. G., *op. cit.*, *passim.*
20. Lewinson, A., *Soap, Perfumery & Cosmetics* **19**, 1036 (1946).
21. Pickthall, J., *Soap, Perfumery & Cosmetics* **22**, 1338 (1949); *J. Soc. Cosmetic Chemists* **2**, 141 (1951).
22. Hollenberg, I. R., *Drug Cosmetic Ind.* **59**, 334, 422, 644, 722 (1946); **60**, 50, 128 (1947).
23. Hollenberg, I. R., *Soap, Perfumery & Cosmetics* **21**, 474 (1948).
24. Davies, J. T., *Perfumery Essent. Oil Record* **43**, 338, 365 (1952).
25. Rothemann, K., *Seifen-Öle-Fette-Wachse* **76**, 119 (1950).

26. King, G. J., *Am. Perfumer* **63,** 349 (1954).
27. Hollenberg, I. R., *Am. Perfumer* **59,** 29 (1952).
28. Ruemele, T., *Perfumery Essent. Oil Record* **45,** 198 (1954).
29. Schweisheimer, W., *Seifen-Öle-Fette-Wachse* **80,** 707 (1954).
30. Wells, F. V., *Am. Perfumer* **59,** 19, 107 (1952).
31. Hilfer, H., *Drug & Cosmetic Ind.* **68,** 322, 393 (1951).
32. Currie, C. C. and Francisco, D. M., *Am. Perfumer* **64,** 421 (1954).
33. Macias-Sarria, J., *Am. Perfumer* **48,** 63 (1946).
34. Federici, N. J., *Soap Sanit. Chemicals* **23,** No. 5, 35, 90 (1947).
35. Ruemele, T., *Perfumery Essent. Oil Record* **44,** 214 (1953).
36. Collier, S. L., *Soap, Perfumery & Cosmetics* **27,** 388 (1954).
37. Winter, F. S., *Perfumery & Cosmetics* **21,** 42 (1948).
38. Janowitz, H. C., *Seifen-Öle-Fette-Wachse* **80,** 619 (1954).
39. Janowitz, H., *Seifen-Öle-Fette-Wachse* **80,** 647, 672 (1954).
40. Janowitz, H., *Seifen-Öle-Fette-Wachse* **80,** 703 (1954).
41. Peel, N. S., *Soap, Perfumery & Cosmetics* **25,** 503 (1952).
42. Peel, N. S., *Soap, Perfumery & Cosmetics* **25,** 615, 729 (1952).
42a. Hilfer, H., *Drug & Cosmetic Ind.* **74,** 498, 596 (1954).
42b. Keithler, W. R., *Drug & Cosmetic Ind.* **75,** 767, 860 (1954).
43. Peel, N. S., *Soaps, Perfumery & Cosmetics* **26,** 275 (1953).
44. *Schimmel Briefs* **199,** (Oct., 1951).
45. Sollazzo, G., *Boll. chim. farm.* **91,** 323 (1952).
46. Avis, K. E., *Am. J. Pharm.* **119,** 271 (1947).
47. Swallow, W., *Pharm. J.* **147,** 226 (1941).
48. Cessna, O. C., Ohlmann, E. O., and Roehm, L. S., *Am. Perfumer* **49,** 369 (1947).
49. Griffin, W. C., Behrens, R. W., and Cross, S. T., *J. Soc. Cosmetic Chemists* **3,** 5 (1952).
50. Vasic, V., *Soap, Perfumery & Cosmetics* **28,** 1131 (1955).
51. *Schimmel Briefs* **227,** (Feb., 1954).
52. De Navarre, M. G. and Bailey, H. E., *J. Soc. Cosmetic Chemists* **7,** 427 (1956).
53. Wynne, W., *Am. Perfumer* **54,** 381 (1949).
54. Pickthall, J., *Soap, Perfumery & Cosmetics* **28,** 69 (1955).
55. Hilfer, H., *Drug & Cosmetic Ind.* **72,** 464, 554 (1953).
56. Bruening, C. F., *J. Assoc. Official Agr. Chem.* **24,** 889 (1941); **25,** 903 (1942).
57. Atlas Powder Co., "A Guide to Formulation of Industrial Emulsions"; "Span 85" and "Tween 81" are trade-marked products of the Atlas Powder Co.
58. Armour & Co., Chemical Div., "Armour Etho-chemicals." "Etho-fat 60/15" and "Ethofat 60/20" are trade-marked products of Armour & Co.
59. Armour & Co., Chemical Div., *op. cit.* "Ethomeen S/12" and "Arquad 2C" are trade-marked products of Armour & Co.; Silicone DC-200", and "Silcone 300-500 cps" are products of the Dow-Corning Co., "Silicone SF-96" of the General Electric Co.
60. Kroner, A., *Seifen-Öle-Fette-Wachse* **80,** 115, 139 (1954).
61. Frump, J. A., *Soap & Chem. Specialties* **31,** No. 2, 153 (1955).
62. Atlas Powder Co., *op. cit.* "Tween 80" is a trade-marked product of the Atlas Powder Co.
63. Eaton, J. L. and Hughes, F. A., *Proc. Chem. Specialties Mfrs. Assoc.,* Dec. **1950,** 197.
64. Trusler, R. B., *Proc. Chem. Specialties Mfrs. Assoc.,* Dec. **1951,** 127.
65. Behrens, R. W. and Griffin, W. C., *Soap Sanit. Chemicals* **27,** No. 11, 128 (1951).

66. Schoenholz, D. and Kimball, E. S., *Soap Sanit. Chemicals* **23**, No. 8, 131 (1947).
67. Welch, H., *Soap Sanit. Chemicals* **28**, No. 1, 117 (1952).
68. Kselik, G., *Seifen-Öle-Fette-Wachse* **82**, 525 (1956).
69. Clark, R. E., *Soap & Chem. Specialties* **32**, 141 (1956).
70. Treffler, A., *Soap Sanit. Chemicals* **28**, No. 4, 147 (1952).
71. Figliolino, A., *Paint & Varnish Production* **31**, No. 1, 14 (1951).
72. Lesser, M. A., *Soap Sanit. Chemicals* **28**, No. 5, 107 (1952).
73. Wasserman, K. J., *Paint & Varnish Production* **46**, No. 8, 36 (1956).
74. Camperio, G., *Ind. vernice* (Milan) **5**, 38, 55 (1951).
75. Werthan, S., in "Emulsion Technology," pp. 214–229, Brooklyn, Chemical Publishing Co., 1946.
76. Allen, A. O., in Bennett, H. (ed.), "Practical Emulsions," pp. 229–236, Brooklyn, Chemical Publishing Co., 1947.
77. Levesque, C. L., in von Fischer, W. (ed.), "Paint and Varnish Technology," pp. 127–144, New York, Reinhold Publishing Corp., 1948.
78. Ferguson, C. S., and Sellers, J. E., in Mattiello, J. J. (ed.), "Protective and Decorative Coatings," **1**, pp. 338–359, New York, John Wiley & Sons, Inc., 1941.
79. Cheetham, H. C. and Pearce, W. T., in Mattiello, J. J. (ed.), "Protective and Decorative Coatings," **3**, pp. 488–496, New York, John Wiley & Sons, Inc., 1943.
80. Elm, A. C. and Werthan, S., *Offic. Dig. Federation Paint & Varnish Production Clubs*, No. 223, 35 (1943).
81. Iddings, C., "Symposium on Paint," p. 13, Philadelphia, American Society for Testing Materials, 1943.
82. Payne, H. F., *Paint, Oil, Chem. Rev.* **112**, No. 25, 12 (1949).
83. Musgrave, J., *Paint Manuf.* **17**, 121 (1947).
84. Paxon, L. A., *Paint Manuf.* **16**, 368 (1946).
85. Burr, W. W. and Matvey, P. R., *Offic. Dig. Federation Paint & Varnish Production Clubs*, No. 304, 347 (1950).
86. Steig, F. B., *Offic. Dig. Federation Paint & Varnish Production Clubs*, No. 301, 94 (1950).
87. Chatfield, H. W., *Paint, Oil & Colour J.* **122**, 1279 (1952).
88. McLean, A., *Paint Manuf.* **23**, 252 (1953).
89. Davis, J. P., *Paint, Oil & Chem. Rev.* **117**, No. 1, 22 (1954).
90. Cogan, H. D. and Clarke, D. F., *Offic. Dig. Federation Paint & Varnish Production Clubs* **28**, 764 (1956).
91. Palmer, J. F., Jr. and Cass, R. A., *Offic. Dig. Federation Paint & Varnish Production Clubs* **28**, 869 (1956).
92. Agabeg, R. C., *Offic. Dig. Federation Paint & Varnish Production Clubs* **28**, 890 (1956).
93. Hand, J., *Offic. Dig. Federation Paint & Varnish Production Clubs* **28**, 896 (1956).
94. Hurd, R., *Offic. Dig. Federation Paint & Varnish Production Clubs* **28**, 883 (1956).
95. Ford, D. S., *Paint Manuf.* **24**, 186 (1954).
96. Simpson, G. K., *J. Oil & Colour Chemists' Assoc.* **32**, 60 (1949).
97. Campbell, B., *Paint Manuf.* **18**, 356 (1948).
98. Armour & Co., Chemical Div., *op. cit.* "Ethofat 142/15" is a trade-marked product of Armour & Co.
99. Armour & Co., Chemical Div., *op. cit.* "Ethofat 142/20" and "Arquad 2C" are trade-marked products of Armour & Co.

100. Smith, C. M. and Goodhue, L. D., *Ind. Eng. Chem.* **34**, 490 (1942).
101. Woodman, R. M., in "Emulsion Technology," pp. 127–175, Brooklyn, Chemical Publishing Co., 1942.
102. Felber, I. M., *J. Agr. Research* **71**, 231 (1945).
103. Jones, H. A. and Fluno, H. J., *J. Econ. Entomol.* **39**, 735 (1946).
104. Plante, E. C., *Australian J. Sci.* **8**, 111 (1946).
105. Hackman, R. H., *J. Council Sci. Ind. Research* **19**, 77 (1946).
106. Sparr, B. I. and Bowen, C. V., *U. S. Dept. Agr., Bur. Entomol. Plant Quarantine* **E866**, (1953).
107. Brown, G. L. and Riley, G. C., *Agr. Chemicals* **10**, No. 8, 34 (1955).
108. Anon., *Chem. Week*, Oct. 2, 1954, p. 93.
109. Clayton, W., "Theory of Emulsions," 4th Ed., pp. 318–323, Philadelphia, Blakiston, 1943.
110. Corran, J. W., in "Emulsion Technology," pp. 176–192, Brooklyn, Chemical Publishing Co., 1946.
111. Jooste, M. E. and Mackey, A. E., *Food Research* **17**, 185 (1952).
111a. Skovholt, O. and Dowdle, R. L., *Cereal Chem.* **27**, 26 (1950).
112. Pyler, E. J., "Baking Science and Technology," **1**, pp. 301–308, Chicago, Siebel Publishing Co., 1952.
113. Coppock, J. B. M., *J. Sci. Food Agr.* **1**, 125 (1950).
114. Longnecker, H. E., *Natl. Research Council-Natl. Acad. Sci. (U.S.), Publ., No.* **280** (1953).
115. Longnecker, H. E., *Natl. Research Council-Natl. Acad. Sci. (U.S.), Publ.,* **No. 251** (1952).
116. Coppock, J. B. M. and Cookson, M. A., *Chemistry & Industry* **1953**, No. 1, 12.
116a. Sager, C. A., *Biochem. Z.* **321**, 44 (1950); *Monatsschr. Kinderheilk.* **99**, 78 (1951).
117. Clayton, W., *op. cit.*, pp. 313–314.
118. Pattison, E. S., *Food Eng.* **24**, No. 5, 112 (1952).
119. Cox, H. E., *Chemistry & Industry* **1952**, 72.
120. Ferri, C., *Food Inds.* **19**, 784 (1947).
121. Easton, N. R., Kelly, D. J., Bartron, L. R., Cross, S. T., and Griffiin, W. C., *Food Technol.* **6**, 21 (1952).
122. Snyder, W. E., *Milk Plant Monthly* **38**, No. 6, 30 (1949).
123. Potter, F. E. and Williams, D. H., *Milk Plant Monthly* **39**, No. 4, 76 (1950).
124. Redfern, R. B. and Arbuckle, W. S., *Southern Dairy Products J.* **46**, No. 3, 32 (1949).
125. Saul, E. L., *Am. Perfumer* **56**, 311 (1950).
126. Faivre, R., *Chimie & industrie* **56**, 373 (1946).
127. Romanoff, A. L. and Yushok, W. D., *Food Research* **13**, 331 (1948).
128. *J. Am. Med. Assoc.* **152**, 1468 (1953).
129. Geyer, R. P., Olsen, F. R., Andrus, S. B., Waddell, W. R., and Stare, F. J., *J. Am. Oil Chemists' Soc.* **32**, 365 (1955).
130. Berkman, S. and Egloff, G., "Emulsions and Foams," pp. 396–444, New York, Reinhold Publishing Corp., 1941.
131. Clayton, W., *op. cit.*, pp. 281–301.
132. Nellensteyn, F. J., *Chem. Weekblad* **24**, 54 (1927); **35**, 283 (1938); in Alexander, J. (ed.), "Colloid Chemistry," III, pp. 535–546, New York, Reinhold Publishing Corp., 1931.
133. Keppler, G., Blankenstein, P., and Borchers, K., *Angew. Chem.* **47**, 223 (1934).
134. Lyttleton, D. V. and Traxler, R. N., *Ind. Eng. Chem.* **40**, 2115 (1948).

135. Traxler, R. N., in Alexander, J. (ed.), "Colloid Chemistry," **VII**, pp. 507–510, New York, Reinhold Publishing Corp., 1950.
136. Nüssel, H., *Bitumen* **11**, 45 (1941).
137. Becker, W., *Bitumen* **11**, 37 (1941).
138. Rick, A. W., *Seifen-Öle-Fette-Wachse* **75**, 273, 293 (1949).
139. Gabriel, L. G., in "Emulsion Technology," pp. 253–282, Brooklyn, Chemical Publishing Co., 1946.
140. Avetikyan, S. M. and Gol'dberg, D. O., *Kolloid. Zhur.* **13**, 159 (1951); *C.A.* **46**, 8753a.
141. Avetikyan, S. M. and Gol'dberg, D. O., *Kolloid. Zhur.* **12**, 401 (1950).
142. Arriano, R., *Riv. ing.* **3**, 115, 280 (1953).
143. Klinkman, G. H., *Asphalt u. Teer Strassenbautech.* **40**, 85, 95, 105, 125 (1940).
144. Gabriel, L. G. and Peard, W. L., *J. Soc. Chem. Ind. (London)* **60**, 78 (1941).
145. Harsch, R. and Spotswood, E. H., *Proc. Assoc. Asphalt Paving Tech.* **12**, 184 (1940).
146. Evans, A. and Keiller, C. Q., *J. Soc. Chem. Ind. (London)* **69**, Suppl. No. 1, S1 (1950).
147. Blakely, L. E., *J. Am. Water Works Assoc.* **40**, 873 (1948).
148. Grader, R. (to Ebano Asphalt-Werkw A.-G., Ger. Pat. 703, 599, Feb. 6, 1941 (Cl. 80b. 25.06); Ger. Pat. 706,355, April 17, 1941 (Cl. 80b. 25.06).
149. Seidenbusch, M. and Grader, R. (to Ebano Asphalt-Werke A.-G.), Ger. Pat. 722,650, June 4, 1942 (Cl. 23c. 2).
150. Buckley, W. D. (to American Bitumuls Co.), U.S. Pat. 2,256,886, Sept. 23, 1941.
151. Chadder, W. J., Spiers, H. M., and Arnold, E. (to Thermal Industrial and Chemical Research Co. Ltd.), U.S. Pat. 2,247,722, July 1, 1941.
152. Barth, E. J. (to Sinclair Refining Co.), U.S. Pat. 2,243,519, May 27, 1941.
153. Carr, D. E. (to Union Oil Co. of Calif.), U.S. Pat. 2,298,612, Oct. 13, 1942.
154. Johnson, J. M. and Brown, E. C. (to Nostrip, Inc.), U.S. Pat. 2,317,959, April 27, 1943.
155. Mayfield, E. E. (to Hercules Powder Co.), U.S. Pat. 2,328,481, Aug. 31, 1943.
156. Gabriel, L. G. and Rawlinson, J. A., U.S. Pat. 2,327,882, Aug. 24, 1943.
157. Borglin, J. N. (to Hercules Powder Co.), U. S. Pat. 2,326,610, Aug. 10, 1943.
158. Mayfield, E. E. (to Hercules Powder Co.), U. S. Pat. 2,370,911, Mar. 6, 1945.
159. Mayfield, E. E. (to Hercules Powder Co.), U. S. Pat. 2,374,776, May 1, 1945.
160. Arnold, E. (to Woodall-Duckham (1920) Ltd.), U. S. Pat. 2,372,924, Apr. 3, 1945.
161. Fratis, J. E. and Oakley, E. H. (to American Bitumuls Co.), U. S. Pat. 2,406,823, Sept. 3, 1946.
162. Allen, W. W. (to West Bank Oil Terminal, Inc.), U. S. Pat. 2,416,134, Feb. 18, 1947.
163. Johnson, J. M. (to Nostrip, Inc.), U. S. Pat. 2,419,404, April 22, 1947.
163a. Worson, L. (to The Patent and Licensing Corp.), U. S. Pat. 2,468,533, April 26, 1949.
164. Azienda Nazionale Idrogenazione Combustibili (A. N. I. C.), Ital. Pat. 420,771, May 6, 1947.
165. McCoy, P. E. (to American Bitumuls Co.), U. S. Pat. 2,585,336, Feb. 12, 1952.
166. Craig, W. G. (to Lubrizol Corp.), U. S. Pat. 2,652,341, Sept. 15, 1953.
167. Cross, S. T. (to Atlas Powder Co.), U. S. Pat. 2,690,978, Oct. 5, 1954.
168. ASTM Standards, Part 3, *passim*, Philadelphia, American Society for Testing Materials, 1955.

169. Harkins, W. D., *J. Am. Chem. Soc.* **69,** 1428 (1947); "Physical Chemistry of Surface Films," Ch. 5, New York, Reinhold Publishing Corp., 1952.
170. Smith, W. V. and Ewart, R. H., *J. Chem. Phys.* **16,** 592 (1948).
171. Smith, W. V., *J. Am. Chem. Soc.* **70,** 3695 (1948).
172. Flory, P. J., "Principles of Polymer Chemistry," pp. 203–217, Ithaca, Cornell University Press, 1953.
173. Williams, H. L., in Schildknecht, C. E. (ed.), "Polymer Processes," pp. 111–174, New York, Interscience Publishers, Inc., 1956.
174. Bennett, H., "Practical Emulsions," 2nd Ed., p. 396, Brooklyn, Chemical Publishing Co., 1947.
175. Clayton, W., *op. cit.*, pp. 301–311.
176. McLaughlin, G. D. and Theis, E. R., "Chemistry of Leather Manufacture," Ch. 24, New York, Reinhold Publishing Corp., 1945.
177. Roux, M., *Assoc. franc. chim. inds. cuir*, Conf. No. 7, 16 pp. (1949).
178. Smith, P. I., *Leather and Shoes* **119,** No. 3, 18 (1950).
179. Poujade, A. and Poujade, L., Fr. Pat. 865,127, May 14, 1941.
180. Kroch, F. H., Lankro Chemicals, Ltd., and Universal Emulsifiers, Ltd., Brit. Pat. 564,316, Sept. 22, 1944.
181. Sisley, J. P., *Am. Dyestuff Reporter* **38,** 513, (1949).
182. Smith, P. I., *Soap Sanit. Chemicals* **23,** No. 9, 40, (1947).
183. Sittig, M., "Sodium," p. 133, New York, Reinhold Publishing Corp, 1956.
184. Sittig, M., *op. cit.*, pp. 133–142.
185. Hansley, V. L. (to E. I. duPont de Nemours), U. S. Pat. 2,394,608, Feb. 12, 1946.
186. Hansley, V. L. (to E. I. duPont de Nemours), U. S. Pat. 2,487,333, Nov. 8, 1949.
186a. Hansley, V. L. (to E. I. duPont de Nemours), U. S. Pat. 2,487,334, Nov. 8, 1949.
187. Hansley, V. L. and Hilts, W. J. P. (to E. I. duPont de Nemours), U. S. Pat. 2,578,257, Dec. 18, 1951.
188. Bryce, G., Williams, V. H., and Imperical Chemical Industries, Ltd., Brit. Pat. 543,593, March 4, 1942; U. S. Pat. 2,368,638, Feb. 6, 1945.
189. Schutte, A. H. (to The Lummus Co.), U. S. Pat. 2,421,968, June 10, 1947.
190. Larsen, D. H., in Alexander, J. (ed.), "Colloid Chemistry," **VI,** pp. 509–530, New York, Reinhold Publishing Corp., 1946.
191. Parkins, H. W., *Petroleum Engr.* **23B,** No. 4, 60 (1951).
192. Van Dyke, O. W., *World Oil* **131,** No. 6, 101 (1950).
193. Lummus, J. L. (to Stanolind Oil & Gas Co.), U. S. Pat. 2,661,334, Dec. 1, 1953.
194. Rosano, H. L. and Weill, M., *Mém. services chim. état. (Paris)* **37,** No. 3, 219 (1952).
195. Adam, N. K. and Stevenson, D. G., *Endeavour* **12,** 25 (1953).
196. Niven, W. W., Jr., "Fundamentals of Detergency," Ch. 11, New York, Reinhold Publishing Corp., 1950.
197. Milne, D., *Sewage and Ind. Wastes* **22,** 326 (1950).
198. Beuerlein, P., *Oel u. Kohle* **38,** 209 (1942).
199. Marke, R., *Oel u. Kohle* **38,** 176 (1942).
200. Baum, G., *Oel u. Kohle* **40,** 220 (1944).
201. Powell, D., *Safety Eng.* **93,** No. 3, 16 (1947).

202. Dey, B. B., Govindarchi, T. R., and Rajagopalan, S. C., *J. Sci. Ind. Research (India)* **4**, 559 (1946).
203. Pollack, A., *Chem. Ztg.* **76**, 141 (1952).
204. Kadmer, E. H., *Petroleum Refiner* **25**, No. 7, 351 (1946).
205. Zentner, R. (to Hoffmann-LaRoche Inc.), U. S. Pat. 2,628,930, Feb. 17, 1953.
206. Wallenmeyer, J. C., McDonald, F. G., and Henry, R. L. (to Mead Johnson & Co.), U. S. Pat. 2,650,895, Sept. 1, 1953.

CHAPTER 9

Demulsification

In previous chapters, and especially in Chapter 8, the chief concern has been with the production of more-or-less stable emulsions. The present chapter concerns itself with the opposite process, i.e., demulsification, or emulsion breaking. In many laboratory situations or industrial processes, unwanted emulsions are formed, very often of considerable stability, and the breaking of these emulsions represents a problem of considerable complexity.

The discussion of demulsification is divided into three parts. First, the manufacture of butter (which is actually a combined inversion and breaking) will be considered briefly. Following this, the more general methods of breaking W/O and O/W emulsions will be discussed.

Butter

In the discussion of cosmetic emulsions, it was pointed out that these represent a form of emulsion technology which has its roots in antiquity. In a similar vein, it may be considered that the formation of butter represents one of the earliest examples of a practical demulsification process, the demulsification of milk.

Manufacture of Butter. Milk, as it occurs in nature, is a fairly dilute O/W emulsion, the oil phase consisting of the materials known as butter fats. Small quantities of protein and sugar, and some inorganic salts, are also found. The emulsion is presumably stabilized by the presence of a mixed phosphatide-albumin complex at the oil-water interface. Cow's milk contains an average of slightly less than 4 per cent of fat.

Milk, of course, is only a moderately stable emulsion. The phenomenon of creaming, as has been pointed out, takes its name from the tendency of milk to separate into oil-rich and oil-poor fractions. (As is well-known, milk may be completely stabilized by homogenization, cf. p. 230).

However, the phenomenon of creaming is also used to produce a concentrated emulsion. Either by standing or centrifugal separation, the cream is separated from milk, producing an emulsion which contains some 36 per cent of butter fat. This more concentrated emulsion, when subjected to gentle agitation, particularly at moderately low temperatures (13° to 18° C), instead of becoming further stabilized, undergoes a combined breaking and inversion process to form butter, which is a W/O emulsion containing approximately 85 per cent of butter fat.

In the actual processing, the cream becomes steadily more viscous as a result of the aggregation of the fat particles. The incorporation of air in in the mixing process leads to the formation of a pronounced foam. It is noted that just as this foam collapses, the butter "comes," i.e., the little granules or corns of butter become visible to the naked eye. These lumps, when subjected to further processing, become the butter of commerce.

Theory of Butter Formation. Clayton[1] has discussed two principal theories of butter formation, which he designated the phase-inversion theory and the foam theory. According to the first of these theories, a simple phase inversion occurs in the churning process. Clayton feels that this cannot be correct; that the butter, when it "comes," does not represent an inverted emulsion, but a coarse suspension of the fatty granules. It is only on further processing that the water is worked into the fat to form the final W/O emulsion.

On the other hand, Clayton believes that the foaming causes desorption of the emulsifier from the oil-water interface, with consequent breaking of the original cream emulsion.

A recent discussion of butter formation by Mohr[2] agrees that the final emulsion is produced in the finishing process, but argues that the emulsion breaks because the low temperatures cause the "protective sheath" of stabilizer to crystallize, while the fat is still fairly fluid.

Additional light has been thrown on the process by the work of Sandelin.[3] This worker investigated the separation in "synthetic" butters formed by emulsifying butter fat in aqueous solutions of a number of emulsifying agents. Each emulsion was skimmed with a separator, and the resulting cream churned. Titrimetric data show that the fat globules carry more acid than alkali into the cream and butter, from solutions of sodium and potassium oleate, sodium abietate, gum arabic, and saponin. This seems to indicate hydrolytic adsorption with anions predominating in the interfacial layer. The opposite effect is found with solutions of sodium caseinate and egg albumin. Sandelin concludes that breaking of fat emulsions by agitation may be due to the discharge of adsorbed anions by frictional electricity, while, in agreement with Clayton, he points out that desorption by foaming may be significant.

Petersen[4] has reviewed the theories of butter formation in a paper which contains some interesting photomicrographs.

Butter Fat. For some applications the complete separation of the butter fat or oil from the cream is desirable. This is most simply done by heating melted butter until the water boils off, producing clarified butter or *ghee*, widely used in the East as a cooking fat.

The separation may also be accomplished by less stringent techniques. Patton[5] has reported on an extensive survey of chemical demulsifiers. In studying the effectiveness of various organic compounds to destroy the normal emulsion of cream during heating, the acids, alcohols, and amines were found to be most effective; the aldehydes, ketones, and esters least effective. Maximum demulsifying power in a given group appeared to occur in the vicinity of the compound containing four carbon atoms. Based on these findings, an aqueous reagent containing butylamine and butanol was developed. It is capable of demulsifying a number of liquid dairy products, including pasteurized and homogenized milks, light and heavy creams, ice cream mix, etc.

In another study, Stine and Patton[6] studied the effect of surface-active compounds. Ninety-seven commercially available agents were screened, and twenty-six of them were found to be of value. Various amounts of the surface-active agent were added to cream samples, the mixtures heated on a water bath, and then centrifuged. Most of the successful agents showed a clear layer before centrifuging. One of the effective agents (*Tergitol* 7) was studied intensively in regard to the variables of agent concentration, fat content of cream, and heat treatment.

Water-in-Oil Emulsions

Probably the largest literature on the subject of demulsification concerns itself with the breaking of W/O emulsions. This is no accident, for the industrially important oil field emulsions fall into this class, and their demulsification is an economically important technique.

Oil Field Emulsions. As crude petroleum rises from the fissures of the earth, its passage through narrow openings, accompanied by water and gases, and agitation by pumping, give rise to conditions favorable to formation of W/O emulsions. It has been estimated that fully 25 per cent of all the crude petroleum produced in the United States reaches the surface in emulsion form.[7] Such oil-field emulsions cannot be processed without first removing the major part of the water.

Morrell and Egloff[8] have listed seven basic methods by which such emulsions may be dehydrated. They are:

1. Settling.
2. Heating or distilling at atmospheric pressure.
3. Heating or distilling at elevated pressures.
4. Electrical dehydration.
5. Use of chemicals.

6. Centrifuging.
7. Filtration.

All of these methods, and combinations of these methods, are used extensively at the present time, and the literature is extensive. Probably the most used method is the chemical separation technique developed by a pharmacist, William S. Barnickel.[8] The electrical technique, invented by F. G. Cottrell,[10] is probably of equal value.

In general, the chemical methods are based on the technique of introducing an agent which counteracts the stabilizing influence of naturally-present emulsifying agents. The original chemical treatment of Barnickel involved the addition of ferrous sulfate, and numerous formulations involving inorganic salts followed. However, the modern techniques involve the use of more-or-less complex organic surface-active agents. Their action may often be conveniently understood in terms of their effect on the effective HLB of the stabilizers (cf. pp. 189–196).

Monson and Stenzel[11] have given a schematic drawing of a chemical demulsification plant (in which other than chemical methods are applied). The plant is reproduced as a block-diagram in Figure 9-1. The authors point out that a practical plant would probably not contain all the units illustrated, but might contain almost any combination. In this plant the chemical demulsifier is metered into the flow line immediately at the well-head. The liquid then passes into a gas separator for the removal of dissolved or occluded gases. Following this, it is passed into a heating unit, which presumably accelerates the chemical action somewhat, as well as adding the beneficial effect of the heat itself. (Indeed, a great many such emulsions are separable by no more than heating following by settling.) The emulsion then passes to a settling tank, in which the major separation takes place, and then is passed through the hay column (so-called because

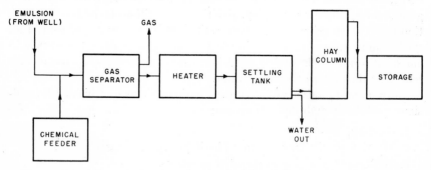

FIGURE 9-1. Schematic drawing of a chemical demulsification plant for the resolution of oil-field emulsions.[11] Not all of the units shown here would be included in a practical plant.

it is basically a column packed with hay or excelsior) in which further separation may take place, presumably because of the large surface offered. The dehydrated petroleum is then pumped to storage.

A petroleum emulsion subjected to such treatment will contain less than 1 per cent of water. Similar results can be achieved by the application of high-frequency alternating voltages. High-speed motion pictures have demonstrated that the effect of the voltage is to cause flocculation of the water droplets, and that the primary effect occurs within a fraction of a second following the application of current.[12]

Extension discussion of the older literature on these methods is given by Berkman and Egloff,[13] Morrell and Egloff,[14] and Monson and Stenzel.[15]

A more recent technique in the application of chemical demulsifiers is the so-called "down-the-hole" method, in which the demulsifying chemicals are pumped down into the well, to prevent the emulsification from taking place. This is discussed briefly by Cardwell.[16]

Literature on Oil-Field Emulsions. In addition to the summaries quoted above, recent publications include the following: Shaskin[17] describes the techniques used in certain Russian oil fields. These involve chemical treatment with black petroleum sulfonaphthenic acids, which brings the water content down to 0.5 per cent. Other Russian techniques are discussed by Aslanova.[18]

Zwick[19] has treated the problem of resolving emulsions formed in Colombian crude oil. This involves the use of heating, chemical treatment, and passage through an excelsior-packed column. Passler[20] discusses a technique of electric dehydration, and concludes that the energy requirements for twenty-four hour operation of a plant which will demulsify 80 to 100 cubic meters of an emulsion containing 40 per cent water is 15 to 20 kw-hrs. In the operation, the original droplet size of about 0.0001 mm increased to 0.01 mm after four minutes of treatment, and to 0.05 mm after eight minutes.

Fussteig[20a] has described the technique of heating under pressures of 3 to 4 atmospheres.

Blair[21] gives an extensive evaluation of the various methods used, and concludes that chemical treatment is the most economical. Such chemicals are most efficient when added near the well-head. Beaver[22] has discussed the use of Cottrell electrical demulsification equipment.

Monson[23] has also discussed the economics of chemical treatment, and, in another paper[24] has treated of the laboratory techniques which go into the evaluation of the correct demulsifier for a particular application. Davis and Crippen[25] have discussed the same problem, and have shown that the effectiveness of a demulsifier can be estimated by observing the color change of the emulsion.

Pettefer[26] has recently discussed the use of high-voltage electrical techniques in heavy Californian crudes. Ackelsburg[27] has suggested the use of ultrasonics in the resolution of oil-field emulsions.

In addition, to these more widely-used methods, others, such as centrifugation and filtration, are employed. For example, centrifugation at very high rates (i.e., 17,000 to 40,000 rpm) will usually break emulsions of this sort. Materials containing 70 to 80 per cent of water may often be reduced to a water concentration of 0.5 per cent.

Filtration through fritted glass or filter cakes of infusiorial earth will reduce the water-content of oil-field emulsions to as little as 0.2 per cent. In order to effect separation by this method it is necessary that the filter medium be selectively wetted by water.

Patents on Petroleum Demulsification. As has been indicated above, the patent literature on the resolution of oil-field emulsions is extensive. A number of recent patents of interest will, however, be cited.

Campbell[28] has revealed a method of separation involving contacting the emulsion with an aqueous brine of the same composition as the aqueous phase of the emulsion, thus inducing agglomeration of the water particles.

Oehler[29] has discussed the use of porous materials such as excelsior for the purpose. Kirkbride[30] has suggested filtration through fibrous glass at elevated temperatures, while Adams, Barlow, and Shapiro[31] have revealed the use of an apparatus in which the emulsion is contacted with fragmented glass. Glass wool as a filtration medium is also described in the use of an apparatus patented by Barton.[31a]

Schiel[32] has revealed a technique for breaking oil-field emulsions by filtration under pressure and at temperatures of 60 to 140° C.

Hatfield[33] has patented the use of a porous mass of sized carbon particles bonded with carbon as a filtration medium for demulsification. In another patent,[34] he describes filtration through such an aggregate, while at the same time a high voltage is impressed between this and another electrode.

Recent patents on electrical demulsification include a method due to McDonald[35] which is suitable for the demulsification of emulsions containing small amounts of water. Perkins has described treatment of the emulsion with alkaline[36] and acid[37] agents prior to electrical demulsification.

Small[38] has described the use of high-frequency vibration in conjunction with the use of electrical discharge maintained at the point of maximum compression. In this method, the voltage and current requirements are dependent on the specific gravity of the oil.

Deutsch[39] has described the use of intermittent direct current superimposed on alternating current at intervals of several seconds. In another patent,[40] the emulsion is made the dielectric of a condenser and subjected to the alternate charge and discharge of a high-voltage source.

A combination of electrical and gravitational separation methods is the subject of a patent by Eddy.[41]

Even more numerous than those describing particular types of apparatus are the patents describing the preparation of chemical demulsifiers. This is largely due to the fact, alluded to earlier, that oil-field emulsions vary in their characteristics, and a chemical agent which will work very well with one emulsion may be totally ineffective against another. Some indication of the wide variety of compounds which have been proposed for this purpose is given below.

De Mering[42] has described the use of sulfonated mineral oils. Wayne[43] has revealed the use of modified alkyd resins of various types, while Blair[44] has patented the use of subresinous high molecular weight materials derived from maleic anhydride.

Bonnet[45] has revealed the use of nitrosophenols or nitrosoaromatic carboxylic acids; in another patent the use of picric acid or other nitro compounds is described.[46]

Tranoski and Uhlmann[47] uses a compound such as acetone in combination with a sulfonated fatty acid to effect demulsification. Stagner[48] uses a mixture of sodium silicate and sodium hydroxide in such proportions that the ratio of Na_2O to SiO_2 is not less than 0.7. As an additive in electrical dehydration, Hanson[49] has recommended the use of complexing agents such as alkali metal hexametaphosphates or pyrophosphates for emulsions in which the aqueous phase contains appreciable amounts of calcium ions.

Goodloe and Berger[50] suggest the use of a water-soluble, oil-insoluble mahogany sulfonate, preferably the sodium salt. Blair[51] has patented the use of alkanolamine esters of condensation products of α-ethylenic acids or anhydrides with unsaturated fatty acids, while Bond and Savoy[52] reveal the use of a soap formed by the reaction of a mineral acid with the complex amine formed by reaction between formaldehyde, phenol, and a non-aromatic secondary amine.

Savoy[53] recommends the use of alkali metal salts of sulfonated mono- and dicyclic terpenes, such as pinene and limonene.

De Groote and Keiser[54] have revealed the use of water-soluble, organic solvent-insoluble polymers of various types.

A large number of patents include the use of various nitrogen derivatives. For example, De Groote, Keiser, and Blair[55] have described the use of derivatives of hydroxylamine, while De Groote and Keiser[56] have revealed the use of diamine compounds.

Clayton[57] suggests the use of salts of basic amines derived from the combination of a water-soluble petroleum sulfonic acid and a heat-polymerized basic hydroxyamine; in another patent,[58] amine salts of alkylated naphthalene sulfonic acids are described.

De Groote and Keiser[59] have also suggested the use of derivatives of quinoline and pyridine for this application, while Wayne[60] has recommended such compounds as the N-alkylated sulfonamides of a water-soluble petroleum sulfonic acid.

Polymerized aminoalcohols are the subject of a patent by Monson, Anderson, and Jenkins.[61] Monson[62] has also revealed the use of substitute imidazolines, while Hughes[63] has suggested the use of azolidine derivatives.

A number of patents cover the use of alkylene oxide condensation products of various sorts. Moeller[64] has described a number of such compounds, including ethylene oxide derivatives of modified castor oils. Boedeker[65] has patented the use of polyoxyethylene derivatives of an isomeric mixture of alkyl phenols. De Groote and Keiser[66] have patented a large number of such compounds. Salathiel[67] has discussed the use of compounds formed by heating linseed oil and nonaethylene glycol together.

De Groote[68] has revealed the use of propylene oxide derivatives of polyhydric compounds, in particular, of sugars.

Two interesting patents relate to the use of water-swellable cellulosic compounds. A French patent[69] describes the use of starch, while Burnam[70] has suggested the use of ethylcellulose.

Other Water-in-Oil Emulsions. Undesirable water-in-oil emulsions result from several industrial practices. An important type is the so-called gas-tar emulsion which arises in the production of illuminating gas by passing oil through heated brick, the emulsions forming in the wash-boxes and scrubbers of the plant. Aside from the disposal problem, the tars which may be separated from such emulsions have some commercial value as road-paving material, etc.

The techniques used in resolving such emulsions are similar to those employed for oil-field emulsions. Recently, Pearce[71] has discussed the use of electrical methods for this purpose. Batchelder[72] has patented a process for resolving such emulsions by solution of at least a portion of the hydrocarbon in a liquified hydrocarbon, normally gaseous at atmospheric pressure, e.g., propane. A complete discussion of the problem of the prevention and resolution of such emulsions has been given by Linden and Parker.[73]

A similar problem arises in the formation of emulsions of water in fuel oil, which is sometimes encountered on ships. Obviously, these emulsions are also amenable to treatments of the type used for oil-field emulsions. Preventive measures, involving the inclusion of the demulsifying agent in the oil at the time it is pumped into the fuel tanks, are often used. Lawrence[74] has recently considered the theoretical problems involved in the separation of this type of emulsion.

Oil-in-Water Emulsions

Of equal importance to the W/O emulsions discussed in the previous section are the O/W emulsions which also arise in various industrial practices. However, the quantity of material which has to be treated for separation is much less, and the variability in treatments which have to be used for a given system is smaller.

Wool Scouring Wastes. The most important unwanted industrial emulsion of the O/W type is probably the one which results from the process of scouring wool. According to Clayton,[75] this effluent contains emulsified wool wax (lanolin), soap residues, free fatty acids, higher alcohols, dirt, bacteria, and proteins. A typical effluent will contain 0.5 to 4.0 per cent wool wax, 0.1 to 0.4 per cent soap, and about 1 per cent of dirt.[76]

According to Truter,[77] there are four major methods of separating the grease component from the emulsions. These are:

Chemical processes:

1. Acid cracking (a) *per se.*
    (b) using ion-exchange resins.
2. Acid cracking as a subsidiary to calcium salt precipitation.

Mechanical processes:

3. Centrifugal separation.
4. Aeration or froth flotation.

In the regular acid-cracking process (which is probably the most widely-used), the liquor is first passed through a coarse filter to remove gross matter, and is then run into a large tank where enough sulfuric acid is added to reduce the pH (initially 9 to 10.5) to 4. After addition of the acid, the liquid is allowed to stand for a few hours, whereupon the grease settles out in the form of a sludge or "magma," which still contains about 90 per cent water. This is pumped to hot storage tanks from which it is fed continuously to a battery of filter-presses; following filtration, the liquor passes into a vessel in which complete separation takes place. A flow chart of this process is shown in Figure 9-2. Truter[77] should be referred to for a more complete discussion of this process, as well as the operating losses involved. An extended description of the Bradford acid-cracking process by Truter[78] is also extremely valuable.

Acidification can also be carried out by passage of the liquor through a bed of ion-exchange resin (e.g., *Zeo-Karb 225*), resulting in an exchange of sodium ions of the soap with hydrogen ions of the resin.

Addition of various heavy metal salts, e.g., magnesium, calcium, etc., has been proposed. Although careful control of the relative concentrations

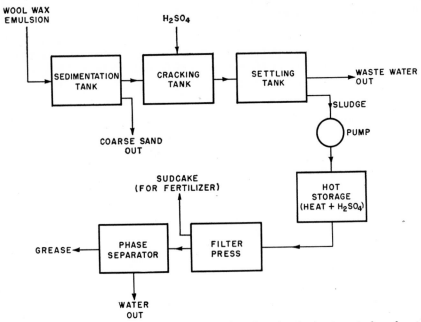

F<small>IGURE</small> 9-2. Flow diagram of an acid-cracking plant for the treatment of wool wax emulsions.[77]

of cations should make this a satisfactory method, such control is difficult to achieve in practise. However, the combination of acid-cracking with such chemical treatment has proven satisfactory.

The technique of centrifugation may be used for the separation of the wool-scouring waste emulsion, by taking advantage of the fact that the maximum difference in density of the wool wax and water occurs at 70° C. Numerous types of high speed centrifugal separators are used for this purpose.[79]

The other mechanical technique which may be employed is that of froth flotation. On agitation, the wool-scouring liquor forms an extremely stable froth, which is quite rich in wax content. Continued skimming of the froth gives a fairly satisfactory separation.

The earliest methods of using froth separation involved simple vigorous agitation, or "battage," with hand paddles. More elaborate versions of this, using mechanical agitation, have been proposed. Figure 9-3 illustrates a simple froth separation unit.

The more modern adaptation of this technique involves the use of jets of air or other gas in the form of a stream of fine bubbles. This forms an extremely active froth, and highly efficient separations are possible.[80]

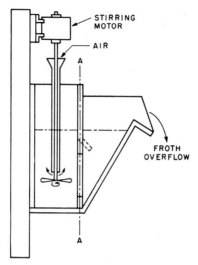

FIGURE 9-3. A simple froth-flotation cell, used for the demulsification of wool scouring wastes.

Other Oil-in-Water Emulsions. In contradistinction to the water-in-oil emulsions which occur in fuel tanks of ships, are the so-called "ballast water" emulsions of the oil-in-water type which form, as the name indicates, in the ballast tanks of ships. The disposal of these wastes is very often a problem when ballast has to be discharged in port. An aeration method, similar to that employed for wool wax, has been described.[81] Usually, the addition of a surface-active agent is required to assist in the separation.[82]

A similar undesirable emulsion arises in the operation of steam engines. The condensate from the cylinders of such engines is often found to form stable O/W emulsions with lubricating oil; the resolution of these emulsions is necessary to avoid the fouling of the boiler tubes.

In separation of these emulsions, electrolysis with iron electrodes has been used. The charged droplets of oil discharge on the electrodes; the discharge is further accelerated by the iron ions formed by the electrolysis. Hydrated iron oxide is formed, the whole mass precipitates, and is readily removed by filtration. Such a coagulate will consist of nearly equal quantities of the iron compound and the oil, with a trace of water.

The addition of the salts of multiply charged cations, e.g., aluminum, iron, barium, etc. is often successful in resolving these emulsions, as is the use of filtration through filter discs of calcium carbonate. The oil is retained on the disc, while the water passes through.

The general methods of emulsion resolution which have been described in this chapter are, of course, of general application. Suitable variations will recommend themselves to the individual worker. Many methods which have been used empirically for years may be readily understood, and possibly improved, by the application of the theoretical considerations relating to emulsion stability outlined in Chapters 4 and 5.

Bibliography

1. Clayton, W., "Theory of Emulsions," 4th ed., pp. 428–435, Philadelphia, The Blakiston Co., 1943.
2. Mohr, W., *Fette u. Seifen* **57,** 925 (1955).
3. Sandelin, A. E., *Finska Kemistamfundets Medd.* **54,** 53 (1945).
4. Petersen, N., *Fette u. Seifen* **51,** 59 (1944).
5. Patton, S., *J. Dairy Sci.* **35,** 324 (1952).
6. Stine, C. M. and Patton, S., *J. Dairy Sci.* **35,** 665 (1952).
7. Berkman, S. and Egloff, G., "Emulsions and Foams," p. 219*n*, New York, Reinhold Publishing Corp., 1941.
8. Morrell, J. C. and Egloff, G., in Alexander, J. (ed.), "Colloid Chemistry," **III,** p. 505, New York, Reinhold Publishing Corp., 1931.
9. Monson, L. T. and Stenzel, R. W., in Alexander, J. (ed.), "Colloid Chemistry," **VI,** p. 538, New York, Reinhold Publishing Corp., 1946.
10. Cottrell, F. G., U.S. Pat. 987,115, March 21, 1911.
11. Monson, L. T. and Stenzel, R. W., *loc. cit.*, pp. 538–542.
12. Monson, L. T. and Stenzel, R. W., *loc. cit.*, pp. 542–545.
13. Berkman, S. and Egloff, G., *op. cit.*, pp. 219–395.
14. Morrell, J. C. and Egloff, G., *loc. cit.*, pp. 503–507.
15. Monson, L. T. and Stenzel, R. W., *loc. cit.*, pp. 535–552.
16. Cardwell, P. H., in Alexander, J. (ed.), "Colloid Chemistry," **VII,** pp. 455–475, New York, Reinhold Publishing Corp., 1950.
17. Shashkin, P. I., *Vostochnaya Neft* **1940,** No. 5–6, 55.
18. Aslanova, M. A., *Vostochnaya Neft* **1940,** No. 5–6, 59.
19. Zwick, B. F., *Oil Weekly* **101,** No. 10, 107 (1941).
20. Passler, W., *Oel u. Kohle* **37,** 194 (1941).
20a. Fussteig, R., *Teer u. Bitumen* **40,** 127 (1942).
21. Blair, C. M., Jr., *Oil Gas J.* **44,** No. 11, 116 (1945).
22. Beaver, C. E., *Trans. Am. Inst. Chem. Engrs.* **42,** 251 (1946).
23. Monson, L. T., *Petroleum World* **43,** No. 4, 41 (1946).
24. Monson, L. T., *Petroleum Engr.* **18,** No. 2, 67 (1946).
25. Davis, C. M. and Crippen, R. G., *Oil Gas J.* **50,** No. 9, 64 (1941).
26. Pettefer, R. L., *World Oil* **134,** No. 1, 168 (1952).
27. Ackelsberg, M. R., *City Coll. Vector* (N.Y.) **13,** 92 (1949).
28. Campbell, S. E., U.S. Pat. 2,270,412, Jan. 20, 1942.
29. Oehler, C. C., U.S. Pat. 2,257,244, Sept. 30, 1941.
30. Kirkbride, C. G. (½ to Owens-Corning Fiberglass Corp. and ½ to Standard Oil Co. of Indiana), U.S. Pat. 2,522,378, Sept. 12, 1950.
31. Adams, G. L., Barlow, R. G., Shapiro, A. (to Socony-Vacuum Oil Co.), U.S. Pat. 2,224,624, Dec. 10, 1941.
31a. Barton, P. D., U.S. Pat. 2,588,794, Mar. 11, 1952.
32. Schiel, R., Ger. Pat. 802,344, Feb. 8, 1951 (Cl. 23*b*. 105).
33. Hatfield, M. R. (to National Carbon Co.), U.S. Pat. 2,336,482, Dec. 14, 1943.

34. Hatfield, M. R. (to National Carbon Co.), U.S. Pat. 2,336,542, Dec. 14, 1943.
35. McDonald, L. E., (to Petrolite Corp., Ltd.), U.S. Pat. 2,366,565, June 5, 1945.
36. Perkins, R. B., Jr. (to Petrolite Corp., Ltd.), U.S. Pat. 2,447,529, Aug. 24, 1948.
37. Perkins, R. B., Jr. (to Petrolite Corp., Ltd.), U.S. Pat. 2,447,530, Aug. 24, 1948.
38. Small, A. D., U.S. Pat. 2,420,687, May 20, 1947.
39. Deutsch, W. I. (to Siemens-Lurgi-Cottrell-Elektrofilter G.m.b.h.), Ger. Pat. 688,135, Jan. 25, 1940 (Cl. 12d. 29/01).
40. Deutsch, W. I. (vested in Alien Property Custodian), U.S. Pat. 2,382,697, Aug. 14, 1945.
41. Eddy, H. C. (to Petrolite Corp., Ltd.), U.S. Pat. 2,315,051, March 30, 1943.
42. De Mering, B. S. (to Standard Oil Development Co.), U.S. Pat. 2,209,445, July 30, 1940.
43. Wayne, T. B., U.S. Pats. 2,214,783–4, Sept. 17, 1940.
44. Blair, C. M., Jr. (to Petrolite Corp., Ltd.), U.S. Pat. 2,216,310, Oct. 1, 1940.
45. Bonnet, C. F. (to American Cyanamid Co.), U.S. Pat. 2,260,798, Oct. 28, 1941.
46. Bonnet, C. F. (to American Cyanamid Co.), U.S. Pat. 2,301,609, Nov. 9, 1942.
47. Tranoski, P. T. and Uhlmann, E. H., U.S. Pat. 2,269,134, Jan. 6, 1942.
48. Stagner, B. A., U.S. Pat. 2,284,106, May 26, 1942.
49. Hanson, G. B. (to Petrolite Corp., Ltd.), U.S. Pat. 2,325,850, Aug. 3, 1943.
50. Goodloe, P. M. and Berger, H. G. (to Socony-Vacuum Oil Co.), U.S. Pat. 2,317,050, April 20,1943; U.S. Pat. 2,355,778, Aug. 15, 1944.
51. Blair, C. M., Jr. (to Petrolite Corp., Ltd.), U.S. Pat. 2,423,365, July 1, 1947.
52. Bond, D. C. and Savoy, M. (to Pure Oil Co.), U.S. Pat. 2,457,634, Dec. 28, 1948.
53. Savoy, M. (to Pure Oil Co.), U.S. Pat. 2,457,735, Dec. 28, 1948.
54. De Groote, M. and Keiser, B. (to Petrolite Corp., Ltd.), U.S. Pat. 2,499,369, Mar. 7, 1950; U.S. Pat. 2,499,370, Mar. 7, 1950.
55. De Groote, M., Keiser, B., and Blair, C. M., Jr. (to Petrolite Corp., Ltd.), U.S. Pat. 2,216,312, Oct. 1, 1940.
56. De Groote, M. and Keiser, B. (to Petrolite Corp., Ltd.), U.S. Pat. 2,241,011, May 6, 1941.
57. Clayton, E. E. (to Petrolite Corp., Ltd.), U.S. Pat. 2,300,972, Oct. 23, 1942.
58. Clayton, E. E. (to Petrolite Corp., Ltd.), U.S. Pat. 2,309,935, Feb. 2, 1943.
59. De Groote, M. and Keiser, B. (to Petrolite Corp., Ltd.), U.S. Pat. 2,335,262, Nov. 30, 1943; U.S. Pat. 2,334,390, Nov. 16, 1943.
60. Wayne, T. B., U.S. Pat. 2,335,554, Nov. 30, 1943.
61. Monson, L. T., Anderson, W. W., and Jenkins, F. E. (to Petrolite Corp., Ltd.), U.S. Pat. 2,407,895, Sept. 17, 1946.
62. Monson, L. T. (to Petrolite Corp., Ltd.), U.S. Pat. 2,589,198, Mar. 11, 1952.
63. Hughes, W. B. (to Cities Service Oil Co.), U.S. Pat. 2,638,451, May 12, 1953.
64. Moeller, A. (to I. G. Farbenindustrie A.-G.), Ger. Pat. 702,012, Jan. 2, 1941 (Cl. 23b. 1.05); (vested in Alien Property Custodian), U.S. Pat. 2,307,058, Jan. 5, 1943.
65. Boedeker, K. and Winnacker, K. (vested in Alien Property Custodian), U.S. Pat. 2,317,726, April 27, 1943.
66. De Groote, M. and Keiser, B. (to Petrolite Corp., Ltd.), U.S. Pat. 2,307,494–5, Jan. 5, 1943; U.S. Pat. 2,338,010, Nov. 16, 1943; U.S. Pat. 2,385,970, Oct. 2, 1945; U.S. Pat. 2,470, 808, May 24, 1949.
67. Salathiel, R. A. (to Standard Oil Development Co.), U.S. Pat. 2,401,966, June 11, 1946.

68. De Groote, M. (to Petrolite Corp., Ltd.), U.S. Pat. 2,552,528–34, May 15, 1951.
69. N.V.W.A. Scholten's chemische Fabrieken, Fr. Pat. 851,549, Jan. 10, 1940.
70. Burnam, T. W., U.S. Pat. 2,327,996, Aug. 31, 1943.
71. Pearce, C. A. R., *Brit. J. Appl. Phys.* **6,** 68, 113 (1955).
72. Batchelder, H. R. (to United Gas Improvement Co.), U.S. Pat. 2,383,362, Aug. 21, 1945.
73. Linden, H. R. and Parker, R., "Prevention and Resolution of Tar Emulsions in High-BTU Oil Gas Production," Chicago, Institute of Gas Technology, 1953.
74. Lawrence, A. S. C., *Chem. and Ind.* **1948,** 615.
75. Clayton, W., *op. cit.*, p. 427.
76. Truter, E. V., "Wool Wax," p. 108, New York, Interscience Publishers, Inc., 1956.
77. Truter, E. V., *op. cit.*, p. 110.
78. Truter, E. V., *op. cit.*, pp. 307–318.
79. Truter, E. V., *op. cit.*, pp. 117–124.
80. Truter, E. V., *op. cit.*, pp. 125–131
81. Monson, L. T. and Stenzel, R. W., *loc. cit.*, pp. 550–551.
82. Roberts, C. H. M. and Niswander, R. V. (to Petrolite Corp., Ltd.), U.S. Pat. 2,260,757, Oct. 28, 1941.

Testing of Emulsion Properties

In the evaluation of emulsions and of the materials which are used to prepare emulsions, there are a number of fairly standardized measurements which can be usefully made. This Appendix describes some of these tests in some detail; more detail will be found in the various treatises to which reference is made in the body of the Appendix.

Measurements of surface and interfacial tension, viscosity, the determination of emulsion type and stability, and, more briefly, the various electrical measurements of interest will be described.

Measurement of Surface and Interfacial Tension

Harkins[1] has catalogued the various methods of measuring surface and interfacial tension with respect to their accuracy and with respect to their applicability to pure liquids or solutions. The classification scheme is as follows:

A. Single Liquids:
    i. Capillary rise.
    ii. Drop weight.
    iii. Ring method.
    iv. Bubble pressure.
    v. Hanging drop.
B. Solutions:
    i. Drop weight.
    ii. Ring method.
    iii. Sessile bubble or drop.

Ferguson[1a] has also devised a classification scheme for the techniques of the measurement of surface tension, based on whether or not the contact angle is critical. This classification is reproduced as Table A-1; the methods discussed in detail in the present section are starred.

The capillary height method is considered to be the one capable of the highest accuracy. However, it is slow for surface tension, and is not as satisfactory for interfacial tension as the drop weight method (but cf. below, p. 303). Harkins considers the drop weight method to be the best

TABLE A-1. METHODS FOR MEASUREMENT OF SURFACE TENSION[1a]

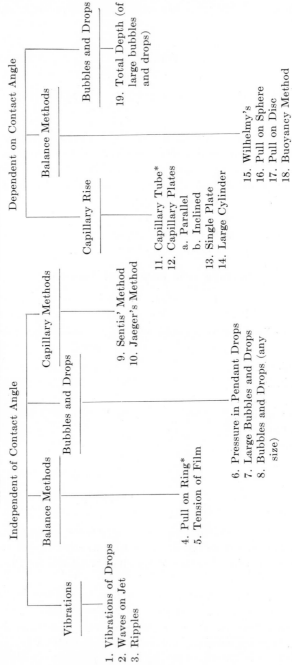

Independent of Contact Angle

Vibrations
1. Vibrations of Drops
2. Waves on Jet
3. Ripples

Balance Methods

Bubbles and Drops
4. Pull on Ring*
5. Tension of Film

6. Pressure in Pendant Drops
7. Large Bubbles and Drops
8. Bubbles and Drops (any size)

Capillary Methods
9. Sentis' Method
10. Jaeger's Method

Dependent on Contact Angle

Capillary Rise
11. Capillary Tube*
12. Capillary Plates
   a. Parallel
   b. Inclined
13. Single Plate
14. Large Cylinder

Balance Methods
15. Wilhelmy's
16. Pull on Sphere
17. Pull on Disc
18. Buoyancy Method

Bubbles and Drops
19. Total Depth (of large bubbles and drops)

20. (Unclassified) Drop-weight Method*

general method for surface and interfacial tensions when both accuracy and speed are considerations.

The ring method is very fast and quite accurate for surface tension; however, Harkins feels that it has been insufficiently tested for interfacial tension. The bubble pressure method is considered to be moderately good; while the hanging drop is, at present, the least accurate method.

For solutions, the drop weight method is preferred for both surface and interfacial tension, provided long time effects are not involved. The ring method is excellent for surface tension even when time effects are involved, and the sessile bubble or sessile drop methods are extremely good when long-term time effects are involved.

Other methods for the measurement of surface tension have been described, such as, e.g., the method of ripples, and the behavior of oscillating jets and drops.[2]

Capillary-Height Method. As presented in most elementary treatises, the theory of the capillary-height method is somewhat inexact, although the equation which results is quite satisfactory for tubes of small diameter.

If a tube of small radius dips into a vessel containing a liquid, the liquid will rise in the tube (Figure A-1), and the liquid in the tube will be in hydrostatic equilibrium with the liquid in the bulk container. The liquid film at the meniscus will exert a linear pull equal to the circumference of the tube $2\pi r$. If the surface tension of the liquid is γ, and the contact angle at the meniscus is θ, the upward force F_u is equal to $2\pi r \gamma \cos \theta$. In most

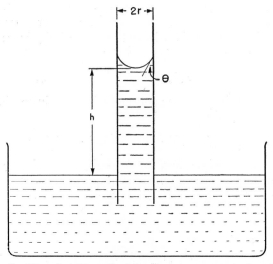

FIGURE A-1. Capillary-height method for the measurement of surface tension.

cases $\vartheta = 0$, or very nearly so, and

$$F_u = 2\pi r\gamma \tag{A.1}$$

The weight of the liquid up to the bottom of the mensicus (i.e., in the distance h) is equal to $\pi r^2 hd$, where d is the density of the liquid. The *total* weight supported by the surface tension of the liquid is this weight *plus* the weight of the liquid in the meniscus, W'. Thus, the force acting downward in the capillary tube is

$$F_d = \pi r^2 hgd + W' \tag{A.2}$$

At equilibrium,

$$F_u = F_d \tag{A.3}$$

or,

$$\gamma = \tfrac{1}{2}rhgd + W'/2\pi r = \tfrac{1}{2}rhgd + mg/2\pi r, \tag{A.4}$$

where m is the unknown, but very small, mass of the liquid in the meniscus. For very narrow capillaries, this quantity is negligible and Eq. A.4 reduces to

$$\gamma = \tfrac{1}{2}\, rhdg, \tag{A.5}$$

which is the commonly quoted result.

When the angle of contact is not zero it must be determined, and the right-hand side of Eq. A.5 multiplied by the cosine.

Harkins[3] has pointed out that when the vapor of the liquid around the tube has an appreciable density (i.e., a high vapor pressure), then a part of the weight of the liquid is balanced hydrostatically by the vapor. Eq. A.2 then becomes

$$F_d = \pi r^2 hg(d - d_0) + gv(d - d_0), \tag{A.6}$$

where d_0 is the density of the vapor, and v is the volume of the meniscus. Equation A.4 then becomes

$$\gamma = \tfrac{1}{2}rhg(d - d_0) + gv(d - d_0)/2\pi r \tag{A.7}$$

The volume of the meniscus may not be negligible and it may be necessary to correct for it. For tubes of quite small diameter, it is possible to write the equation in the form

$$\gamma = \tfrac{1}{2}rh'g(d - d_0); \tag{A.8}$$

where h' is a corrected height

$$h' = h + r/3 \tag{A.9}$$

This method of correction is not always satisfactory, in which case the quantity v has to be estimated by the method of Bashforth and Adams, described in detail by Adam.[4]

The capillary-height method may also be used for the measurement of interfacial tension, using the equation

$$\gamma_i = \tfrac{1}{2}rhg(d_1 - d_2), \tag{A.10}$$

where d_1 and d_2 are the densities of the two liquid phases, respectively. Since the density difference may be quite small, large values of h are often encountered, even for large capillaries. Harkins and Humphery[5] found, for example, that the tension at a benzene-water interface gave a rise of 78 mm in a tube of 1.4 mm diameter. An elegant procedure for making this measurement is described by these authors; however, it should be noted that the meniscus angle at liquid-liquid interfaces is seldom zero, and this interposes a problem.

Capillary-Height Apparatus. Apparatus for measurement of surface tension by capillary height may vary from the extremely simple devices described in undergraduate physical chemistry laboratory manuals (and available at small cost from most supply houses) to quite elaborate arrangements.

The simplest type of capillary rise apparatus consists of a capillary tube mounted, through a stopper, in a large diameter test-tube. Tubing from broken thermometers is often employed for this purpose, since the diameter of a fairly good thermometer capillary is sufficiently small and sufficiently uniform to be used. The calibrations which are already on the tube may often be used, and permit the substitution of a hand lens for a cathetometer in measuring the capillary rise.

Much more elaborate systems have been described by Richards and Carver,[6] Young, Gross, and Harkins,[7] and Harkins and Jordan.[8]

A variation of the method is due to Ferguson and Dowson,[9] and makes use of the measurement of the pressure which has to be imposed in order to force the meniscus in the tube back to the level of the liquid in the outside tube. Ferguson[10] has also reported on methods for the measurement of the surface tension of small quantities of liquids, as have Mouquin and Natelson[11] and Nevin.[11a] Methods depending on the rise between parallel sheets of glass have also been described.[12]

In all capillary-rise measurements, the cleanliness of the capillary is most important. Harkins and Humphery[5] recommend steaming out prior to use.

It is necessary to know the radius of the capillary in order to apply Eq. A.5. Adam[13] points out that perfect circularity, although frequently in-

TABLE A-2. SURFACE TENSIONS OF PURE LIQUIDS FOR CALIBRATION
(*In Air, 20° C*)

Liquid	Harkins & Brown[15]	Richards & Carver[6]
Water	72.80	72.73
Benzene	28.88	28.88
Toluene	—	28.43
Chloroform	—	27.14
Ethyl Ether	—	16.96

sisted on, is not necessary; a 6 per cent ellipticity gives rise to an error of less than 0.1 per cent.

If a capillary of uniform bore is obtained, the radius may be measured microscopically with a filar micrometer. Harkins[14] describes an elaborate method, involving the weighing of the capillary filled to various extents with mercury, which permits the determination of the per cent deviation from the average diameter. An experimentally derived curve, correlating the average deviation with a correction for the capillary height, may be constructed.

For most routine work, a usable diameter can be obtained by calibration with liquids of known purity. Young, Gross, and Harkins[7] give the following relation between the surface tension of water and temperature:

$$\gamma_{H_2O} = 75.680 - 0.138t - 0.0_3356t^2 + 0.0_647t^3 \qquad (A.11)$$

Values for the surface tension of a number of pure liquids at 20° C, as reported by Harkins and Brown[15] and Richards and Carver[6] are given in Table A-2.

When employed with all possible precautions, the capillary-rise method is the most precise method for determining surface tensions. It is less useful for the determination of interfacial tension, but within the limitations and the techniques described by Harkins and Humphery[5] it could yield useful results.

Drop-Weight Method. If a drop of liquid is imagined as emerging from a capillary tube in the form of a cylinder of radius equal to that of the tube, the maximum weight of liquid which could be supported by surface forces would be given by

$$W = mg = 2\pi r\gamma; \qquad (A.12)$$

this is known as Tate's Law.[16]

The mass of the drop which is detached from such a capillary tip is, however, found to be less than that predicted by Eq. A.12. There are two reasons for this: first, and obviously, the edge of the drop is only rarely vertical; secondly, the whole drop is not detached. On detachment, both a large and a small drop are detached, while a small portion of the original drop remains on the tip.

The mathematical theory of this behavior has been discussed by Lohnstein[17] and Harkins and Brown.[15] The latter authors found that the weight of the drop was a function of the quantity r/a,

$$W = 2\pi r \gamma f(r/a), \tag{A.13}$$

where r is the capillary radius, and a is a quantity known as the capillary constant,[4] and defined as

$$a = (r/h)^{1/2}, \tag{A.14}$$

where h is the capillary rise for a given liquid in a tube of radius r. Obviously, a has a characteristic value for each liquid.

In general, the quantity a is not known, but Harkins and Brown point out that $f(r/a)$ is also a function of $r/V^{1/3}$, where V is the volume of the drop. Thus, Eq. A.13 may be written

$$W = 2\pi r \gamma \phi(r/V^{1/3}), \tag{A.15}$$

where, although the function ϕ is different from the function f, it has the same numerical values.

If a new function F is defined, such that

$$F = \tfrac{1}{2}\pi\phi(r/V^{1/3}), \tag{A.16}$$

then

$$\gamma = \frac{mg}{r} F \tag{A.17}$$

Harkins and Brown have compiled values of F showing that, if this correction is applied, the drop-weight method is capable of great accuracy. Values of F, according to Harkins and Brown,[15, 18] are given in Table A-3.

Drop-Weight Apparatus. The basis for determining surface tension by the drop-weight method is the determination of the weight of a drop obtained from a capillary of known radius (or, what is essentially the same thing, the drop volume). This is done experimentally by weighing a large number of drops, and calculating the average weight. Substitution into Eq. A.17 (using Table A-3) gives the surface tension.

In order to use the corrections, the volume V of the drop must be known. This is, of course, calculated from the weight by use of the density. This last quantity need not be known with a high degree of precision, since the correction is not very sensitive to density differences.

TABLE A-3. VALUES OF F FOR DROP-WEIGHT CORRECTIONS[15, 18]

V/r^3	F	V/r^3	F	V/r^3	F
5000	0.172	2.3414	0.26350	0.729	0.2517
250	.198	2.0929	.26452	.692	.2499
58.1	.215	1.8839	.26522	.658	.2482
24.6	.2256	1.7062	.26562	.626	.2464
17.7	.2305	1.5545	.26566	.597	.2445
13.28	.23522	1.4235	.26544	.570	.2430
10.29	.23976	1.3096	.26495	.541	.2430
8.190	.24398	1.2109	.26407	.512	.2441
6.662	.24786	1.124	.2632	.483	.2460
5.522	.25135	1.048	.2617	.455	.2491
4.653	.25419	0.980	.2602	.428	.2526
3.975	.25661	.912	.2585	.403	.2559
3.433	.25874	.865	.2570		
2.995	.26065	.816	.2550		
2.637	.26224	.771	.2534		

N.B.: The experimental error in these corrections is within 0.1 per cent when V/r^3 lies between 2.637 and 1.2109, and is within 0.2 per cent between 10.29 and 0.865.

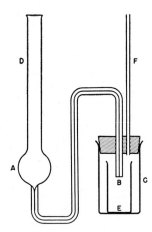

FIGURE A-2. Simple form of drop-weight apparatus for the determination of surface tension.

The simplest type of drop-weight apparatus is shown schematically in Figure A-2. The liquid whose surface tension is to be determined is drawn into the tube through the tip, B, until it is filled to just above the bulb, A. The tared weighing bottle, E, is placed in the chamber, C, and the whole apparatus immersed in a constant temperature bath so that only tubes D and F emerge from the bath liquid. Slight vacuum is applied to the system through F, until the droplet has *almost* reached its full size. It is then allowed to drop off the tip by itself. This is repeated until some convenient number (e.g., twenty-five) of drops have been so collected. The average weight and volume of a single drop is then readily determined.

In the construction of all drop-weight equipment the nature of the capillary tip is most important. According to Harkins,[19] it should be made of straight, heavy-walled Pyrex capillary tubing, with the edge ground absolutely plane, and the circumference machined with the aid of a lathe to absolute circularity.

A convenient, well-built, drop-weight unit capable of high precision, and constructed according to the designs of Harkins, is available from many laboratory supple houses. Many more elaborate units have been described in the literature. For example, Harkins and Harkins[20] have described a unit which is suitable for the determination of the surface tension of volatile liquids.

The drop-weight method is also applicable to the determination of interfacial tension. An apparatus suitable for this purpose, according to Harkins and Humphery[5] is shown in Figure A-3. In using this device, the aqueous phase is placed in the pipette *ABD*, and allowed to drop into the oil phase, held in a vessel (not shown), in the container *F*. For interfacial tension,

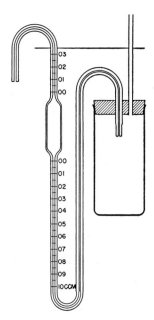

FIGURE A-3. Drop-weight apparatus suitable for the measurement of interfacial tensions, after Harkins and Humphery.[5]

Eq. A.17 takes the form

$$\gamma = V(d_1 - d_2)g/2\pi r\phi(r/V^{1/3})$$
$$= [V(d_1 - d_2)g/r]F,$$

(A.18)

where d_1 and d_2 are the densities of the respective phases, V is the *volume* of the drop, and F is the correction function derived from Table A-3.

More complex drop weight units have been devised for interfacial tension by Harkins and co-workers.[21]

Philippovich[22] has described a semimicro drop-weight device which can be used to make measurements on as little as 0.5 ml of liquid, but which, in some cases, showed disagreement of the order of 9 per cent with other methods.

McGee[23] has described a technique in which the drop-weight is measured as a function of the period of drop fall. The results so obtained are extrapolated to an infinite time of fall. This method should be of value in systems where aging effects are significant, and is probably capable of high precision.

Ring Method. Methods depending on the detachment of a wire or plane surface have been known for a long time. Thus, Wilhelmy determined surface tension by the measurement of the force required to detach a flat plate (such as a glass microscope slide) while Timberg introduced the idea of using a ring of wire. The use of the ring is basic to most modern detachment devices; the theory of this method is considered below.

In using the ring method, the ring is placed in the liquid surface (or interface) and the force required to detach the ring from the surface is measured. If one assumes that as the ring is pulled upward it supports a cylinder of liquid, the total pull P required to detach the ring is equal to the mass of the liquid in the cylinder, and

$$P = Mg = 2\pi\gamma R' + 2\pi\gamma(R' + 2r) = 4\pi\gamma(R' + r) = 4\pi\gamma R, \quad \text{(A.19)}$$

where M is the mass of liquid supported by the ring, R' is the *inside* radius of the ring, r is the radius of the wire of the ring (thus the outside diameter of the ring $= R' + 2r$), and R is the average radius of the ring $(R' + r)$.

Actually, of course, Eq. A.19 is not correct, due to the fact that the liquid pulled away from the surface is not a cylinder. The shape that is observed experimentally is shown in Figures A-4A and A-4B, showing the distension of the surface film when the ring displacement is small, and just before detachment, respectively. A similar distension is observed in interfacial measurements (Figure A-5).

Harkins and Jordan[8] and Freud and Freud[24] have shown that the shape of the liquid pulled up by the ring is a function of the ratios R^3/V and

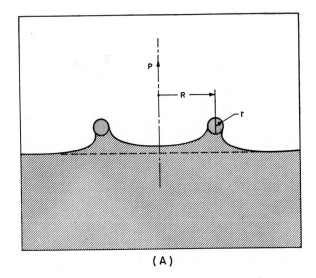

(A)

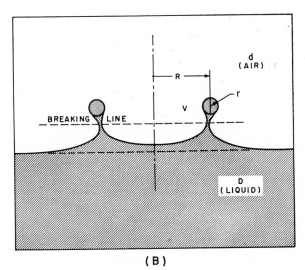

(B)

FIGURE A-4. Distortion of surface film by tensiometer ring. (A) Small displacement. (B) Large displacement, just before detachment. *Courtesy: Central Scientific Co.*

R/r, where V is the volume of the liquid pulled up. The shape is also a function of the surface tension, so that Eq. A.19 becomes

$$\gamma = (Mg/4\pi R) \cdot f(R^3/V, R/r) = (Mg/4\pi R) \cdot F \qquad \text{(A.20)}$$

The value of the correction function F may be obtained from the tables

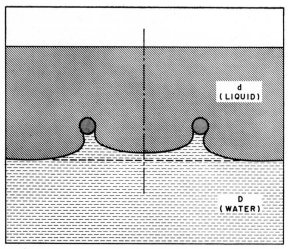

F<small>IGURE</small> A-5. Distortion of interfacial film by ring in interfacial tension measurements. *Courtesy: Central Scientific Co.*

calculated by Freud and Freud, reproduced as Table A-4. It should be noted that, over a long range, the correction term is quite close to unity; for approximate determinations the correction may not be required.

Recently, Fox and Chrisman[25a] have studied the corrections which must be applied in the study of liquids of high density and low surface tension. They have extended the Harkins and Jordan corrections into this range (Table A-5).

An alternate method of determining the correction factor F is by reading it from the graphs calculated by Zuidema and Waters.[25] Figure A-6 reproduces the curves for a ring 6 cm in circumference, giving the values of F as a function of $P/(D - d)$, where $(D - d)$ is the density difference at the interface (where the air/liquid interface is involved, d may be neglected). The curves are given for several values of the ratio R/r.

The need for correction may be obviated entirely by determining the values of P for a number of pure liquids, and constructing a calibration curve. This curve will, of course, be characteristic of the particular ring.[25b]

Ring Method Apparatus. In principle, any balance may be used to measure the pull on the ring. Harkins and Jordan[8] have described a chainomatic balance which can be used for this purpose, and which is capable of high precision. However, at present most ring measurements are carried out with the aid of the apparatus devised by du Nouy,[26] or modifications thereof.

In this device the pull is measured by means of a torsion-wire balance. A commercially available form of the du Nouy tensiometer is illustrated in

R^3/V	$R/r = 30$	32	34	36	38	40	42	44	46	48	50	52	54	56	58	60
0.30	1.012	1.018	1.024	1.029	1.034	1.038	1.042	1.046	1.049	1.052	1.054					
.31	1.006	1.013	1.018	1.024	1.028	1.033	1.039	1.041	1.044	1.046	1.049					
.32	1.001	1.008	1.012	1.019	1.023	1.028	1.033	1.035	1.039	1.041	1.045					
.33	0.9959	1.003	1.008	1.014	1.018	1.024	1.028	1.030	1.035	1.036	1.040					
.34	.9913	0.998	1.003	1.010	1.014	1.019	1.023	1.026	1.031	1.032	1.036					
.35	.9865	.993	0.999	1.006	1.008	1.015	1.019	1.022	1.026	1.027	1.031					
.36	.9824	.989	.995	1.002	1.005	1.010	1.015	1.018	1.022	1.024	1.027					
.37	.9781	.985	.991	0.998	1.001	1.006	1.011	1.014	1.018	1.020	1.024					
.38	.9743	.981	.987	.995	0.998	1.003	1.007	1.010	1.015	1.017	1.020					
.39	.9707	.977	.983	.991	.994	0.9988	1.004	1.007	1.011	1.013	1.017					
.40	.9672	.974	.980	.986	.991	.9959	1.000	1.004	1.008	1.010	1.013	1.016	1.018	1.020	1.021	1.022
.41	.9636	.970	.976	.983	.987	.9922	0.997	1.001	1.005	1.007	1.010	1.013	1.015	1.017	1.019	1.019
.42	.9605	.968	.973	.980	.984	.9892	.994	0.998	1.002	1.004	1.007	1.010	1.013	1.014	1.016	1.017
.43	.9577	.964	.970	.977	.981	.9863	.991	.995	.999	1.001	1.005	1.007	1.010	1.011	1.014	1.014
.44	.9546	.961	.967	.974	.979	.9833	.988	.992	.997	0.998	1.002	1.005	1.007	1.009	1.011	1.011
.45	.9521	.959	.965	.971	.976	.9809	.986	.990	.993	.996	0.9993	1.002	1.004	1.006	1.009	1.009
.46	.9491	.956	.962	.969	.973	.9779	.983	.987	.991	.994	.9968	1.000	1.002	1.004	1.006	1.007
.47	.9467	.954	.960	.966	.971	.9757	.980	.985	.988	.992	.9945	0.998	1.000	1.002	1.004	1.005
.48	.9443	.951	.957	.963	.968	.9732	.978	.983	.986	.989	.9922	.995	0.997	0.999	1.002	1.003
.49	.9419	.949	.955	.961	.966	.9710	.976	.981	.984	.987	.9899	.993	.995	.997	1.000	1.001
.50	.9402	.946	.952	.959	.964	.9687	.973	.978	.981	.985	.9876	.991	.993	.995	.997	.9984
.51	.9378	.944	.950	.956	.961	.9665	.971	.976	.979	.983	.9856	.989	.991	.993	.995	.9965
.52	.9354	.942	.948	.954	.959	.9645	.969	.974	.977	.981	.9836	.987	.989	.991	.994	.9945
.53	.9337	.940	.946	.952	.957	.9625	.967	.972	.975	.979	.9815	.985	.987	.990	.992	.9929
.54	.9315	.938	.944	.950	.955	.9603	.965	.970	.974	.977	.9797	.983	.986	.988	.990	.9909
.55	.9298	.936	.942	.948	.953	.9585	.964	.968	.972	.975	.9779	.981	.984	.986	.988	.9892
.56	.9281	.934	.940	.946	.951	.9567	.962	.966	.970	.974	.9763	.980	.982	.984	.986	.9879
.57	.9262	.932	.939	.944	.949	.9550	.960	.964	.968	.972	.9745	.978	.980	.983	.984	.9861
.58	.9247	.930	.938	.942	.947	.9532	.958	.963	.966	.970	.9730	.976	.979	.981	.982	.9842
.59	.9230	.929	.935	.940	.946	.9515	.956	.961	.965	.968	.9714	.975	.977	.979	.981	.9827

TABLE A-4.—(Continued)

R^3/V	R/r = 30	32	34	36	38	40	42	44	46	48	50	52	54	56	58	60
.60	.9215	.927	.933	.939	.944	.9497	.954	.959	.963	.967	.9701	.973	.976	.978	.979	.9813
.62	.9184	.924	.930	.936	.941	.9467	.951	.956	.960	.964	.9669	.970	.973	.975	.976	.9784
.64	.9150	.921	.927	.932	.938	.9439	.948	.953	.957	.961	.9643	.968	.970	.972	.973	.9754
.66	.9121	.918	.925	.930	.935	.9408	.946	.950	.954	.959	.9614	.965	.967	.969	.971	.9728
.68	.9093	.915	.921	.927	.932	.9382	.943	.948	.951	.956	.9590	.963	.965	.967	.968	.9703
.70	.9064	.912	.919	.924	.929	.9352	.940	.945	.949	.953	.9563	.960	.962	.966	.966	.9678
.72	.9037	.910	.916	.921	.927	.9328	.937	.943	.946	.951	.9542	.957	.960	.962	.964	.9656
.74	.9012	.907	.913	.919	.924	.9303	.935	.940	.944	.949	.9519	.955	.958	.960	.962	.9636
.76	.8987	.905	.911	.916	.922	.9277	.933	.938	.942	.947	.9499	.953	.956	.958	.960	.9616
.78	.8964	.902	.908	.914	.920	.9258	.930	.936	.939	.944	.9475	.951	.954	.956	.958	.9598
.80	.8937	.900	.906	.912	.918	.9230	.928	.933	.937	.942	.9454	.949	.952	.954	.956	.9581
.82	.8917	.898	.904	.909	.915	.9211	.926	.931	.935	.940	.9436	.947	.950	.952	.954	.9563
.84	.8894	.895	.902	.907	.913	.9190	.924	.929	.933	.938	.9419	.946	.949	.951	.953	.9548
.86	.8874	.893	.900	.905	.911	.9171	.922	.927	.932	.936	.9402	.944	.947	.949	.951	.9534
.88	.8853	.891	.898	.903	.909	.9152	.921	.926	.930	.934	.9384	.942	.945	.947	.950	.9517
.90	.8831	.889	.896	.902	.907	.9131	.919	.924	.928	.933	.9367	.940	.943	.946	.948	.9504
.92	.8809	.887	.894	.900	.905	.9114	.917	.922	.926	.931	.9350	.939	.942	.945	.947	.9489
.94	.8791	.885	.892	.898	.904	.9097	.915	.919	.925	.929	.9333	.937	.940	.943	.945	.9476
.96	.8770	.883	.890	.896	.902	.9074	.914	.917	.923	.928	.9320	.936	.939	.942	.944	.9462
.98	.8754	.882	.888	.894	.900	.9064	.912	.916	.922	.926	.9305	.934	.937	.940	.943	.9452
1.00	.8734	.880	.886	.892	.899	.9047	.910	.914	.920	.925	.9290	.933	.936	.939	.941	.9438
1.05	.8688	.875	.882	.888	.895	.9007	.906	.912	.916	.921	.9253	.929	.932	.936	.938	.9408
1.10	.8644	.871	.878	.885	.891	.8970	.903	.908	.913	.917	.9217	.925	.929	.933	.935	.9378
1.15	.8602	.867	.875	.881	.888	.8937	.900	.905	.910	.914	.9183	.922	.926	.930	.933	.9352
1.20	.8561	.864	.871	.878	.885	.8904	.897	.902	.907	.911	.9154	.920	.923	.927	.930	.9324
1.25	.8521	.860	.868	.875	.882	.8874	.893	.899	.904	.908	.9125	.916	.920	.924	.927	.9300
1.30	.8484	.856	.864	.871	.879	.8845	.891	.896	.901	.905	.9097	.914	.917	.921	.925	.9277
1.35	.8451	.853	.861	.869	.876	.8819	.888	.893	.898	.903	.9068	.911	.915	.919	.922	.9253
1.40	.8420	.850	.858	.866	.873	.8794	.885	.891	.896	.900	.9043	.909	.913	.916	.920	.9232
1.45	.8387	.847	.855	.863	.871	.8764	.883	.888	.893	.898	.9014	.906	.910	.914	.918	.9207

R^3/V	$R/r = 30$	32	34	36	38	40	42	44	46	48	50	52	54	56	58	60	65	70	75	80
1.50	0.8356	0.844	0.853	0.861	0.868	0.8744	0.881	0.886	0.891	0.895	0.8995	0.904	0.908	0.912	0.916	0.9190				
1.55	.8327	.841	.850	.858	.866	.8722	.878	.883	.888	.893	.8970	.901	.906	.910	.914	.9171				0.9382
1.60	.8297	.839	.848	.856	.863	.8700	.876	.881	.886	.891	.8947	.899	.904	.908	.912	.9152	0.922	0.928	0.933	.9365
1.65	.8272	.836	.845	.853	.861	.8678	.874	.879	.984	.889	.8927	.897	.902	.906	.910	.9133	.921	.927	.931	.9354
1.70	.8245	.834	.843	.851	.859	.9658	.872	.877	.882	.886	.8906	.895	.900	.904	.909	.9116	.919	.925	.930	.9341
1.75	.8217	.831	.840	.849	.857	.8638	.870	.875	.880	.884	.8886	.893	.898	.902	.907	.9097	.918	.924	.929	.9328
1.80	.8194	.829	.838	.847	.855	.8618	.868	.873	.878	.882	.8867	.891	.896	.900	.905	.9080	.916	.922	.927	.9317
1.85	.8168	.827	.836	.845	.853	.8596	.866	.871	.876	.881	.8849	.889	.895	.899	.903	.9066	.915	.921	.926	.9305
1.90	.8143	.824	.834	.943	.851	.8578	.864	.869	.974	.879	.8831	.888	.893	.897	.902	.9047	.913	.919	.925	.9291
1.95	.8119	.822	.832	.841	.849	.8559	.862	.867	.872	.877	.8815	.886	.891	.895	.900	.9034	.912	.918	.923	.9281
2.00	.8098	.820	.830	.839	.847	.8539	.860	.865	.870	.875	.8798	.884	.890	.893	.899	.9016	.910	.917	.922	.8270
2.10	.8056	.816	.826	.835	.843	.8502	.856	.862	.867	.872	.8768	.881	.886	.890	.895	.8991	.908	.914	.920	.9247
2.20	.8015	.812	.822	.831	.839	.8464	.853	.858	.864	.869	.8738	.879	.883	.887	.892	.8962	.905	.911	.917	.9226
2.30	.7976	.808	.818	.828	.835	.8428	.849	.855	.861	.866	.8710	.876	.880	.884	.890	.8935	.903	.909	.915	.9206
2.40	.7936	.804	.814	.824	.832	.8393	.846	.852	.857	.863	.8680	.873	.878	.882	.887	.8910	.900	.907	.913	.9185
2.50	.7898	.800	.811	.820	.828	.8360	.843	.849	.854	.860	.8651	.870	.875	.879	.885	.8884	.898	.904	.910	.9166
2.60	.7861	.797	.807	.817	.825	.8325	.840	.846	.851	.857	.8624	.868	.872	.877	.882	.8859	.895	.902	.908	.9145
2.70	.7824	.793	.803	.813	.822	.8291	.836	.843	.848	.854	.8598	.865	.870	.874	.880	.8837	.893	.900	.906	.9126
2.80	.7788	.790	.800	.810	.818	.8260	.834	.840	.846	.852	.8570	.862	.867	.872	.877	.8813	.891	.898	.904	.9107
2.90	.7752	.786	.796	.806	.815	.8230	.831	.837	.843	.849	.8545	.860	.865	.870	.875	.8790	.889	.896	.902	.9089
3.00	.7716	.783	.793	.803	.812	.8200	.828	.834	.841	.846	.8521	.858	.863	.868	.873	.8770	.887	.894	.900	.9068
3.10	.7677	.779	.790	.800	.809	.8170	.825	.832	.838	.844	.8494	.855	.860	.866	.871	.8750	.885	.892	.899	.9049
3.20	.7644	.776	.787	.797	.806	.8140	.822	.829	.835	.842	.8472	.853	.858	.864	.869	.8730	.883	.890	.897	.9030
3.30	.7610	.772	.783	.793	.803	.8113	.820	.827	.833	.840	.8449	.851	.856	.862	.866	.8710	.881	.888	.895	.9012
3.40	.7572	.769	.780	.790	.800	.8083	.817	.824	.831	.837	.8424	.849	.854	.860	.864	.8688	.879	.886	.893	.8993
3.50	.7542	.766	.777	.788	.798	.8057	.814	.822	.829	.835	.8404	.847	.852	.858	.862	.8668	.877	.884	.892	.8974

TABLE A-5. CORRECTION FACTORS (F) FOR THE RING METHOD[25a]

R^3/V	$R/r = 40$	50	52	54	56	58	60
3.50	0.8063	0.8407	0.847	0.852	0.858	0.863	0.8672
3.75	.8002	.8357	.842	.848	.853	.858	.8629
4.00	.7945	.8311	.837	.843	.849	.854	.8590
4.25	.7890	.8267	.833	.839	.845	.850	.8553
4.50	.7838	.8225	.829	.835	.841	.847	.8518
4.75	.7787	.8185	.825	.832	.838	.843	.8483
5.00	.7738	.8147	.822	.828	.834	.840	.8451
5.25	.7691	.8109	.818	.825	.831	.837	.8420
5.50	.7645	.8073	.815	.821	.828	.834	.8389
5.75	.7599	.8038	.811	.818	.825	.830	.8359
6.00	.7555	.8003	.808	.815	.821	.827	.8330
6.25	.7511	.7969	.805	.812	.818	.825	.8302
6.50	.7468	.7936	.801	.808	.815	.822	.8274
6.75	.7426	.7903	.798	.806	.813	.819	.8246
7.00	.7384	.7871	.795	.803	.810	.816	.8220
7.25	.7343	.7839	.792	.800	.807	.813	.8194
7.59	.7302	.7807	.789	.797	.804	.811	.8168

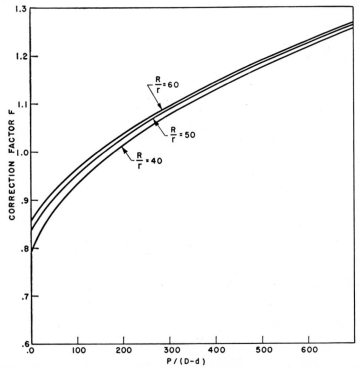

FIGURE A-6. Correction function for ring measurement of surface tension for a 6 cm ring.[25]

Figure A-7. In using the tensiometer, the liquid to be measured is placed in a shallow, wide dish, and the dish elevated until the ring makes contact with the surface. Tension is then applied to the wire by rotating the knurled knob, while the dish is slowly raised at such a rate that the index marker remains zeroed. This reduces the distortion of the film prior to detachment, and thus increases the accuracy. When the tension on the wire is equal to the pull required to detach the ring, the lever arm suddenly springs up. The tension is then read directly on the circular scale (which should be previously calibrated by determination of the tension corresponding to various small weights).

This technique is open to certain objections. The necessity for raising the liquid makes it difficult to avoid the setting up of waves or other vibrations in the liquid surface, which interferes with accuracy. An arrangement which prevents this is to remove the entire stand and screw arrangement and place the liquid container on a solid surface. The entire tensiometer is then mounted on a jack, and the instrument is gradually *lowered* during the determination.

Another objection to the use of this device is the difficulty of arranging for temperature control. For fairly rough measurements, the liquid under test may be brought to temperature in a constant-temperature bath, and rapidly transferred to the dish for measurement. This often permits measurements to be made within a temperature variation of $\pm 0.5°$ C.

For precision work, however, some sort of temperature control is necessary. Dunning and Johansen[27] have developed a combined elevating platform and thermostat for use with either the chainomatic or tension balance. It is also possible to place the entire instrument in an air thermostat, and operate it by remote control attachments.

Another method of temperature control which has been found to be quite effective, involves mounting the instrument above the constant-temperature bath. The liquids to be measured are immersed in the bath contained in an Erlenmeyer flask, or in a fairly large diameter cylinder which is fitted with a narrow neck (to minimize evaporation). The ring is attached to the torsion-wire assembly by means of a stiff wire, long enough to reach down to the liquid. The scale must, of course, be recalibrated to take into account the additional weight of the wire.

In all measurements with the ring method, complete wetting of the ring must be ensured in order to get reproducible and meaningful results. This is ordinarily not a problem with surface tension and most interfacial tension measurements. When making interfacial tension measurements with liquids *denser* than the aqueous phase, however, the fact that platinum is preferentially wetted by water makes it necessary to carry out the measurements

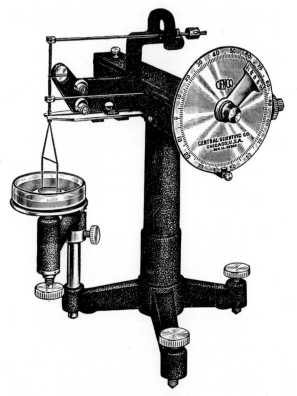

FIGURE A-7. A du Nouy surface-tension apparatus. *Courtesy: Central Scientific Co.*

in a *downward* direction. Certain of the commercial ring apparatuses are constructed so as to make this possible.

Alternatively, the ring could be made of some material other than platinum, such as for example, teflon, or glass coated with some water-repellant material.

Maximum Bubble Pressure Method. One other method of measuring surface tension, the maximum bubble pressure method, will be described, although not in as much detail as the previous methods. This method depends on the fact that if a small bubble of a gas (e.g., air) is blown at the bottom of a capillary dipping into a liquid, the pressure in the bubble increases initially, while the bubble grows and the radius of curvature decreases. If the bubble is small enough to be considered as spherical, the smallest value of the radius of curvature (and the maximum pressure) occurs when the bubble is a hemisphere. Growth beyond this point results in a drop in pressure, liquid rushes in, and the bubble, in effect, bursts. The maximum pressure is given by the relation

$$P = gh_1(D - d) + 2\gamma/r, \qquad (A.21)$$

where the first term on the right-hand side of Eq. A.21 represents that portion of the pressure required to force the liquid down the tube to the level h_1 of the end of the tube below the plane surface of the liquid, and r is the radius of the capillary.

If the capillary tube is larger, Eq. A.21 cannot be applied and correction terms must be used. The method of correction is described by Adam.[4, 28]

Sugden[29] has described a unit, using two tubes of different diameter, which is capable of high precision; he also describes methods of using this technique with semi-micro quantities.

Measurement of Viscosity

The viscosity of an emulsion is often quite possibly an important factor in determining its stability. From the point of view of ultimate use, the viscosity is also important. For example, it is desirable that a mayonnaise be of a certain definite "stiffness," not so viscous that it cannot be easily spread or whipped into other materials, but not so thin that it will actually flow. In other applications, however, a fairly free-flowing emulsion may be desirable. Viscosity measurements may also be a guide to the type of emulsion, or to changes in emulsion form.

Methods of Viscosity Measurement. Three basic methods for the measurement of liquids and plastic fluids are listed by Green and Melsheimer.[30] These are:

1. Flow through a capillary tube or orifice.
2. Rotational methods.
3. a. Falling balls, falling plungers, etc.
   b. Rising bubble.

In one form or another, most of these method are used in making industrial measurements, and most of them are applicable to measurements involving emulsions. The choice of the particular method will depend on convenience, and on the nature of the particular system being examined. Scott Blair[31] has compiled a useful list of the various types of materials that have been studied and the methods used.

Capillary Tube Methods. The flow of a liquid in a capillary is governed by Poiseuille's[32] law:

$$\eta = \frac{\pi P r^4 t}{8VL} \qquad (A.22)$$

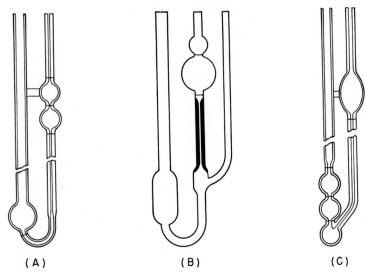

FIGURE A-8. Capillary viscometers. (A) Ostwald type. (B) Ubbelohde type. (C) Viscometer for opaque liquids.[40]

where η is the viscosity, P the pressure driving the liquid (in the case of gravity flow this is the hydrostatic head), r the radius of the capillary, and t the time for a volume V to travel a length L. The equation, as stated, is true only for streamline flow, but this condition is readily met for capillaries which are quite long as compared to their diameters. The commercially available capillary viscometers satisfy this criterion.

Figure A-8(A) illustrates the simplest type of Ostwald viscosity pipette; other versions are available, as shown in (B) and (C) of the illustration. In the Ostwald apparatus the liquid flows through the capillary under gravity, and the time required for the meniscus to pass between the two calibrations noted.

Although Eq. A.22 is perfectly straightforward, certain of the terms are difficult to determine, and it is usual to calibrate the viscometer in terms of a liquid of known viscosity. If this is done, one needs to know only the densities of the two liquids. Since, r, V, and L are functions of the geometry of the pipette

$$\eta_1/\eta_2 = P_1 t_1/P_2 t_2, \tag{A.23}$$

and, since flow is under gravity

$$\eta_1/\eta_2 = hd_1gt_1/hd_2gt_2 = d_1t_1/d_2t_2 \tag{A.24}$$

The units of viscosity are seen to be g/cm-sec, which is called the *poise*, in honor of Poiseuille. A more convenient unit for most measurements is

TABLE A-6. NBS VISCOSITY STANDARDS[33a]

NBS Viscosity Oil Standard	Kinematic Viscosity (cs*)					
	68° F*	77° F	86° F	100° F	104° F	122° F
D	2.5	2.2	—	1.8	—	—
H	9.1	7.7	—	5.4	—	—
I	15	12	—	8.0	—	—
J	25	20	—	12	—	—
K	50	39	—	22	—	—
L	110	84	—	43	—	—
M	390	280	—	130	—	—
N	1600	1100	—	460	—	—
OB	38,000	24,000	—	—	7,000	—
P	—	—	50,000	—	22,000	10,000

* Viscosities are quoted in centistokes (cf. p. 000), and temperature is given in degrees Fahrenheit.

one hundredth of this quantity, the *centipoise* (abbreviated cp). Water is widely used as a reference liquid; its absolute viscosity has recently been quite precisely redetermined and is reported by Swindells[33] to be equal to 1.002 cp at 20° C. It will be noted that the relative viscosity which would be calculated from Eq. A.24, using water as the reference liquid, differs from the absolute viscosity by only a small fraction of a per cent.

For liquids of higher viscosity, viscometers with larger capillaries are commonly employed. With these the use of water as a calibrating liquid is inconvenient. However, standard liquids, covering a wide range of viscosities, may be obtained from the National Bureau of Standards (Table A-6).[33a]

It should be noted that there is a small inexactitude in Eq. A.22, owing to the fact that the liquid leaves the capillary possessing some kinetic energy. The corrected form of Eq. A.22 is then[34]

$$\eta = \pi h d g r^4 t / 8VL - mdV/8L\pi t, \tag{A.25}$$

where m is a complex constant which may be taken as unity in most applications without much error. Equation A.25 can be written in a simplified form

$$\eta = Cdt - Bd/t \tag{A.26}$$

It is obvious that the second term can be rendered negligible by designing the viscometer so that B is small and t is large. In any case, calibration with a standard liquid will suffice to determine the constants. Dorsey[35] has given an interesting discussion of the corrections involved in the use of capillary viscometers.

The validity of Eq. A.24 depends, among other things, on the equality of the volume of liquid in the two determinations. Ordinarily this is not difficult to achieve.

It is necessary to keep the capillary exactly vertical in order to maintain a constant driving head. In any determination of viscosity the control of temperature is also important, since the temperature coefficient is appreciable. The capillary viscometer is especially satisfactory in this regard, since it can readily be used in a constant-temperature bath. The effect of such factors has been considered by Riley and Seymour[36] and by Geist and Cannon.[37]

The capillary method is probably the most satisfactory method of viscosity determination, from the theoretical standpoint. It is not satisfactory for substances of high viscosity, however, e.g., of the order of 10^4 centipoise or greater. It is also difficult to use with opaque liquids, although this can in some measure be obviated. For example, Jones and Talley[38] describe a photoelectric method for observing the passage of the meniscus; more recently, Andrievskii and Karelin[39] have reported a magnetic method. Cannon[40] has devised a special viscometer which can be used for opaque liquids. A discussion of the various types of capillary viscometers is given by McGoury and Mark.[41] The ASTM Standards should also be referred to.[42]

Efflux Methods. Because the Ostwald-type viscometer is useful in the low viscosity region, it is of limited value in the study of emulsions. Other viscometers have been devised in which the rate of efflux through a small orifice is measured; the orifice is, in effect, a very short capillary. The Poiseuille equation, Eq. A.22, no longer exactly applies, but relative measurements are still possible.

However, a generalized form of Eq. A.26 is commonly used:

$$\nu = \eta/d = At - B/t, \tag{A.27}$$

where ν is the so-called *kinematic viscosity*, whose units are defined as the *stoke* and *centistoke* (abbreviated *cs*) by analogy with the corresponding units of absolute viscosity. In many standardized instruments of this type the viscosity is merely reported in terms of the number of seconds required for the efflux of a specified volume of liquid.

One of the most widely used efflux instruments is the so-called "Saybolt" apparatus, which exists in two modifications, known respectively as the "Universal" and the "Furol" types.[43] The basic instrument consists of a jacketed vessel with an overflow gallery at the top and an efflux orifice at the bottom. The vessel contains a bit more than 60 ml, and the viscosity is measured by determining the time required for 60 ml to discharge into a

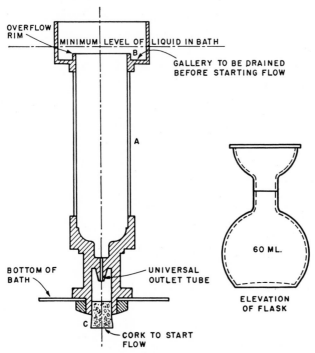

OVERFLOW RIM

MINIMUM LEVEL OF LIQUID IN BATH

B

GALLERY TO BE DRAINED
BEFORE STARTING FLOW

A

60 ML.

BOTTOM OF
BATH

UNIVERSAL
OUTLET TUBE

ELEVATION
OF FLASK

C

CORK TO START
FLOW

FIGURE A-9. Efflux viscometer, Saybolt type.

calibrated vessel under gravity (Figure A-9). The difference between the Universal and Furol models lies in the size of the orifice. In the former case it is 1.225 cm long by 0.1765 cm in diameter, in the latter type the diameter is 0.315 cm. Obviously, the Furol type is intended for use with more viscous materials. The data obtained with the Saybolt instrument are usually reported in "Saybolt" seconds.

A number of similarly designed instruments are also used, the most popular being the so-called "Engler" viscometer.

Much simpler designs for efflux units are also available. For example, in the paint industry viscometers which consist simply of a cylindrical cup with a standardized orifice at the center of the bottom are commonly used. The most popular version of this is the "Ford" cup,[44] which comes fitted with interchangeable orifices of various diameters, and an overflow lip to eliminate bubbles and adjust the volume of the sample.

Falling Ball Methods. All falling-ball methods for viscosity determination depend on Stokes' law:[45]

$$\eta = 2r^2(d_1 - d_2)_g/9v, \qquad (A.28)$$

where d_1 is the density of the sphere, d_2 is the density of the liquid, r is the radius of the sphere, and v is the velocity of the sphere,

Unfortunately, the derivation of this equation contains several assumptions which cannot be justified in any practical instrument.* Ladenburg[46] has investigated the corrections which have to be applied and has deduced the result

$$\eta = \frac{2r^2(d_1 - d_2)g}{9v(1 + 8r/R)(1 + 3r/L)}, \qquad (A.29)$$

where R is the radius of the tube in which the ball is falling, and L is the length of the column of liquid below the falling sphere that is not traversed during the timed fall.

Obviously, the equation can be put in the form

$$\eta = A(d_1 - d_2)/v, \qquad (A.30)$$

where A is a constant characteristic of the particular tube-sphere combination.

A more refined version of the falling ball type is the Hoeppler instrument, in which the sphere actually rolls down an inclined plane through the liquid (Figure A-10). The viscosity is a much more complicated function, and includes the angle of the inclined plane. For a given instrument, however, it reduces to a form analogous to Eq. A.30, except that the constant is now a function of the angle of the instrument, which must therefore be calibrated against this variable as well.

The Hoeppler instrument is superior to simple falling ball techniques in that proper choice of the material of the sphere and the angle can make the rate of fall such that true streamline flow is obtained, and the results are therefore meaningful.

A variation on the falling-sphere technique is the use of a perforated disc attached to a rod. Theoretical analysis of the fall of the disc is extremely difficult, and has not been attempted. However, in the form, e.g., of the "Gardiner Mobilometer" it finds considerable application in the paint and allied industries.

Rotational Methods. The classical rotation methods depend on the measurement of shearing forces between two rotating coaxial cylinders. The so-called "Couette" viscometer is a highly precise device which can be used for research determinations.

---

* Some of these problems might be resolved by the use of other than spherical shapes; however, as Scott Blair points out, only W. S. Gilbert has suggested the use of "elliptical billiard balls."

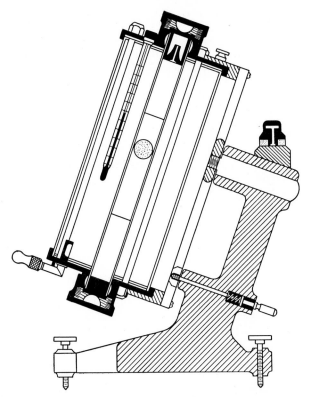

FIGURE A-10. Rolling ball viscometer, Hoeppler type.

More practical devices are the Stormer and MacMichael instruments. In the former device an inner cylinder or bob, submerged in the liquid, is caused to rotate by imposing a torque through a series of gears and a rotating drum around which passes a silk thread, to which is attached a descending weight which exerts the driving force.

A common modification of this instrument uses a forked paddle in place of the rotating bob. The data obtained are generally reported as the time required for a 100 revolutions, at some defined weight. A somewhat better method is to determine the weight required to give a standard rate of rotation.

The MacMichael instrument consists essentially of an outer vessel which contains the sample, jacketed for temperature control, geared to a driving motor. The cup is rotated at a given speed and the torque exerted on a concentrically suspended bob is measured by means of a calibrated torsion wire. The angle through which the wire is twisted is indicated in degrees

MacMichael. Multiplication by an appropriate calibration factor gives the actual torque.

The mathematical theory of both devices is similar and has been given by Reiner.[47] He has derived the relation

$$\Omega = \left(\frac{T}{4\pi h\eta}\right)\left(\frac{1}{R_b} - \frac{1}{R_c}\right) - \left(\frac{f}{\eta}\right)\left[\ln\left(\frac{R_c}{R_b}\right)\right],$$ (A.31)

where T is the angular torque, R_b is the radius of the bob, R_c is the radius of the cup, h is the depth of immersion, Ω is the angular velocity, η is the viscosity, and f is the yield value (of significance in non-Newtonian systems).

Equation A.31 may be conveniently rewritten as two separate equations, i.e.,

$$\eta = (T - T_0)S/\Omega$$ (A.32a)

$$f = T_0 C$$ (A.32b)

where T_0 is the intercept on the T-axis, and S and C are constants whose values depend on the geometry of the instrument. With Newtonian liquids T_0 is, of course, zero.

A recently developed unit, the "Brookfield Synchro-lectric Viscometer," is of the rotational type, dispensing, however, with the necessity for a special container for the liquid to be studied. The unit permits readings over a large range of values, and viscosities as high as 100,000 cps, with a considerable degree of precision. The unit is illustrated in Figure A-11. In this device a low-speed synchronous motor drives a specially shaped spindle through the liquid, and the torque exerted by the viscous drag on the spindle is measured by a calibrated spring. Since a large number of spindle shapes and several speeds of rotation are obtainable, meaningful values can be obtained even on non-ideal liquids. For extremely viscous materials, a special stand is available which moves the spindle through the liquid during the measurement, thus continuously bringing fresh liquid into contact with the spindle.

Roth[48] has described the measurement of viscosity by ultrasonic methods. An extremely useful discussion of the interpretation of flow data, with application to cosmetic formulations, has been given by Weltman.[49]

Measurement of Surface and Interfacial Viscosity

The measurement of surface and interfacial viscosity has not been widely used as a method of studying emulsions. Nonetheless, it is quite probable that these properties, especially interfacial viscosity, are extremely im-

FIGURE A-11. The Brookfield "Synchro-lectric" viscometer. *Courtesy: Brookfield Engineering Laboratories, Inc.*

portant, to the extent that emulsion stability depends on the rigidity of the interfacial film.

Methods of Measurement. With respect to surface viscosity, Adam[50] lists two principal experimental methods. The first of these is the so-called "surface-slit" viscometer, which is, in effect, a two-dimensional capillary. The second method depends on the measurement of the rate of damping of the oscillations of a disc in the surface.

The surface-slit method is much more elegant, although experimentally more difficult. Joly has designed an instrument in which the difference of pressure at the two ends of the long narrow channel, through a barrier in the surface (which constitutes the "surface slit"), is kept constant. It can then be shown that the amount passing through the slit in unit time is equal to

$$Q = Fd^3/12\eta l, \tag{A.33}$$

where F is the applied force, d the width, and l the length of the slit, while η is the coefficient of surface viscosity. Equation A.33 is obviously the two-dimensional analogue of the Poiseuille equation. For very narrow slits, it has been found that the amount passing through is, in fact, quite closely proportional to the cube of the slit width; with wider slits, however, Q does not increase as rapidly as d^3.

The surface slit method cannot be conveniently applied to the measurement of interfacial viscosities, and the oscillating disc or bob method must be employed. A convenient form of this device is described by Wilson and Ries.[51] In this instrument the bob is fastened to a torsion pendulum, which is set into oscillation, and the period and amplitude of the oscillation determined. The extension of devices of this sort to measurement of interfacial viscosity has recently been described by Criddle and Meader.[51a] Figure A.12 shows the design of a disc suitable for interfacial measurements.

It may be shown that the interfacial viscosity is given by the relation

$$\eta = \eta_0(r_t - r_0)/r_0, \tag{A.34}$$

where r_t is the observed damping factor in the interface, r_0 the damping factor in the bulk of the oil, and η_0 the bulk viscosity of the oil.

Determination of Emulsion Type

In many investigations, the determination of the type of emulsion (O/W or W/O) which is produced may be important. Unfortunately, an unequivocal determination may not always be possible, and deductive reasoning based on the data obtained from several methods may be required. Hauser

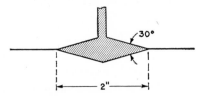

FIGURE A-12. A disc suitable for the measurement of interfacial viscosity by the oscillating bob method of Criddle and Meader.[51a]

and Lynn[52] list five methods by which the desired information may be obtained.

Dye Solubility Method. According to Hauser and Lynn this method was suggested by Robertson.[53] A colored dye soluble in one component, but insoluble in the other, is added to the emulsion, and the mixture gently agitated. If the color spreads through the whole emulsion, the phase in which the dye is soluble is the continuous one; if the color appears in discontinuous spots, it is the disperse phase.

In the past, most descriptions of this test have recommended the use of the oil-soluble "Red Sudan III." However, the certainty of the test is much increased if both water- and oil-soluble dyes are added to different portions of the emulsion. Recently, it has been suggested that "Brilliant Blue FCF" be used as the water-soluble dye, and "Oil Red XO" as the oil-soluble one.[54]

Phase Dilution Method. This method depends on the fact that an emulsion is readily dilutable by the liquid which constitutes the continuous phase. The test is most readily carried out by placing two drops of the emulsion on a glass slide or in the depressions of a spot-plate. A drop of one component is added to each drop and stirred lightly. Whichever component blends easily with the emulsion is considered to be the continuous phase. Observation of the dilution under a low-powered microscope may often be helpful.

Conductivity Method. Since most oils are poor conductors while aqueous systems are, in the main, good conductors, a crude measure of conductance may serve to identify the continuous phase. A simple device which may be used for this purpose, is illustrated in Figure A-13. In water-continuous emulsion, the neon lamp will glow when the electrodes are immersed in the liquid; when the oil phase is continuous, the lamp will not light.

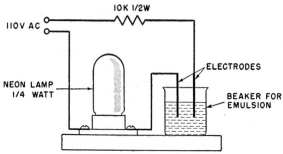

FIGURE A-13. A simple conductance device for the determination of emulsion type.

Unfortunately, there are certain systems where caution must be used. When an O/W emulsion is stabilized by an ionic emulsifying agent, the aqueous phase will have a high conductance. When nonionic emulsifiers are used, however, this may not be the case. It has been suggested that a small amount of sodium chloride be added to the emulsion to improve the conductivity of the aqueous phase; such an addition must be made with caution, however, since the added electrolyte could conceivably have an effect on the emulsion type. It might also be noted that W/O emulsions in which the phase volume of the disperse phase is high (*c.* 60 per cent) may have a moderately good conductivity.

Fluoresence Method. This method utilizes the fact that many oils fluoresce under ultraviolet light. Thus, examination of a drop of the emulsion under fluorescent light microscope may serve to identify the emulsion type. If the whole field fluoresces the emulsion is W/O; if, on the other hand, only a few fluorescing dots are evident the emulsion is O/W.

Wetting of Filter Paper. This method applies to emulsions of heavy oils and water, and depends on their respective abilities to wet filter paper. A drop of the emulsion is placed on a piece of filter paper; if the liquid spreads rapidly, leaving a small drop at the center, the emulsion is O/W. If no spreading occurs, the emulsion is W/O. However, with oils such as benzene, which spread on filter paper, this test is useless.

Other Methods. In addition to these five classic methods, other observations may be made. For example, the type of emulsifying agent, while not by any means controlling, is often an indication of the emulsion type. Similarly, the sign of the zeta potential (as obtained, for example, from electrophoresis measurements) may be a clue, since the droplets of an O/W emulsion are usually negatively charged, and vice versa. The sign of the droplet charge may also be determined without recourse to elaborate measurements, by direct microscopic observation of the direction of migration of the dispersed droplets in a micro-electrophoresis apparatus of simple design.

The phenomenon of creaming may be used as an indication of emulsion type. It is interesting to note that this method apparently has not been described previously. Since it is the disperse phase which creams, observation of the direction of creaming, together with a knowledge of the relative densities of the two phases, suffices to determine emulsion type. In all but the most stable emulsions, creaming may be induced by centrifugation at high speeds (cf. p. 137). This method is inapplicable in very viscous systems, and in complex systems where it is not always entirely possible to decide which component corresponds to which phase. However, in many cases this method can be used unequivocally.*

Determination of Droplet-Size Distribution

In studies of the stability of emulsions, the change in the particle-size distribution with time is often an important datum. There are, in general, three distinct methods by which this information may be obtained, i.e., by direct microscopic observation, by various sedimentation techniques, and by light-scattering measurements.

Direct Microscopic Measurement. This method has been widely used in the past. While it is extremely laborious, it is probably the method in which there is the most certainty in the results. Observation of the emulsion under a microscope fitted with a micrometer eyepiece, and preferably in a haemocytometer cell, permits the tabulation of the numbers of droplets in the various size ranges. From this, a distribution curve may be plotted.

It should be pointed out that, in order to obtain meaningful results, a large number of droplets should be measured. Harkins and Fischer,[55] for example, in work using this technique measured 50,000 droplets for each determination.

With concentrated emulsions, it is necessary to dilute the emulsion in order to be able to carry out the count. This may introduce a serious error, since the very act of dilution may cause changes in the distribution pattern. For O/W emulsions this possibility may be reduced by diluting with a solution of gelatin in water, the high viscosity of this diluent, in effect, "freezing" the emulsion. For the most part, however, this possible source of error has been disregarded.

A more serious limitation on the determination by this method is the impossibility of measurement of droplets in the fractional-micron range. This can introduce a serious error in the overall distribution.

Sedimentation Methods. If an emulsion creams at any sort of observable rate, measurement of the amount creamed per unit time permits the construction of a distribution curve. Most methods which depend on this phenomenon measure the change in density. Hauser and Lynn[55a] describe the specially designed hydrometers of Casagrande which can be used for this purpose, and present a nomogram which can be used for the solution of Stokes' Law on the basis of these determinations.

---

* The writer's attention has been drawn to a method which depends on the sound which the emulsion makes when shaken up and down in a bottle. According to report a W/O emulsion "rings" while an O/W emulsion sounds "flat." This technique should be applied with discretion.

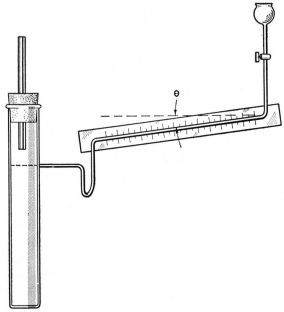

FIGURE A-14. Apparatus for the determination of droplet-size distribution by sedimentation measurements.[56]

A more useful technique is due to Kraemer and Stamm.[56] In this method the large cylinder of the apparatus illustrated in Figure A-14 is filled with the emulsion and the capillary sidearm is filled with the continous phase. As creaming occurs, the change in density at the top of the large column is reflected by a change in the level of disperse phase in the capillary. It can be then shown that

$$P = \frac{d_1 d_2}{d_1 - d_2} [A \sin \theta + s][a - b], \qquad (A.35)$$

where P is the weight of disperse phase which has passed the plane of the column where the sidearm enters, d_1 is the density of the continuous phase, d_2 is the density of the disperse phase, A and s are the cross-sectional areas of the large cylinder and the capillary, respectively, θ is the angle of inclination of the capillary, and a and b are the positions of the capillary meniscus at the beginning and end of the period during which P has accumulated. P may be plotted as a function of time to produce an accumulation curve, which may then be graphically differentiated to produce a distribution curve.

In the case of more stable emulsions, however, this method is not very satisfactory. To be sure, creaming may be accelerated by centrifugation;

indeed, the ultracentrifuge has been used, albeit rarely, for this purpose. Nichols and Bailey[57] have described, in some detail, the determination of the particle-size distribution for a Nujol emulsion by this technique.

Figurovskii[58] considers sedimentometric analysis as the fastest and most reliable method of determining the degree of dispersion and stability of various dispersed systems. A number of methods are detailed in his paper.

Light-Scattering Methods. The methods used by Langlois, Gullberg, and Vermeulen[59] for the determination of the particle-size distribution (in terms of interfacial area) have been described (cf. p. 54). Van der Waarden[60] has used orthodox light-scattering techniques to obtain what is in essence an *average* droplet size, and which is therefore also simply a measure of the total interfacial area (which may quite often be the information of most value).

Sloan[61] has described a method by which angular-dependence light scattering may be used to deduce the particle-size distribution of polydisperse systems. A series of papers by Sloan,[62] Arrington,[63] and Wortz[64] have given further applications of this technique, while a paper by Aughey and Baum[65] describes an instrument suitable for the measurements required.

Determination of Emulsion Stability

In the vast majority of commercial emulsions stability is the desieratum. In a smaller, though no less important, number, instability (total or limited) is required. In either case, it is desirable to have some method of measuring the stability of the emulsion under consideration.

Accelerated Aging Tests. A great deal about the shelf-life of an emulsion can often be learned by means of accelerated aging tests. For example, storage at elevated temperatures (e.g., 100 to 150° F) for periods in the order of two weeks may be instructive. Similarly, storage at lower than normal temperatures may be important. Certain emulsions, for example, emulsion paints, are required to be stable through several freeze-thaw cycles. This is a particularly stringent test.

Levius and Drommond[66] have considered elevated-temperature testing in some detail, and suggest the use of size frequency analysis as a corrolary to the storage. Examples of the use of freeze-thaw cycles in testing may be found, for example, in the papers of Walker[67] on the coagulation of neoprene latices through freezing, and of Fletcher and Mayne[68] on the stability of poly(vinyl acetate) emulsion paints.

Accelerated creaming tests may be induced by centrifugation. Under these circumstances, Stokes' Law becomes:

$$v = \frac{2\omega^2 R r^2 (d_1 - d_2)}{9\eta}, \qquad \text{(A.36)}$$

where the acceleration due to gravity g has been replaced by $\omega^2 R$, ω is the angular velocity of the centrifuge, and R is the distance of the sample from the center of rotation.

The value of $\omega^2 R$ may obviously be increased to many times that of gravity (indeed, to many thousand times); if one assumes that the effect is directly proportional to the gravitational force one may predict long-range behavior. For example, centrifugation at 3750 rpm in a 10 cm radius centrifuge for a period of five hours would be equivalent to the effect of gravity for about one year.

Unfortunately, it is unlikely that any such direct correlation exists, at least for fairly stable emulsions. On the other hand, with less stable emulsions, such that some creaming is evident after about 20 minutes of centrifugation, the correlation may well be quite exact.[65a].

Determination of Extent of Separation. In many cases, the question of stability is a black-and-white case, i.e., the emulsion is stable or it is not. In other cases, however, it may be desirable to measure the extent of breakdown as a function of time. This may be particularly valuable as a guide in formulation.

As has been pointed out above, Levius and Drommond[66] have indicated that particle size distribution is a valid measure of this property. The methods described in the previous section are applicable here. Sanders, Suter, and Garverich[69] have described a hydrometer technique which measures the change in density as separation occurs. Griffin and Behrens[70] have patented a device which may be used for determining the phase separation of emulsions by transmitted light, and which makes possible the direct observation of the relative proportions of the separated layers.

Turbidimetric or nephelometric methods may be used as a measure of emulsion separation; Bolton and Marshall[71] describe the use of simple equipment for this purpose.

Colburn,[72] Davies,[73] and Karas[74] discuss the problem of the stability of cosmetic emulsions in general. Davies points out that observations on model systems may enable the formulator to decide whether a particular system would be more stable as a W/O or an O/W emulsion.

Electrical Mesurements

Since the stability of an emulsion is so much a function of the electrical properties of the interface, methods of measurement of these properties can be quite important in the evaluation of emulsions, and in prediction of their stability or instability.

It has been suggested, in an earlier section, that direct observation of the motion of the droplets under an electric field could serve as an indica-

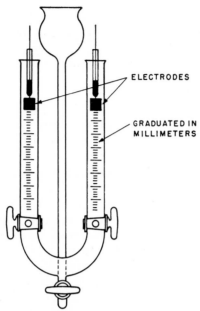

Figure A-15. A simple apparatus for the measurement of electrophoretic mobility by the boundary-layer method.

tion of the emulsion type. Observations of this type, at a slightly higher degree of precision, would also serve as a rough measure of the zeta potential.

A rather simple experimental technique for the determination of the zeta potential uses the moving-boundary method. Figure A-15 represents a practical form of the apparatus. In use, the U-tube is filled with the emulsion to a point below the level of the electrodes. The electrodes are then covered with water or with some electrolyte solution. A direct current of 110 to 250 volts is applied and the rate of motion of the water-emulsion interface is observed. From this, the applied voltage, and the geometry of the cell, the zeta potential may be calculated from

$$\zeta = \frac{6\pi\eta v}{\epsilon E},\qquad\text{(A.37)}$$

where η is the viscosity of the dispersion medium, v the electrophoretic velocity, ϵ the dielectric constant of the dispersion medium, and E the potential gradient. The technique is described in detail by Abramson[75] and Moore.[76]

If only the sign of the surface potential is required, an apparatus due to Chittum[77] may be employed. In this unit, the electric polarity of the aqueous phase of a W/O emulsion is measured in terms of gas evolved from a pair of electrodes, at least one of which is characterized by its ability to liberate gas when brought into contact with the aqueous phase. This is useful in determining the types of reagents required for the breaking of emulsions (*cf.* Chapter 9).

Miscellaneous Measurements

There are a large number of special measurements which may be made on emulsions and the materials which make up the emulsions. Many of these are standard tests in the various industries and are well known to the workers in the field. The various ASTM Methods publications are a mine of such methods, which may often be adapted to particular problems.

Methods of evaluating the efficiency of emulsifying agents have been discussed earlier (pp. 188–205). Behrens and Griffin[78] have discussed the evaluation of emulsifiable concentrates (for agricultural applications), and Russ[79] has described a test for evaluating the water-absorbency of W/O emulsions.

Analysis of emulsions may be carried out by ordinary methods. However, Harker, Heaps, and Horner[80] have proposed the use of adsorption columns for the analysis of soap and detergent-stabilized emulsions, and this technique is extended to emulsions stabilized by nonionic detergents by Green, Harker, and Howitt.[81]

Bibliography

1. Harkins, W. D., in Weissberger, A. (ed.), "Technique of Organic Chemistry," 2nd ed., I (*Physical Methods*), Part I, pp. 363–364, New York, Interscience Publishers, Inc., 1949.
1a. Ferguson, A., *Endeavour* (London) II, 34 (1943).
2. Adam, N. K., "Physics and Chemistry of Surfaces," 3rd ed., pp. 385–387, London, Oxford University Press, 1941.
3. Harkins, W. D., *loc. cit.*, pp. 365–366.
4. Adam, N. K., *op. cit.*, pp. 365–369.
5. Harkins, W. D. and Humphery, E. C., *J. Am. Chem. Soc.* **38**, 236 (1916).
6. Richards, T. W. and Carver, E. K., *J. Am. Chem. Soc.* **43**, 827 (1921).
7. Gross, P. L. K., Dissertation, Chicago, 1926; Harkins, W. D., *loc. cit.*, p. 367.
8. Harkins, W. D., and Jordan, H. F., *J. Am. Chem. Soc.* **52**, 1751 (1930).
9. Ferguson, A. and Dowson, P. E., *Trans. Faraday Soc.* **17**, 384 (1921).
10. Ferguson, A., *Proc. Phys. Soc.* (London) **36**, 37 (1923); **44**, 511 (1932).
11. Mouquin, H. and Natelson, S., *J. Phys. Chem.* **35**, 1931 (1931).
11a. Nevin, C. S., *J. Am. Oil Chemists' Soc.* **33**, 95 (1956).
12. Adam, N. K., *op. cit.*, pp. 371–372.
13. Adam, N. K., *op. cit.*, p. 369.
14. Harkins, W. D., *loc. cit.*, pp. 368–369.
15. Harkins, W. D. and Brown, F. E., *J. Am. Chem. Soc.* **41**, 499 (1919).
16. Tate, —, *Phil. Mag.* **27**, 176 (1864).
17. Lohnstein, T., *Ann. Physik* **20**, 237, 606 (1906); **21**, 1030 (1906).

18. International Critical Tables, Vol. IV, p. 435, New York, McGraw-Hill, 1928.
19. Harkins, W. D., *loc. cit.*, pp. 375–376.
20. Harkins, H. N. and Harkins, W. D., *J. Clin. Investigation* **7**, 263 (1929).
21. Harkins, W. D., *loc. cit.*, pp. 378–382.
22. Philippovich, A., *Erdöl u. Kohle* **5**, 412 (1952).
23. McGee, C. G., *Abstracts*, American Chemical Society, 124th Meeting, September, 1953.
24. Freud, B. B. and Freud, H. Z., *J. Am. Chem. Soc.* **52**, 1772 (1941).
25. Zuidema, H. H. and Waters, G. W., *Ind. Eng. Chem.*, *Anal. Ed.* **13**, 312 (1941).
25a. Fox, H. W. and Chrisman, C. H., Jr., *J. Phys. Chem.* **56**, 284 (1952).
25b. Macy, R., *J. Chem. Ed.* **12**, 573 (1935).
26. Du Nouy, P. L., *J. Gen. Physiology* **1**, 521 (1919); Rodewald, H. J., *Rev. Sci. Instr.* **24**, 229 (1953).
27. Dunning, H. N. and Johansen, R. T., *Rev. Sci. Instruments* **24**, 1154 (1953).
28. Adam, N. K., *op. cit.*, pp. 373–376.
29. Sugden, S., "The Parachor and Valency," pp. 215–216, New York, Knopf, 1930.
30. Green, H. and Melsheimer, L. A., in Matiello, J. J. (ed.), "Protective and Decorative Coatings," **4**, p. 116, New York, John Wiley, 1944.
31. Scott Blair, G. W., "Survey of General and Applied Rheology," pp. 89–101, New York, Pitman Publishing Corp., 1944.
32. Poiseuille, J. L. M., *Mém. Savants Étrangers* **9**, 433 (1846).
33. Swindells, J. F., *Nat. Bur. Standards Tech. News Bull.* **36**, 5 (1952).
33a. ASTM Standards, Part 5, p. 195, Philadelphia, American Society for Testing Materials, 1955.
34. McGoury, T. E. and Mark, H., in Weissberger, A. (ed.), "Technique of Organic Chemistry," 2nd ed., **I** (*Physical Methods*), Part I, pp. 333–334, New York, Interscience Publishers, Inc., 1949.
35. Dorsey, N. W., *Phys. Rev.* **28**, 833 (1926).
36. Riley, J. and Seymour, G. W., *Ind. Eng. Chem.*, *Anal. Ed.* **18**, 387 (1946).
37. Geist, J. M. and Cannon, M. R., *Ind. Eng. Chem.*, *Anal. Ed.* **18**, 611 (1946).
38. Jones, G. and Talley, S. K., *Physics* **4**, 215 (1933).
39. Andrievskii, A. I. and Karelin, N. N., *Priborostroenie*, No. 4, 24 (1956); *C.A.* **50**, 12580h.
40. Cannon, M. R., *Ind. Eng. Chem.*, *Anal. Ed.* **16**, 708 (1944).
41. McGoury, T. E. and Mark, H., *loc. cit.*, pp. 331–336.
42. ASTM Standards, Part 5, pp. 192–230, Philadelphia, American Society for Testing Materials, 1955.
43. ASTM Standards, Part 5, pp. 17–22, Philadelphia, American Society for Testing Materials, 1955.
44. Green, H. and Melsheimer, L. A., *loc. cit.*, p. 124.
45. Stokes, G. G., *Trans. Cambridge Phil. Soc.* **9**, 8 (1851).
46. Ladenburg, R., *Ann. Physik* **22**, 287 (1907).
47. Reiner, M., *J. Rheology* **1**, (1929).
48. Roth, W., *J. Soc. Cosmetic Chemists* **7**, 553 (1956).
49. Weltmann, R. N., *J. Soc. Cosmetic Chemists* **7**, 599 (1956).
50. Adam, N. K., *op. cit.*, pp. 394–397.
51. Wilson, R. E. and Ries, E. D., "Colloid Symposium Monograph," **I**, p. 145, Baltimore, Williams and Wilkins, 1923.
51a. Criddle, D. W. and Meader, A. L., Jr., *J. Appl. Phys.* **26**, 838 (1955).
52. Hauser, E. A. and Lynn, J. E., "Experiments in Colloid Chemistry," pp. 129–131, New York, McGraw-Hill, 1940.

336   EMULSIONS: THEORY AND PRACTICE

53. Robertson, T. B., *Kolloid-Z.* **7,** 7 (1910).
54. *Schimmel Briefs*, **247,** (October, 1955.)
55. Fischer, E. K. and Harkins, W. D., *J. Phys. Chem.* **36,** 98 (1932).
55a. Hauser, E. A. and Lynn, J. E., *op. cit.*, pp. 144–145.
56. Kraemer, E. O. and Stamm, A. J., *J. Am. Chem. Soc.* **46,** 2709 (1924).
57. Nichols, J. B. and Bailey, E. D., in Weissberger, A (ed.), *op. cit.*, pp. 673–679.
58. Figurovskii, N. A., *Trudy Vsesoyuz. Konferentsii Anal. Khim.* **2,** 399 (1943); also, Somov, V. S. and Zhukovskaya, I. Ya., *Zavodskaya Lab.* **16,** 1130 (1950); *C.A.* **45,** 1823a.
59. Langlois, G. E., Gullberg, J. E., and Vermeulen, T., *Rev. Sci. Instr.* **25,** 360 (1954).
60. Van der Waarden, M., *J. Colloid Sci.* **9,** 215 (1954).
61. Sloan, C. K., *J. Phys. Chem.* **59,** 834 (1955).
62. Sloan, C. K., *Abstracts*, American Chemical Society, 125th Meeting, April, 1954.
63. Arrington, C. H., Jr., *Abstracts*, American Chemical Society, 125th Meeting, April, 1954.
64. Wortz, C. G., *Abstracts*, American Chemical Society, 125th Meeting, April, 1954.
65. Aughey, W. H. and Baum, F. J., *J. Optical Soc. Am.* **44,** 833 (1954).
65a. Merrill, R. C., Jr., *Ind. Eng. Chem., Anal. Ed.* **15,** 743 (1943).
66. Levius, H. P. and Drommond, F. G., *J. Pharm. Pharmacol.* **5,** 743 (1953).
67. Walker, H. W., *J. Phys. & Colloid Chem.* **51,** 451 (1947).
68. Fletcher, A. C. and Mayne, J. E. O., *Paint Manuf.* **25,** 116 (1955).
69. Sanders, H. L., Suter, H. R., and Garverich, E. R., *Soap & Chem. Specialties* **30,** No. 5, 99 (1954).
70. Griffin, W. C. and Behrens, R. W., *Anal. Chem.* **24,** 1076 (1952); (to Atlas Powder Co.), U. S. Pat. 2,673,484, Mar. 30, 1954.
71. Bolton, M. E. and Marshall, A. W., *Soap, Sanit. Chemicals* **25,** No. 9, 129 (1949).
72. Colburn, W., *J. Soc. Cosmetic Chemists* **2,** 193 (1951).
73. Davies, J. T., *Perfumery Essent. Oil Record* **43,** 338 (1952).
74. Karas, S., *Am. Perfumer* **62,** 113 (1953).
75. Abramson, H., "Electrokinetic Phenomena," *passim*, New York, Reinhold Publishing Corp., 1934.
76. Moore, D. H., in Weissberger, A. (ed.), "Technique of Organic Chemistry," 2nd ed., I (*Physical Methods*), Part II, pp. 1685–1712, New York, Interscience Publishers, Inc., 1949.
77. Chittum, J. E. (to California Research Corp.), U. S. Pat., 2,678,911, May 18, 1954.
78. Behrens, R. W. and Griffin, W. C., *Agr. Food Chem.* **1,** 720 (1953).
79. Russ, A., *Seifen-Öle-Fette-Wachse* **75,** 254 (1949).
80. Harker, R. P., Heaps, J. M., and Horner, J. L., *Nature* **173,** 634 (1954).
81. Green. T.. Harker, R. P., and Howitt, F. O., *Nature* **174,** 659 (1954).

APPENDIX B

Commercially Available Emulsifying Agents

In the following pages an attempt has been made to list all the commercial emulsifying agents, by their trade-names, which are available at the present time. It must be recognized that, in spite of all precautions, such a list must be incomplete. It has been compiled with the assistance of the compendious listings of Sisley and Wood,[1] McCutcheon,[2] and Price,[3] and by reference to the published data of the various manufacturers.

The agents are listed alphabetically by trade-name, followed by the name of the manufacturer (usually somewhat abbreviated), and the chemical composition of the agent (when the information is available). The last column gives the applications of the agent: the type of emulsions formed (*e.g.*, "W/O emulsions") and/or specific materials for which the agent is considered particularly suitable (*e.g.*, "mineral oils"; following a semicolon, the broad areas of application of the agent are indicated (*e.g.*, "agricultural sprays"). In many cases there are no entries under the first of these; the context is usually clear.

It has not been felt worthwhile to indicate that the particular agent is anionic, cationic, or nonionic. This is usually clear from the chemical description in column two. When it is not clear, the agent cannot readily be classified.

The use of the work of Sisley and Wood as a source has enabled the writer to include a number of agents of European manufacture. A listing of the full names of the manufacturers and their addresses (when available) follows the listing of agents.

Emulsifying Agent (trade-name)	Manufacturer	Composition	Applications
Abietate of potash or soda		As given	
Acco Emulsifier #5	Glyco	Polyethylene glycol ester of fatty acids	Cosmetics
Actinol L	Amalgamated SF	Ethanolamine salt of polyalkyl naphthalene sulfonic acid	Textiles
Acto 450 500 550W 600 630 700	Esso	Alkyl aryl sulfonates (M.W. 465–480)	Rust preventives, cutting oils, emulsion breaking
Actusol	DuBois	Not given	Greases; metal industry
Advawet 10 25 33 33-Na-6	Advance	Alkyl aryl polyether alcohol Sodium salt of sulfonated mineral oil Polyglycol ester Ethylene oxide condensation product with fatty acid esters	Latex paints, water base inks Agricultural sprays O/W; emulsion paints Paints
Aerosol AY MA NAL NAO OT 100% 18 22	Cyanamid	Diamyl sulfosuccinate Dihexyl sodium sulfosuccinate Monolauryl sodium sulfosuccinate Monoöleyl sodium sulfosuccinate Sodium dioctyl sulfosuccinate N-octadecyl disodium sulfosuccinate N-octadecyl tetrasodium (1,2-dicarboxyl ethyl) sulfosuccinate	W/O Emulsion polymerization Emulsion polymerization
Agent S-489	Glyco	Amine type condensation product	Waxes, polishes
Agent TR BPE	Cyanamid		O/W O/W
Agrimul 70-A C GM T improved	Nopco	Alkyl aryl polyether alcohol	DDT, Lindane Aldrin, Dieldrin, Endrin Chlordane, herbicidal esters Chlordane, Toxaphene, Strobane, herbicidal esters

Trade name	Manufacturer	Composition	Applications
Ahco Base Oil W-100	AH	Sulfonated petroleum derivatives	Mineral lubricants, cutting oils, textile softeners
Ahcol 575	AH	Sulfonated red oil	
Albasol BF	Nopco	Potassium soap, solubilized with glycol	
Albatex BD	Ciba	Sodium m-nitrobenzene sulfonate	Textiles
POK paste		Benzimidazole compound	Textiles
Alcolec DS HS-2 HS-3 MS-10 M-12 4135 WDR	American Lecithin	Lecithin	Cosmetics, pharmaceuticals
Aldo 25 28 33 40	Glyco	Propylene glycol monostearate Glyceryl monostearate	Foods Baking, ice cream, cosmetics
		Glyceryl oleostearate	Cosmetics, foods
Alipal CO-433	Antara	Sodium salt of alkyl phenoxy polyoxyethylene ethanol	Cosmetics
CO-436		Ammonium salt of above	Copolymerization of vinyl monomers, cosmetics
Aliphatic ester sulfate	Onyx	As given	Textiles
Alkanol DW solution 189-S	DuPont	Sodium alkyl aryl sulfonate Long-chain hydrocarbon sulfonate	
Alkaterge A C E T	CSC	Substituted oxazolines	Textiles, emulsion paints
Alrodyne 315 6104 8020	Geigy	Polyethylene glycol fatty esters	Agricultural sprays
Alrolene 70 65	Geigy	Not given	Metal cleaning

Emulsifying Agent (trade-name)	Manufacturer	Composition	Applications
Alromine RA			
S-50	Geigy	Fatty alkylolamine condensate	Cosmetics
S-25			Cosmetics
G-50			Cosmetics
G-25			Textiles
			Textiles
Alrosene PD	Geigy	Alkyl aryl amide sulfonate	
Alrosept	Geigy	Quaternary ammonium compound	
Alrosol O	Geigy	Fatty alkylolamine condensates	DDT
S			Cosmetics
Alrosperse 11P	Geigy	Fatty amine condensate	
100		Blend of nonionic and cationic agents	
Amascour W	American Aniline	Complex cationic agent	Wool scouring
Amine O	Geigy	Oleic (?) tertiary amine	Mineral oils; herbicidal sprays, textiles
C		Coco (?) tertiary amine	
S		Stearic (?) tertiary amine	
Amine 220	Carbide	1-hydroxyethyl, 2-heptadecenyl glyoxalidine	Mineral oils; tar emulsion breaker
ES		Amine ester	Waxes
Aminostearin	Glyco	Modified amine ester of stearic acid	
Ammonyx AO	Onyx	Lauryl dimethyl amine oxide	Mineral oils
CO		Cetyl dimethyl amine oxide	
OO		Alkenyl dimethyl amine oxide	
Ammotate	Beacon	Ammonium oleoresinate	
Amoa A2	Amoa	Casein hydrolyzed by alkyl naphthalene sulfonate	
A3			
A5			
BN		Alkyl naphthalene sulfonate	
DN			
Falco WA		Lauryl alcohol sulfate	Cosmetics
Antarate K	Antara	Alkyl carboxylate	Mineral oils, hydrocarbon solvents

Trade name	Manufacturer	Composition	Uses
L520		Alkyl aryl polyethylene glycol sulfonate	Cosmetics
R155		Sodium oleyl methyl taurate	Emulsion paints, textiles
S870		Sodium alkyl naphthalene sulfonate	Paints, printing inks
Antarox A200, A201, A400	Antara	Sulfonated red oil	
A401, A402, A403, A404	Antara	Ethylene oxide condensate with alkyl phenol	Aromatic solvents, chlorinated solvents, vegetable oils
B200, B201, 290		Polyethyleneglycol oleate Alkyl polyglycol ester ether Polyethylene glycol ricinoleate	Cosmetics, absorption bases Herbicidal sprays, oils, fats, waxes Casein and oleoresinous paints, cosmetics
B390, B590, D100		Polyethylene glycol stearate Alkyl polyoxyethylene glycol ester Polyglycol oleyl ether	Insecticides, inks, solvents Non-mineral oils, fats, waxes; emulsion paints
D300		Alkyl polyoxyethylene glycol thioether	
April Wetting Agent	Essential	Alkyl aryl sulfonate	
Aquatergent	CSC	Sulfated boro-fatty acid amide	
Arctic Syntex O36 L	Colgate	Ethylene oxide condensation product with nonyl phenol Sodium H-coco monoglyceride sulfate	
Areskap 50, 100	Monsanto	Monobutyl phenyl phenol sodium sulfonate	Metals processing
Aresket 300	Monsanto	Monobutyl diphenyl sodium monosulfonate	
Aresklene 375, 400	Monsanto	Dibutylphenyl phenol sodium disulfonate	
Arlacel 20, 40, 60, 80, 83, 85	Atlas	Sorbitan monolaurate Sorbitan monopalmitate Sorbitan monostearate Sorbitan monoöleate Sorbitan thioöleate Sorbitan sequioleate	Cosmetics, pharmaceuticals
Arlacel A, B, C	Atlas	Mannide monoöleate Mannitan monoöleate Sorbitan sesquioleate	W/O; cosmetics

Emulsifying Agent (trade-name)	Manufacturer	Composition	Applications
Armac 8	Armour	n-Octyl ammonium acetate	
10		n-Decyl ammonium acetate	
12		n-Lauryl ammonium acetate	
16		n-Palmityl ammonium acetate	
18		n-Octadecyl ammonium acetate	
C		Cocoyl ammonium acetate	
S		Stearyl ammonium acetate	
T		Tallow ammonium acetate	
Arquad 12	Armour	n-Lauryl ammonium chloride	
16		n-Palmityl ammonium chloride	
18		n-Octadecyl ammonium chloride	
C		Cocoyl ammonium chloride	
S		Stearyl ammonium chloride	
2C		Dicocoyl ammonium chloride	O/W
2HT		Di-H-tallow ammonium chloride	O/W
Atlox 81	Atlas	Polyoxyethylene sorbitan monoöleate	Agricultural sprays
672		Polyoxyethylene sorbitan laurate	
1045		Polyoxyethylene sorbitol laurate	
1045A		Polyoxyethylene sorbitol oleate-laurate	
1086, 1087, 1276		Polyoxyethylene fatty esters	
2065, 2085		Sulfonated oil with polyoxyethylene esters of mixed fatty and resin acids	
Atmul 82 122	Atlas	Mixed lard mono- and di-glycerides	Foods
Atpet	Atlas	Fatty acid ester	Petroleum
A-2-Z	Planetary	Built sodium alkyl aryl sulfonate	
Avasol 114 114A 153 153A 66 66A	Alframine	Fatty acid derivatives of tertiary amine plus aldehyde reaction products	Insecticidal sprays, petroleum emulsions
Base 301-R	Drew	Coconut fatty acid amine condensate	W/O; paraffin wax
401-M		Polyoxyethylene esters and sulfonates	Insecticidal sprays
Bemul	Beacon	Glyceryl monostearate	Cosmetics, shortenings, lubricants

Trade name	Chemical description	Applications
Betanol 152 401 520 540 560 564 701	High molecular weight esters	Cosmetics, paints, pharmaceuticals
Beacon		
Belgotex		Textiles
Betepon A AS T	Sodium oleyl-stearyl isothionate Sodium sulfonate of fatty amide Sodium oleyl methyl taurate	
Blendene	Terpene fatty acid salt complex	O/W, oils and solvents
Brij 30, 35	Ethylene oxide condensation product with lauryl alcohol	
Burkester	Polyethylene glycol fatty ester	Textiles
Burkart-Schier		
C-430	Complex fatty acid ester	Dry cleaning preparations
Calpex	Sulfated ester	Textiles
Hall		
Apex		
Carbowax 6000 monoöleate	Polyethylene glycol monoöleate	
Glyco		
Carbowax 1000 monoöleate	Polyethylene glycol monoöleate	
Glyco		
Cardene	Fatty alkanolamide	Textiles
Carlisle		
Cationic Agent C	Laurylamine salt of lauryl amido ethyl phosphoric acid	Textiles
D	Stearylamine salt of stearyl amido ethyl phosphoric acid	
Victor		
Cationic Amine 220	Alkyl trimethylammonium chloride	Oils; agricultural sprays, printing inks, elastomers, resins
Carbide		
Catol 2	Alkyl trimethylammonium chloride	Textiles
Carbide		
Catylex	Salt of tertiary amine	Textiles
Hart		
Catylon C D	Salt of tertiary amine	Textiles
Hart		
Cemulsol 64 A B	Ethylene oxide condensation product Ethylene oxide condensation product with fatty acids	Textiles Textiles, cosmetics, insecticides
CPI		

Emulsifying Agent (trade-name)	Manufacturer	Composition	Applications
Ceramol	Aceto	Partly sulfated fatty alcohol	Pharmaceuticals, cosmetics
Cerasynt S SA SE SN M MN DM IP		Glyceryl monostearate	Cosmetics, pharmaceuticals
	Van Dyk	Ethylene glycol monostearate	
		Diethylene glycol monostearate	
		Glycol amido stearate	Cosmetics
Cetalvon	ICI	Cetyl trimethylammonium bromide	
Chemoil 412	Standard	Sulfonated castor oil	Textiles
Chemsol 990	Synthron	Alkyl aryl sulfonate	
Chlorsol	Drew	Polyoxyalkylene ester	Chlordane
Cire di Sipol #2	Sinnova	Ethylene oxide condensation product with cetyl alcohol	Textiles, cosmetics
Clavanol Conc.	Dexter	High molecular weight ethylene oxide condensation product	Textiles
Clearate Special Extra WD LV	Naftone	Soya lecithin Soya lecithin derivatives	Emulsion paints
Collex	SPCMC	Casein hydrolysate	Fats, animal, vegetable, mineral oils
Condensate NI S	Carbide	Ethylene oxide condensation product with fatty acid amine Fatty acid condensate	
Cordon 810 700/75 PB 870 NB 870 LB 870 LB 890 LB 900	Finetex	Sulfonated red oil Sulfonated olive oil Sulfonated tall oil Sulfonated tall oil Sulfonated tall oil Sulfonated castor oil Sulfonated sperm oil	Textiles, leather finishing

Trade name	Manufacturer	Composition	Use
Corikal A} B}	General Dyestuff	Alkyl naphthalene sulfonate	
Crillex 1 to 13 17, 19 20 to 23	Croda	Polyethylene oxide esters of fatty acids Ethylene oxide condensate with propylene glycol monostearate Polyethylene oxide esters of stearic acid	Cosmetics O/W emulsions Essential oils, O/W emulsions
D-Spers-O-Ac} D-Spers-W}	Planetary	Condensation product of polyglycol fatty acid and sodium sulfosuccinic acid	Agricultural, insecticidal sprays
Daxad #21 #23 #27	Dewey & Almy	Calcium salt of alkyl aryl sulfonic acid Sodium salt of alkyl aryl sulfonic acid Sodium salt of alkyl aryl sulfonic acid plus inorganic colloid	Finely dispersed emulsions of sulfur-in-water; other fungicides and insecticides
Dehydag Wax E	Fallek	Sodium salt of higher saturated fatty alcohol sulfates	O/W emulsions
Demal-14	Emulsol	Polyglycol ester of oleic acid	Cosmetics, pharmaceuticals
Desephor	Despé	Ethylene oxide condensation product with cetyl alcohol	Waxes; cosmetics
Detergent O-245	Wolf	Fatty acid amide condensate	Textiles
Deterpon A} 2A}	A	Sodium sulfonate of a fatty amide	Textiles
Diglycol laurate oleate stearate}	Carlisle; Glyco; C. P. Hall; Kesler; Van Dyk	As given	Cosmetics, foods, textiles
Dipex SP 702} 703}	Stanco	Sulfonated mineral oil	
Dispersant NI-O	Oronite	Ethylene oxide condensation product with alkyl phenol	W/O emulsions; dry cleaning
Doittau 14 32	Doittau	Sulfonated butyl ester of ricinoleic acid Sulfonated monoglyceride of ricinoleic acid	Textiles

Emulsifying Agent (trade-name)	Manufacturer	Composition	Applications
DO 8	IGF	Ethylene oxide condensation product with fatty alcohol	Mineral oil
Drumul 30W	Drew	Purified ethylene oxide condensation product with fatty acid	Naphthol dyeing
Druterge NCR	Drew	Blend of coconut fatty acid amine condensate and polyoxyethylene alkyl aryl ether	Textiles
NT		Ethylene oxide derivative of tall oil	
ON		Alkyl aryl polyoxyethylene ether	
TG Conc.		Oleyl methyl tauride sodium salt	
NCE		Ethylene oxide condensation product with vegetable oil	
Duponol C	DuPont	Sodium lauryl sulfate USP	
G		Fatty alcohol amine sulfate	Fungicidal sprays
L-144-WDG		Sodium fatty alcohol sulfate	Cosmetics
L-142		Sodium fatty alcohol sulfate	
ME Dry		Technical sodium lauryl sulfate	
OS		Higher fatty alcohol sulfate, amine salt	O/W emulsions
RA		Alcohol ether sodium sulfate	Textiles
WS		Higher fatty alcohol sulfate, amine salt	Buffing compounds
80		Sodium octyl sulfate	Electroplating, textiles, leather
Ecconol A} 5}	Essential	Fatty acid alkanolamide	Liquid floor cleaners
E 607-40 Spec.	Emulsol	Quaternary ammonium compound of pyridine-betaine, prepared from selected fatty acids	O/W emulsions
Emargol	Emulsol	Sulfoacetyl compound of oleomargarine mono- and diglycerides	Foods

Trade name	Composition	Use
Emcol 14	Fatty acid polyglyceride ester	Baking
3812	Fatty amine derivative; anion-active	
EMS	Sodium salt of fatty acid ester of polyhydric alcohol	
H-50A	Fatty acid condensation product with polyhydric alcohol	
H-65C	Blend of polyalcohol carboxylic acid esters and oil-soluble sulfonates	Gives 2% clear dispersion of chlordane-in-water
H-77	Blend of polyalcohol carboxylic acid esters and sulfonated oil	Agricultural emulsions
H-83T	Blend of polyoxyethylene ethers and oil-soluble sulfonates	Chlordane, toxaphene
H-140	Blend of polyoxyethylene ethers and oil-soluble sulfonates	Malathion
H-400		
H-500	Blend of polyoxyethylene ethers and oil-soluble sulfonate	Agricultural emulsifiable concentrates
H-600		
K-8300	Complex aliphatic sulfonate	Polymer emulsions
MS	Glyceryl monostearate	Foods
MST		
P10569	Amine salt of sulfonic acid	Oil-soluble
RDCD	Diethylene glycol laurate	
RGL	Fatty acid mono- and diglycerides	Cake mixes
RHT	Blend of fatty acid mono- and diglycerides	Foods, cosmetics, pharmaceuticals
TL	Fatty alkyl amine	Oils, fats, waxes, resins
Emkabase	Naphthenic, amino sulfonate blend	Mineral oil
Emkagen	Amide amino condensate	
Emulgade F	Mixture of higher fatty alcohols and alcohol sulfates	O/W emulsions; cosmetics, pharmaceuticals
Emulgane CT	Polyoxyamine ester of fatty acids	Vegetable oils, greases, mineral oils
Emulgobel CAN	Hydrolyzed glue	Greases, oils, waxes
Extra	Ethylene oxide condensation product with a fatty acid	Wool oils
Antara		
Belgotex	Ethylene oxide condensation product with tallow alcohols	Rubber, textiles, leather
Emulphogene AM-870		

Emulsifying Agent (trade-name)	Manufacturer	Composition	Applications
Emulphor A	IGF	Ethylene oxide (12 moles) condensation product with oleic acid	Textiles
AG Oil		Ethylene oxide condensation product with a commercially pure organic acid	
A Extra		Ethylene oxide (6.5 moles) condensation product with di(isohexylisoheptyl)-phenol	Leather
EL		Ethylene oxide (20–40 moles) condensation product with castor oil	Castor oil, waxes, organic solvents, resins
ELA		Ethylene oxide condensation product with soap-free fatty acid	
ELN		Ethylene oxide (20 moles) condensation product with di(isoheptyl)-phenol	Olein
EL-620 } EL-719 }	Antara	Ethylene oxide condensation product with vegetable oil	Oils, solvents, agricultural, leather, emulsion paints; Animal, vegetable fats and oils, herbicides, insecticides
FM Oil soluble FFO	IGF	Triethanolamine monooleate	Oils
MW	IGF	Ethylene oxide (8 moles) condensation product with hexyl-heptyl-naphthol; Ethylene oxide (7.5–8 moles) condensation product with di(isohexylisoheptyl)-phenol	Mineral oil; textiles
O		Ethylene oxide (20 moles) condensation product with oleyl alcohol	Olein or mixtures containing olein
OL		Ethylene oxide (40 moles) condensation product with abietinol	Olein, neutral oils
ON		Condensation product of polyethylene glycol	
ON-870		Ethylene oxide condensation product with fatty alcohol	Mineral oil, fatty acids, waxes, latex emulsions, leather
STH } STX }		Condensation product of long-chain aliphatic chloride with ammonia, then monochloracetic acid	Mineral oils; cutting and wire drawing oils
VN-430	Antara	Ethylene oxide condensation product with a fatty acid	Mineral oils, liquid fatty oils, pesticides; cutting oils, cosmetics, textiles

Product	Company	Composition	Application
Emulsamin	Ciba	Menthol diurethane	Dyeing (as the hydrochloride)
Emulsan K D	RWW	Polyethylene glycol laurate Polyethylene glycol oleate	Wool degreasing
Emulser CF	Arkansas	Synthetic fatty ester	O/W emulsifier; agricultural sprays
Emulside 680B	Van Dyk	Polymerized higher glycol fatty acid ester	Emulsifiable chlordane concentrates
Emulsifier 803M	Drew	Blend of petroleum sulfonates, polyglycol ester, and glycols	
Emulsifier G-21 G-61 G-261 G-861	Kessler	Blend of mono- and diglycerides	Baking
Emulsifier H R L M	Monsanto	Blended surface-active agents	Insecticides, herbicides
Emulsifier L-32 L-34 L-45B L-34A L-45 O-205 O-141	Wolf	Fatty acid ester of higher polyalcohol	O/W emulsions, vegetable, animal, mineral oils Leather Vegetable, mineral, animal oils
Emulsifier P-8610 X-15	Kessler	Polyhydric alcohol ester of edible saturated fatty acids Fatty acid ester	Cosmetics, foods
Emulsifier W-763	Wolf	Sulfonated naphthenic acid, sodium salt	Aliphatic, aromatic solvents, textiles, leather, insecticides
Emulsifier 2 3 M-O-1	Kessler	Polyglycol ester of fatty acids	O/W emulsifiers; degreasing compounds, agricultural sprays Leather, textiles
Emulsifier 202	Remsen		Toxicants, weedkillers
Emulsifier 610-A	Van Dyk	Polyethylene glycol fatty acid ester	Pigment dispersions, cosmetics

Emulsifying Agent (trade-name)	Manufacturer	Composition	Applications
Emulsionnant LLO	IGF	Mepasine amide	
Emulsol 607 609	Emulsol	Alkyl pyridinium chloride Lauryl ester of amino-butyric acid	
Emultex C O	Adjubel	Phosphonate of diisopropyl naphthalene Emultex C plus hydrolyzed albuminous matter	Waxes, natural greases, paraffin Vegetable, animal oils
S		Ethylene oxide condensation product of fatty materials plus fats	Mineral, animal, vegetable oils, waxes
Essential 40	Essential	Alkyl aryl sulfonate	
Estol T-50% T-75% T-Extra	Drew	Sulfonated tallow	Textiles
Ethofat C/15 C/25		Ethylene oxide condensation product with coco fatty acids	
60/15 60/20 60/25		Ethylene oxide condensation product with stearic acid	
142/15 142/20 242/15 242/20 242/15 242/60	Armour	Ethylene oxide condensation product with rosin and fatty acid combinations N.B.: Number following bar minus ten equals moles ethylene oxide added	
Ethomeen C/12 C/15 C/20 C/25 T/12 T/15 T/25 S/12 S/15 S/20 S/25 18/12 18/15 18/20 18/60	Armour	Tertiary amines; ethylene oxide condensation products of primary fatty amines. Code letter indicates fatty acid sources, ethylene oxide content as above	

Trade name	Manufacturer	Composition	Uses
Ethomid C/15 RO/15 HT/15 HT/25 HT/60	Armour	Ethylene oxide condensation products of fatty acid amides. Coded as above.	
Emulgin M8	Fallek	Fatty alcohol preparation	Insecticides, mineral oil; textiles
Excelsior Oil A	General Dyestuff	Sulfonated vegetable oil	Textiles
Finish KB	Sandoz	Sulfonated fat	Textiles
Formula L-32 L-34A L-45	Wolf	Cf. Emulsifier L-32, etc.	
G-1425 -1431 -1441 -1493	Atlas	Ethylene oxide condensation product with sorbitol and lanolin	Cosmetics (*1425* water-dispersible; *1431, 1441* water-soluble; *1493* oil-soluble)
G-1702 -1704 -1706 -1725 -1726 -1727 -1734	Atlas	Ethylene oxide condensation product with sorbitol and beeswax	O/W emulsions; cosmetics (*1702, 1704, 1706* oil-dispersible; *1725 1726, 1727, 1734* water-dispersible)
G-2000	Atlas	Mixture of mannitan monopalmitates	Cosmetics, wax polishes, insecticidal sprays
G-2149 -2151 -2152 -2153	Atlas	Polyoxyethylene stearate	Drugs, cosmetics
G-2162	Atlas	Polyoxyethylene propyleneglycol monostearate	
G-2800 -2854 -2855 -2859	Atlas	Polyoxyethylene sorbitol oleate	Vitamin oils

Emulsifying Agent (trade-name)	Manufacturer	Composition	Applications
G-7076H -7426N	Atlas	Polyoxyethylene sorbitan dilaurate Polyoxyethylene sorbitan monopalmitate	
G-8916P -8916T	Atlas	Polyoxyethylene sorbitan esters of mixed fatty and rosin acids (tall oil)	
G-9446N	Atlas	Polyoxyethylene sorbitan monoöleate	
Glaurin	Glyco	Diethylene glycol monolaurate	Electrolytic emulsions
Glucaterge 12 28	CSC	Fatty acid condensate of N-methyl glucamine and coconut or cottonseed acids	
Glyceryl monolaurate monoöleate monostearate monoricinoleate	Alrose, Carlisle, Colgate, Glyco, Goldschmidt, Kessler, etc.	As given	Foods, baking
Glycol esters	As above		
Goremul A	Glyco	Polyhydric alcohol ether fatty acid condensate	
Hallco CPH-123	Hall	Complex fatty acid derivative	Agricultural sprays, insecticides
Hartofol C	Hart	Alkyl aryl sulfonate	
HHS	Publicker	Akyl aryl sulfonate	
Hydroterge B	Hydrocarbon		Oils, greases
Hymolon K	Hart	Fatty acid amine condensate	Textiles
Hymotol	Hart	Polyethylene glycol ester of fatty acid	Textiles
Hyponate L-50	Sonneborn	Petroleum sulfonate of M.W. 415/430	Synthetic rubbers, resins
Hytergen BM	Hart	Modified alcohol sulfate	Textiles

Igepal CA-630 -710 CO-430 -530 -610 -630 -710 -730 -850 -880 DM-710		Ethylene oxide condensation product with isoöctyl phenol Ethylene oxide condensation product with nonyl phenol Ethylene oxide condensation product with higher alkyl phenol	Cosmetics, textiles, agricultural sprays
Igepon T-33 T-43 T-51 T-73 T-77	Antara	Sodium N-methyl-N-oleyl taurate	
IN-181 -438 -531 -2503	Grasselli (Du Pont)	Lauryl alcohol sulfate Oleyl alcohol sulfate	Wool scouring Agricultural sprays
Intracol OA O S M	Synthetic	Long-chain fatty acid amide containing multiple amine groups	Textiles, rubber, pigments
Intral 222 224 229 384 231	Synthetic	Ethylene oxide condensation product with a fatty acid Ethylene oxide condensation product with unsaturated fatty acid Sulfonated ester of an unsaturated fatty acid	Textiles, rubber
Intramine WK	Synthetic	Sodium salt of sulfated lauryl alcohol and myristyl collamide	
Invadine C N	Ciba	Sodium alkyl naphthalene sulfonate Sodium sulfonate of the amyl ester of phthalic anhydride	Textiles Textiles

Emulsifying Agent (trade-name)	Manufacturer	Composition	Applications
Iragol M, P	Geigy	Fatty acid sulfonate	Mineral oils, fats, waxes; Animal, vegetable oils
Isonol DL1	Onyx	Dilauryl dimethyl ammonium bromide	Leather, textiles
Janusol	Synthetic	Mixture of lauryl and myristyl esters containing both primary amino and sulfated groups	Textiles, rubber
Jordonal	Jordan	Alkyl aryl sulfonate	
Katapol PN-340	Antara	Alkyl polyoxyethylene amine	Mineral oil; agricultural sprays
Kemulsion Base AA, CC	Kem	Fatty derivative of polyalcohol	Cosmetics; Coal tar disinfectants
Kessco DC-5B, -10, -11	Kessler	Complex polyoxyethylene fatty acid esters	Dry cleaning
Kessco Wax AC, A-21, A-33	Kessler	Fatty acid ester of polyhydric alcohol	Cosmetics, pharmaceuticals
Kessco X-10	Kessler	Fatty acid ester of polyhydric alcohol	Margarine
Kessco 23201, 18303R	Kessler	Fatty acid ester of polyglycol	Leather, textiles, paper, agricultural sprays
Kyro EO	P&G	Ethylene oxide condensation product with alkyl phenol	Textiles
Lamepon K	Maywood	Condensation product of fatty acid chloride with degraded protein	Textiles
Lamepon Tech. O, S, 4C	Kalide	Fatty acid condensate with a protein base	
Lanafin	Stand. Prod.	Blend of fatty acid ester and soap	Textiles
Larapal AD, PT	Jordan	Fatty alcohol sulfates	Textiles

Lavapon WS-1	Rohm & Haas	Ethylene oxide condensation product with alkyl phenol	
Lightening Penetrator X	Commonwealth	Fatty alcohol sulfate	
Lignosite	Puget	Salts of lignosulfuric acid	O/W emulsions
Lipal 4L } 4P } 6ML 60	Drew	Polyoxyethylene fatty acid ester Methoxy polyoxyethylene fatty acid ester Ethylene oxide condensation product with alkyl phenol	
Lissapol LD	ICI	Lauryl alcohol sulfate in the presence of acetic anhydride	
Lubritan P Conc.	Doittau	Sulfonated marine animal oil	Leather
Lupomin	Wolf	Alkylolamide	Textiles
M 885	DuPont	Ethylene oxide condensation product with isooctylene and phenol	
MP-189-S	DuPont	Sodium salt of aliphatic hydrocarbon sulfonate in aqueous solution	Fats, oils
Mapro Degum & Scour	Onyx	Sodium salt of the boro-sulfate of cetyl and stearyl alcohols	Textiles
Mapromol Powder	Onyx	Sodium salt of the boro-sulfate of stearyl alcohol	Textiles
Marasperse C CE} CB	Marathon	Highly purified calcium lignosulfonate Highly purified partially desulfonated sodium lignosulfonate	Sulfur, clay pigments, drilling muds, insecticides Carbon blacks, pigments
N		Highly purified sodium lignosulfonate	Insecticides
Maypon OW	Maywood	Triethanolamine salt of protetin condensation product with fatty acids	DDT
Mentor Beads	Colgate	Alkyl aryl sulfonate	
Merpol F	DuPont	Long-chain hydrocarbon sodium sulfonate	Textiles
Michelene DMA	Michel	Alkyl amido sulfate plus alkyl aryl sulfonate	Textiles
DNI DS		Ethylene oxide condensation product Alkyl amido alcohol	Textiles Textiles

Emulsifying Agent (trade-name)	Manufacturer	Composition	Applications
Miragene S Conc.	Miranol	Fatty acid amide ether derivative	
Monamine AA-100 AD-100 ADD-100 ACO-100 6-92 3-89 3-111 5-19 6-133	Mona	Fatty acid alkanolamides	
Monopole oil	Wolf	Sulfonated castor oil	Textiles
Monosulph	Nopco	Highly sulfonated castor oil	Textiles, latex
Morosol 4L SL 3F 31 DS	Moretex	Polyoxyethylene laurate Polyoxyethylene sorbitol stearate Polyoxyethylene oleate Polyoxyethylene sorbitol stearate Polyoxyethylene stearate	Textiles Waxes, mineral oil; textiles Oils, insecticides W/O emulsions
Morpel X-803	Morton-Withers	Synthetic petroleum sulfonate, water cut, M.W. 320	Textiles
X-912		Synthetic petroleum sulfonate, water cut, M.W. 425	
X-914		Synthetic petroleum sulfonate, oil cut, M.W. 450	
Mulsirex	Turco	Petroleum sulfonates	
Mulsor 3CW #8 #461 224 229 384 K 487 373 433	Synthetic	Fatty acid glycol ester Fatty acid polyoxyethylene ester Blend of fatty acid esters and polyglycols with complex amino bases Ethylene oxide condensation product with fatty acid	Cutting oils Textiles Metal cleaning, latexes Textiles

Trade Name	Manufacturer	Composition	Use
Myrj 45/52	Atlas	Polyoxyethylene stearate	O/W emulsions; cosmetics, pharmaceuticals
Myvatex 5-20/7-40/7-85	DPI	Monoglycerides in various bases	Foods
Myverol Type 18-00/18-05/18-30/18-70/18-85/18-90	DPI	Monoglycerides purified by molecular distillation	Foods
Naccolene A. Conc.	National Aniline	Modified petroleum sulfonate	Organic solvents
Naconal NR/NRSF	National Aniline	Alkyl aryl sulfonate	
Nagamine 142A	Synthetic	Amine ester of long-chain fatty acid	Textiles
Nekal BA-75	Antara	Sodium alkyl naphthalene sulfonate	Textiles, leather, paint, rubber
Neutronyx 268/330	Onyx	Polyethylene glycol (300) distearate; Polyethylene glycol (600) soybean fatty acid ester	Cosmetics, textiles
331/332/333/834		Polyethylene glycol fatty acid ester	Cosmetics, insecticides
		Polyethylene glycol (800) oleate	O/W emulsions
Nilo EM/EMC/S	Sandoz	Casein hydrolyzed by alkyl naphthalene sulfonate	Oils, fats, waxes; textiles
		Sulfonated olive oil plus olein and ammonium soap	Textiles
VO-51		Ethylene oxide condensate	Highly refined mineral oils
Ninate 402/411	Ninol	Calcium alkyl aryl sulfonate; Amine alkyl aryl sulfonate	Insecticides; Kerosene

Emulsifying Agent (trade-name)	Manufacturer	Composition	Applications
Ninol 200, 201, 400, 501, 713			Textiles; Leather, cosmetics
E-66, E-67, E-79, E-88, E-101, HA-10, MC-14	Ninol	Complex nonionic fatty ester amines	Cutting oils, insecticides
737		Condensation product of diethanolamine with fatty acids	Mineral oils; Paraffin wax, mineral oils; Isopropyl ester of 2,4-D; W/O emulsions; Metal cleaning compounds; Textiles
X		Condensation product of diethanolamine with stearic acid	Chlordane, isopropyl ester of 2,4-D
Ninox BDO, BFO, BJO	Ninol	Ethylene oxide condensation product with alkyl phenol	W/O emulsions
Nionine PG	Amalgamated	Polyethylene glycol ester	Textiles
Nonic 218, 234, 260, 261	Sharples	Polyethylene glycol tert-dodecyl thio-ether	Textiles, cosmetics
300		Polyethylene glycol alkyl phenyl ether	Textiles
Nonionic LG	Emulsol	Ethylene oxide condensation product with purified fatty acid ester	Cosmetics, pharmaceuticals
Nonisol 100, 110	Geigy	Polyethylene glycol (400) laurate	Cosmetics, textiles; Fats
200		Polyethylene glycol (400) laurate possibly plus 600	
		Polyethylene glycol (400) oleate	Kerosene; cosmetics, insecticidal sprays
210, 300		Polyethylene glycol oleate; Polyethylene glycol (400) stearate	Kerosene

Trade name	Manufacturer	Chemical composition	Use
Nopco 1-L, 6-L, 10-L	Nopco	Polyoxyethylene monolaurate	Textiles
1-0, 6-0		Polyoxyethylene monoöleate	Textiles
1-S		Polyoxyethylene monostearate	Cosmetic creams
BL-13, BL-18		Polyoxyethylene lauric amido amine	
1186		Sulfonated alkyl ester	Latex
1408		Sulfated castor oil soap	Pine oil, solvents
1471		Highly sulfated vegetable oil	Mineral, vegetable oils
1525		Ethylene oxide condensation product with alkyl phenol	Textiles
2031		Sulfated hydroxystearic acid	Textiles
SHCO		Sulfonated H-castor oil	Cosmetics, ointments
Nopcogen 11-0	Nopco	Oleic alkylolamide	Mineral, vegetable oils
14-L		Lauric alkylolamine condensate	Textiles
16-L		Lauric polyamine condensate	W/O emulsions, mineral, vegetable oils
16-0		Oleic polyamine condensate	Asphalt
20-0		Oleic polyamine condensate	Asphalt
No-Strip	Maguire	Tall oil amine condensate	Textiles, electroplating
NSAE Powder, Paste	Onyx	Sodium alkyl naphthalene sulfonate	Synthetic rubber
Nytron	Solvay	Sodium salt of substituted alkyl sulfonic acid from petroleum	Fats, waxes, pectins; textiles
Orethal 44	Despé	Sodium salt of sulfates of oleyl, stearyl, and cetyl alcohols	Textiles
E		Sulfonated palmitic acid amide condensed with ethylene oxide	Textiles
LL		Sodium lauryl sulfate	Textiles, agricultural sprays
Oronite Wetting Agent, Wetting Agent S	Oronite	Sodium alkyl aryl sulfonate	Orthodichlorobenzene
Orthemul	Fine Organics		

Emulsifying Agent (trade-name)	Manufacturer	Composition	Applications
Orvus AB Granules	P&G	Sodium alkyl aryl sulfonate	Textiles
ES Paste		Modified sodium alkyl sulfate	Textiles, cosmetics
K Liquid		Modified alkyl sulfate	Textiles, cosmetics
WA Paste		Sodium alkyl sulfate	Cosmetics
Peg-42	Glyco	Polyoxyethylene stearate	Foods
Pegmol	Stesen-Reuter	Polyethylene glycol (400) monolaurate	Latex paints
Pegy	Stand. Prod.	Polyglycol ester of fatty acids	Oils, solvents
Penetrant 100	Proven		Leather
Penetrant 120	Apex	Sulfated fatty ester	Textiles
Penetrator WH-9	Apex	Amine condensate	Textiles
Pentoterge 100	Jordan	Fatty amide condensate	Textiles
Perclo X-180 / X-390	Perry		Non-edible use
Perkolloid B / KG / O	AH	Casein degraded by treatment with soda ash	Textiles
Permalene A-100 / A-118 / A-120 / A-122 / A-134	Refined	Fatty amide condensate, fatty esters, and salts of fatty alcohol sulfates	Textiles
Perminal ELM	ICI	Glue hydrolysate	Oils, wool oils
Persol 40	Perkins	Igepon type	
Petrobase 1 / 2 / 3 / 210	Pennsylvania Refining	Petroleum sulfonate base	Petroleum oils
Petromix-9	Sonneborn	Modified petroleum sulfonate plus soap	
Petrosul 500 / 545 / 742 / 745 / 750	Pennsylvania Refining	Petroleum sulfonate	Oils

Petrowet WN	DuPont	Sodium alkyl sulfate	Acid solutions
Phil-O-Sol	Onyx	Sulfonated fatty acid ester	Textiles
Pluronic L44, L61, L62, L64, P75, F68	Wyandotte	Ethylene oxide condensate with hydrophobic base of polypropylene oxide	Textiles, etc.
Polyethylene glycol (400) (di-tri)ricinoleate, dilaurate, monolaurate, monooleate, monoricinoleate, monostearate	Glyco	As given	W/O emulsions; W/O emulsions; Insecticides, paints, textiles; Insecticides; Lubricant; Cosmetics, textiles
Polyethylene glycol (600) monolaurate	Glyco	As given	Insecticides, leather
Polyoxyethylene oleate 480	Van Dyk	As given	Pigments
Polyoxyethylene laurate 1060	Van Dyk	As given	Pigments
Polymene G-24	Quaker	Fatty amide condensate	Textiles
Prestabil Oil V	Antara	Sulfonated fatty acid	Textiles
Proflex	Glyco	Modified protein	Water paints
Promulgen	Robinson-Wagner	Polyethylene glycol fatty acid ester	Cosmetics, pharmaceuticals
Propylene glycol monococoate, monolaurate, monostearate	Colgate; Colgate, Glyco; Colgate	As given	Baking, foods; Foods, cosmetics; Foods, cosmetics
Protean W	Protean	Protein hydrolysate	Textiles
Protenol KX	Continental	Protein condensate	Textiles
Protex Gel	Ciba	Sulfonated laurylamide of alkyl benzamidazol	Textiles

Emulsifying Agent (trade-name)	Manufacturer	Composition	Applications
RN 200 } 200A }	Riches-Nelson	Alkyl aryl sulfonate	
Renex 20	Atlas	Ethylene oxide condensation product with mixed fatty and rosin acids	Textiles
Repcol Emulsifier	Refined	Modified fatty amine condensate	Agricultural sprays, textiles
Rexobase PW	Emkay	Naphthenic derivative	Mineral oil, paraffin
Rueterg 97-S 57-M 40-U 40-T	Finetex	Alkyl aryl sulfonates from "Neolene 400"	Mineral oil, solvents
Santol I } T }	Charlotte	Fatty alcohol sulfates	Textiles
Santolube 374	Monsanto	Mixed base	W/O emulsions
Santomerse 43	Monsanto	Amine salt of alkyl aryl sulfonate	
Sapamine MS	Ciba	Quaternary ammonium compound	
Saponin	Various	Natural origin	Fatty matter
Sarkosyl NL-100 NL-30 LC L S O	Geigy	Sodium lauryl sarcosinate Sodium lauryl sarcosinate Cocoyl sarcosine Lauryl sarcosine Stearyl sarcosine Oleyl sarcosine	
Savon Ampho 18	IGF	Mixture of sodium octadecyl sulfonates	Emulsion polymerization
Sellogen 0-141	Wolf	Mixture of anionic and nonionics	Textiles
Sinnopon A	Sinnova	Sodium or ammonium alkyl aryl sulfonate derived from a diphenyl oxide	
Sipex A SB	Alcolac	Ammonium lauryl sulfate Sodium lauryl sulfate	Latex, PVC
Sipon LT/6	Alcolac	Triethanolamine lauryl sulfate	Cosmetics
Siponic AP BC AAC }	Alcolac	Alkyl phenol ether Branched-chain alcohol ether	

Trade name	Manufacturer	Composition	Use
Soapotol	Commonwealth	Sodium salt of sulfonated fatty acid amide	
Solasol	Aceto	Sodium lauryl sulfate	
Soloil Base H-60, H-60	Oil States	Sodium petroleum sulfonate	Petroleum oils
Soluble Base 11, 600	Carlisle	Sodium alkyl sulfonate plus other emulsifiers	O/W emulsifiers, for soluble oils
Soluble Oil J	Sonneborn	Blend of petroleum sulfonates	
Soluble Oil Base 13-W	Carlisle	Oil-soluble petroleum sulfonate	Cutting oils
Solvadine BL Conc.	Ciba	Sodium alkyl naphthalene sulfonate	Textiles
G		Alkyl aryl sulfonate	
Solvit-A	Emulsol	Diacetyl tartaric acid ester of fatty acid mono- and diglycerides	Oil-soluble vitamins
Sorapon SF-78	Antara	Sodium alkyl aryl sulfonate	Textiles
Sorbit AC, P	Geigy	Mono- and dibutyl naphthalene sodium sulfonates	Textiles
Sotex N, NC, CW, 3CW	Synthetic	Ethylene oxide condensation product with fatty acids	Paints, etc.
Sotex C, WO, CX, 487			W/O emulsions
SP 315, 717	Stanco	Sulfonated petroleum derivatives	Paints, pigments
Span 20, 40	Atlas	Sorbitan monolaurate Sorbitan monopalmitate	Foods
Span 60, 62, 80, 85		Sorbitan monostearate Sorbitan monoöleate Sorbitan trioleate	W/O emulsions
Standapol	Standard	Sulfonated vegetable oil	Textiles
Stearonix	Onyx	Quaternary ammonium chloride	Textiles, agricultural

Emulsifying Agent (trade-name)	Manufacturer	Composition	Applications
Sterox SE	Monsanto	Polyoxyethylene thioether	Textiles
Sulfanole AN, ANO, S Gel, No. 9, No. 93	Warwick	Fatty amine condensate	Textiles
		Ethylene oxide condensation product with alkyl phenol	
Sul-Fon-Ate-OA-5	Tennessee	Sulfonated oleic acid	Textiles
Sulfonate S-40	Sinclair	Sodium mahogany sulfate (M.W. 461)	Cutting oils
Sulfo Turk S	Glyco	Sodium salt of sulfonated hydrocarbon	
Sulphopet	Sonneborn	Petroleum sulfonate	
Sulphosol	Beacon	Naphthenic soaps	
Superamide L9, GC, GR	Onyx	High purity fatty acid alkanolamides	
Surfax 505	Houghton	Fatty ester	
Surfynol 82, 102, 104	Air Reduction	Ditertiary acetylenic glycols	Emulsion paints
Syn-o-tol AV	Drew	Modified coconut oil amine condensate	
Synthetics AD 50	Hercules	Polyethylene glycol ether of hydroabietyl alcohol	W/O emulsions
AD160			O/W emulsions
AD400			O/W emulsions
AF40		Polyethylene glycol ether of alkyl phenol	W/O emulsions
AF80			O/W emulsions; textiles
AF100			O/W emulsions; waxes
AF150			O/W emulsions; textiles
AF200			W/O emulsions; agricultural sprays
AR 50			
AR100		Polyethylene glycol ether of rosin	W/O emulsions
AR150			O/W emulsions
AR200			O/W emulsions

Trade name	Manufacturer	Chemical composition	Application
Synthonon 100	Synthron	Ethylene oxide condensation product with an alkyl phenol	Textiles
Tegin 515 Tegacid	Goldschmidt	Glyceryl monostearate	Cosmetics, pharmaceuticals
Tegin P	Goldschmidt	Propylene glycol monostearate	Cosmetics, pharmaceuticals
Tenlo 400 420	Griffin	Polyhydric alcohol sulfonic acid derivative	Herbicides
Tensol	Synthetic	Sulfonated alkyl naphthalene ether	Textiles, leather
Tergitol 08 4		Sodium 2-ethyl-hexanol-1-sulfate Sodium 7-ethyl-2-methyl-undecanol-4 sulfate	Emulsion polymerization
7	Carbide	Sodium 3,9-diethyl-tridecanol-6 sulfate	Emulsion polymerization, leather
P-28		Sodium di(2-ethylhexyl) phosphate	Textiles
NPX NP-14 NP-27 NP-35 NP-40		Ethylene oxide condensation product with an alkyl phenol	Emulsion polymerization, agricultural concentrates, leather, etc.
TMN XD		Alkyl ether of polyethylene glycol	
Texsoft	Armour	Quaternary ammonium compound	Textiles
Textrafoam	Tex-Chem	Amine condensate	
Titamine TCP	Titan	Derivative of sulfonated alcohol	Textiles, leather
Toximul 100 200 300 400 500	Ninol	Anionic-nonionic blends	Insecticides
Trem 014 024 615 616 618	Griffin	Polyhydric alcohol esters	Agricultural sprays

Emulsifying Agent (trade-name)	Manufacturer	Composition	Applications
Trepenol T-100	Treplow	Triethanolamine salt of sulfated alkyl phenoxy polyoxyethylene ether	
Trepoline 505	Treplow	Non-sulfonated, hydrotropic amine condensate	Textiles, cosmetics
Triton B-1956	Rohm & Haas	Modified phthalic glycerol alkyd resin	Insecticides, fungicides
E-79		Ammonium alkyl aryl polyether sulfonate	Textiles, insecticides
GR-7		Sulfated alkyl esters	Agricultural emulsions
K-60		Stearyl dimethyl benzyl ammonium chloride	Textiles
WR-1339		Polymeric alkyl aryl polyether alcohol	Pharmaceuticals
Triton X-45, X-67, X-100		Alkyl aryl polyether alcohol	Textiles, insecticides
X-151, X-171		Blend of alkyl aryl polyether alcohols with organic sulfonates	Agricultural
X-155		Alkyl aryl polyether alcohol	Insecticides, herbicides
X-177		Blend of alkyl aryl polyether alcohol and modified resin	Aromatic solvents
X-188		Alkyl aryl polyether alcohols with additional stabilizer	Insecticides, herbicides
X-200, X-300		Sodium alkyl aryl polyether sulfonate	
X-301		Sodium alkyl aryl polyether sulfate	
X-400		Stearyl dimethyl benzyl ammonium chloride	Cosmetics
Tween 20, 21	Atlas	Polyoxyethylene sorbitan monolaurate	Essential oils, vitamins
40		Polyoxyethylene sorbitan monopalmitate	
60, 61		Polyoxyethylene sorbitan monostearate	Foods
65		Polyoxyethylene sorbitan tristearate	
80, 81		Polyoxyethylene sorbitan monoöleate	
85		Polyoxyethylene sorbitan trioleate	

Trade name	Manufacturer	Composition	Application
Twitchell 7231 Oil / 7240 Oil / 7250 Oil		Sulfonated mineral oil / Sulfonated fatty acid derivative	Textiles / Textiles
8262 Base / 8266 Base	Emery	Sulfonated mineral oil	Mineral oils; cutting oils
Ultrapole S / DL	Ultra	Fatty acid amine condensate / Lauryl diethanolamide	Textiles
Ultravon JF	Ciba	Polyether alcohol	
Ultrawet 30DS	Atlantic	Alkyl benzene sulfonate	Emulsion polymerization
Veripon	Protean	Protein hydrolysate condensed with fatty acid	Textiles
Victamine C / D	Victor	Substituted amide of alkyl phosphate	Textiles
Victamul 20 / 24C / 27 / 89 / 116C	Victor	Organic phosphate esters	Petroleum
Wettal	Emulsol	Sodium salt of sulfo-acetylated derivative of fatty acid monoglyceride	
Wetting Agent S	Oronite	Sodium alkyl aryl sulfonate	
Witco DGL / DGO / GMO / 77	Witco	Diethylene glycol laurate / Diethylene glycol oleate / Glyceryl monoöleate / Blend of polyalcohol carboxylic acid esters and oil-soluble sulfonate	Emulsion paints / Resin paints
Xyno Cation RO	Onyx	High molecular weight tertiary amine	Agricultural emulsions

Manufacturers of Emulsifying Agents

Aceto	Aceto Chemical Co., Flushing, N. Y.
Adjubel	Adjubel S.A., Lambecque lez Hal, Belgium
Advance	Advance Solvents and Chemical Corp., New York, N. Y.
A H	Arnold, Hoffman & Co., Providence, R. I.
Air Reduction	Air Reduction Chemical Co., New York, N. Y.
Alcolac	American Alcolac Corp., Baltimore, Md.
Alframine	Alframine Corp., Paterson, N. J.
Alrose	Alrose Chemical Co., Providence, R. I.
Amalgamated	Amalgamated Chemical Corp., Philadelphia, Pa.
American Lecithin	American Lecithin Co., Woodside, N. Y.
Amoa	Amoa Chemical Co., Hinckley, England
Antara	Antara Chemicals, New York, N. Y.
Apex	Apex Chemical Co., New York, N. Y.
Arkansas	Arkansas Co., Inc., Newark, N. J.
Armour	Armour and Co., Chicago, Ill.
Atlantic	Atlantic Refining Co., Philadelphia, Pa.
Atlas	Atlas Powder Co., Wilmington, Del.
Beacon	Beacon Chemical Co., Cambridge, Mass.
Belgotex	S.A. Belgotex, Brussels, Belgium
Burkart-Schier	Burkart-Schier Chemical Co., Chattanooga, Tenn.
Carbide	Carbide and Carbon Chemicals Co., New York, N. Y.
Carlisle	Carlisle Chemical Works, Inc., Reading, Ohio
Charlotte	Charlotte Chemical Laboratories, Charlotte, N. C.
Ciba	Ciba Co., Basle, Switzerland, and New York, N. Y.
Colgate	Colgate-Palmolive Co., New York, N. Y.
Commonwealth	Commonwealth Color and Chemical Co., New York, N. Y.
Continental	Continental Chemical Co., New York, N. Y.
CPI	Compagnie Française des Produits Industriels, Asnières, France
Croda	Croda Ltd., Goole, England
CSC	Commercial Solvents Corp., New York, N. Y.
Cyanamid	American Cyanamid Co., New York, N. Y.
Despē	Établissement Despé, Haat, Belgium
Dewey and Almy	Dewey and Almy Chemical Co., Cambridge, Mass.
Dexter	Dexter Chemical Co., New York, N. Y.
Doittau	Établissements Doittau (Solar), Corbeil, France
DPI	Distillation Products Industries, Rochester, N. Y.
Drew	E. F. Drew and Co., Boonton, N. J.
DuBois	DuBois Co., Cincinnati, Ohio
DuPont	E. I. duPont de Nemours and Co., Wilmington, Del.
Emery	Emery Industries, Inc., Cincinnati, Ohio
Emkay	Emkay Chemical Co., Elizabeth, N. J.

Emulsol	Emulsol Chemical Corp., Chicago, Ill.
Essential	Essential Chemicals Co., Milwaukee, Wis.
Esso	Esso Standard Oil Co., New York, N. Y.
Fallek	Fallek Products Co., New York, and Duesseldorf, Germany
Fine Organics	Fine Organics, Inc., New York, N. Y.
Finetex	Finetex, Inc., Pompton Plains, New Jersey
Geigy	Geigy Chemical Co., New York, N. Y. and Basle, Switzerland
General Dyestuff	General Dyestuff Corp., New York, N. Y.
Glyco	Glyco Products Co., New York, N. Y.
Goldschmidt	Goldschmidt Chemical Co., New York, N. Y.
Grasselli	*See* duPont
Griffin	Griffin Chemical Co., San Francisco, California
Hall	C. P. Hall Co., Chicago, Ill.
Hart	Hart Products Corp., New York, N. Y.
Hercules	Hercules Powder Co., Wilmington, Del.
Houghton	E. F. Houghton and Co., Philadelphia, Pa., and Puteaux, France
Hydrocarbon	Hydrocarbon Chemicals Co., Newark, N. J.
ICI	Imperial Chemical Industries, London, England
IGF	I. G. Farben Industries, Ludwigshafen, Germany
Jordon	W. H. and F. Jordon Jr., Mfg. Co., Philadelphia, Pa.
Kalide	Kalide Corp., Lawrence, Mass.
Kem	Kem Products Co., Newark, N. J.
Kessler	Kessler Chemical Co., Philadelphia, Pa.
Maguire	Maguire Industries, Inc., Jamaica, L. I., N. Y.
Marathon	Marathon Corp., Rothschild, Wis.
Maywood	Maywood Chemical Works, Maywood, N. J.
Michel	M. Michel and Co., New York, N. Y.
Miranol	Miranol Chemical Co., Irvington, N. J.
Mona	Mona Industries, Inc., Paterson, N. J.
Monsanto	Monsanto Chemical Co., St. Louis, Mo.
Moretex	Moretex Chemical Products, Spartanburg, S. C.
Morton-Withers	Morton-Withers Chemical Co., Greensboro, N. C.
Naftone	Naftone Inc., New York, N. Y.
National Aniline	National Aniline Div., Allied Chemical and Dye Corp., New York, N. Y.
Ninol	Ninol Laboratories, Inc., Chicago, Ill.
Nopco	Nopco Chemical Co., Harrison, N. J.
Oil States	Oil States Petroleum Co., New York, N. Y.

Onyx	Onyx Oil and Chemical Co., Jersey City, N. J.
Oronite	Oronite Chemical Co., New York, N. Y.
Pennsylvania Refining	Pennsylvania Refining Co., Butler, N. J.
Perkins	Perkins Soap Co., Springfield, Mass.
Perry	Perry Brothers, Inc., Woodside, N. Y.
P & G	Proctor and Gamble Co., Cincinnati, Ohio
Planetary	Planetary Chemicals Co., Creve Coeur, Mo.
Protean	Protean Chemical Co., New York, N. Y.
Proven	Proven Products, Peabody, Mass.
Publicker	Publicker Industries, Inc., Philadelphia, Pa.
Puget	Puget Sound Pulp and Timber Co., Bellingham, Wash.
Quaker	Quaker Chemical Products Corp., Conshohocken, Pa.
Refined	Refined Products Co., Lyndhurst, N. J.
Remsen	Remsen Chemicals, Inc., Oceanside, N. Y.
Riches-Nelson	Riches-Nelson Co., New York, N. Y.
Robinson-Wagner	Robinson-Wagner Co., New York, N. Y.
Rohm & Haas	Rohm and Haas Co., Philadelphia, Pa.
RWW	Reilly-Whiteman-Walton Co., Conshohocken, Pa.
Sandoz	Sandoz Chemical Works, Inc., New York, N. Y.
S.F.	Savonnerie Fournier et Fournier Cimag, Marseilles, France.
Sharples	Sharples Chemicals, Inc., Philadelphia
Sinclair	Sinclair Chemicals Co., New York, N. Y.
Sinnova	Societē Sinnova Sadic, Trilport, France
Solvay	The Solvay Process Co., New York, N. Y.
Sonneborn	L. Sonneborn Sons, Inc., New York, N. Y.
SPCMC	Societé des Produits Chémiques et Matières Colorantes de Mulhouse, France
Stand. Prod.	Standard Chemical Products, Inc., Hoboken, N. J.
Standard	Standard Chemical Co., Philadelphia, Pa.
Stesen-Reuter	Fred'k A. Stessen-Reuter, Inc., Chicago, Ill.
Synthron	Synthron, Inc., Ashton, R. I.
Synthetic	Synthetic Chemicals, Inc., Paterson, N. J.
Tennessee	Tennessee Corp., Atlanta, Ga.
Tex-Chem	Tex-Chem Co., Fairlawn, N. J.
Titan	Titan Chemical Products, Inc., Jersey City, N. J.
Treplow	Treplow Products, Inc., Paterson, N. J.
Turco	Turco Products, Inc., Los Angeles, Calif.
Ultra	Ultra Chemical Works, Inc., Paterson, N. J.
Van Dyk	Van Dyk and Co., Belleville, N. J.
Victor	Victor Chemical Works, Chicago, Ill.
Warwick	Warwick Chemical Co., Wood River, R. I.

Witco	Witco Chemical Co., New York, N. Y.
Wolf	Jacques Wolf and Co., Passaic, N. J.
Wyandotte	Wyandotte Chemicals Corp., Wyandotte, Mich.

Bibliography

1. Sisley, J. P., "Enclyopedia of Surface-active Agents," (trans. from the French and revised by P. J. Wood), New York, Chemical Publishing Co., 1952.
2. McCutcheon, J. W., "Synthetic Detergents and Emulsifiers—Up to Date III," *Soap & Chem. Specialties*, July, Aug., Sept., Oct., 1955 (available as paperbound reprint from the author).
3. Price, D., *Chem. Week*, Oct. 22, 1955, p. 41 *et seq.*

Index

A

Abietic acid, 172
Absorption base, 247
Acetylated beeswax, effect on stability, 248
  emulsifier efficiency, 205
Acid-soap ratio, and inversion, 145, 148
Adhesion, 11
  work of, 12, 13
Aerosols, 176
Agent-in-oil method, 209
Agent-in-water method, 209
Aging tests, accelerated, 331–332
Agricultural sprays, 259–261
  oil type, 261
Agricultural spray oil, self-emulsifying, 260
Alcohol sulfates, 173–175
Alginates, 185
Alginic acid, structure, 186
Alkane sulfonates, 175–177
Alkyl aromatic sulfonates, 177
Alkyl naphthalene sulfonates, 177
Alkylol amines, 182
Alkyl phenol-ethylene oxide compounds, 179
Alternate addition method, 210
Amine emulsifiers, in floor polishes, 255
Amine salts, 178
Amine soaps, 170
2-amino-2-methyl-1,3-propanediol, 171
Amphipathic molecules, 20
Amphiphilic compounds, 24
Amphiphilic molecules, 20
Anionic emulsifying agents, 169–177
Antonoff's rule, 12
Apparent viscosity, definition, 56
Area, cross-sectional, of molecule, 18
  per molecule, 19
Asphalt emulsions, 268–272
  distribution of droplet size, 46, 48
  effect of bitumen on stability, 269
  effect of emulsifier, 269

effect of water hardness, 269
mechanism of breaking, 269
patents, 271–272
production, 268
stability, 268
test methods, 270
viscosity, 269

B

Baked goods, 265–266
  effect of emulsifier, 265
Balance, hydrophil, 16
Balanced molecules, and emulsifier efficiency, 188
Ballast water emulsions, demulsification, 295
Bancroft's rule, 87, 93
Beeswax-borax creams, emulsion type, 248
Bingham plastics, definition, 55
Block copolymers, 179
Breakability, of asphalt emulsions, 269
Brookfield Synchro-lectric viscometer, 324, 325
Brownian motion, effect on stability, 53
  in emulsions, 52
Bulk properties, solutions of surface-active compounds, 29–33
Butter, 285–287
  production, 285
Butter-fat, separation, 286–287
  stabilization in milk, 285
Butter formation, role of foaming, 286
  role of protective sheath, 286
  theories, 286

C

Calcium-sensitive emulsifiers, 171
Capillary-height method, surface tension, 301–303
  apparatus, 303–304
  corrections, 302

373